ROUTLEDGE LIBRARY EDITIONS:
WATER RESOURCES

Volume 1

WATER RESOURCES IN THE ARID REALM

WATER RESOURCES IN THE ARID REALM

CLIVE AGNEW
AND
EWAN ANDERSON

LONDON AND NEW YORK

First published in 1992 by Routledge

This edition first published in 2024
by Routledge
4 Park Square, Milton Park, Abingdon, Oxon OX14 4RN

and by Routledge
605 Third Avenue, New York, NY 10158

Routledge is an imprint of the Taylor & Francis Group, an informa business

British Library Cataloguing in Publication Data
A catalogue record for this book is available from the British Library

ISBN: 978-1-032-74502-2 (Set)
ISBN: 978-1-032-73273-2 (Volume 1) (hbk)
ISBN: 978-1-032-73382-1 (Volume 1) (pbk)
ISBN: 978-1-003-46391-7 (Volume 1) (ebk)

DOI: 10.4324/9781003463917

Publisher's Note
The publisher has gone to great lengths to ensure the quality of this reprint but points out that some imperfections in the original copies may be apparent.

WATER RESOURCES IN THE ARID REALM

Clive Agnew and Ewan Anderson

London and New York

First published 1992
by Routledge
11 New Fetter Lane, London EC4P 4EE

Simultaneously published in the USA and Canada
by Routledge
a division of Routledge, Chapman and Hall, Inc.
29 West 35th Street, New York, NY 10001

Typeset in [Scantext September] by
Leaper & Gard Limited
Printed and bound in Great Britain by
Biddles Ltd, Guildford and King's Lynn

British Library Cataloguing in Publication Data

A catalogue record for this title is available from the British Library.

Library of Congress Cataloging in Publication Data

Agnew, Clive, 1955–
Water resources in the arid realm / Clive Agnew and Ewan Anderson.
p. cm. – (Routledge physical environment series)
Includes bibliographical references and index.

ISBN 0–415–04346–8. – ISBN 0–415–07969–1

1. Water-supply–Management. 2. Arid regions ecology.
3. Arid regions–Middle East. 4. Arid regions–Africa.
5. Hydrology.
I. Anderson, Ewan W. II. Title. III. Series.
TD365.A7 1992 91–38302
333.91′00915–dc20 CIP

And God made a firmament, and divided the waters that were under the firmament, from those that were the firmament.

(Genesis 1: 7)

And we send the fecundating winds, then cause the rain to descend from the sky, therewith providing you with water (in abundance), though we are not the guardian of its stores.

(Al-Hujurat, 22)

Then shall the lame man leap as a hart, and the tongue of the dumb shall be free: for the waters are broken out in the desert, and streams in the wilderness. And that which was dry land shall become a pool, and the thirsty land, springs of water.

(Isaiah 35: 6–7)

For among the rock, there are some from which rivers gush forth; others there are which when split asunder send forth water.

(Al-Baqara, 74)

I have given waters in the wilderness, rivers in the desert, to give drink to my people, to my chosen.

(Isaiah 43: 20)

CONTENTS

PLATES

FIGURES

TABLES

ACKNOWLEDGEMENTS

First and foremost, our special thanks to Kath and Sian for their encouragement and fortitude. We would also like to express our appreciation to the Public Authority for Water Resources and in particular the Director General, William Doyel, Robert Dingman and John Kay for their suggestions for the book and especially for the use of material from Oman. The decision to write this book can be traced to an expedition to the Wahiba Sands, Oman and we thank the Royal Geographical Society for that opportunity and in particular Nigel Winser, the expedition organiser. Support was gratefully received from the following: The Government of Turkey and Bilkent University supported travel to the Ataturk Dam site, the US Geological Survey and the Fulbright Commission for assistance to visit research institutions in the USA. Sahelian climatic data was provided by the AGRHYMET Centre, Niamey and the Climatic Research Unit of the University of East Anglia. The figures used in this book were drawn by the drawing office staff in the Geography departments at University College London and the University of Durham, whom we thank for their patience and skill. Permissions to use particular figures and photographs are listed below but we are grateful to the RGS for the use of some of their slides from the Wahiba Sands Expedition and the Geographical Magazine for permission to reproduce the figures used in Chapter 10.

We would also like to thank the following for granting permission to reproduce the figures listed below in this book:

Figure I.1 Professor D.A. Wilhite, Institute of Agriculture and Natural Resources, University of Nebraska.

Figure I.2 Map from the explanatory note, MAB Technical Notes 7, copyright UNESCO 1977. Reprinted by permission of UNESCO, Paris.

Figure 1.1 World Meteorological Organisation, Geneva.

Figure 1.11 University of Wisconsin Press.

Figure 1.13 Permission granted by Professor R.U. Cooke, University College London.

Figure 1.14 Jacaranda Wiley Ltd., Queensland.

Figure 1.15 Professor D.L. Johnson, Clark University.

Figure 2.1 Professor George Rumney.

Figures 2.6 and 2.7 Author: Dr. G. Farmer, AID FEWS Project, Arlington (and Climatic Research Unit, University of East Anglia) and report produced by: IUCN, Gland, Switzerland.

Figure 3.1 American Association for the Advancement of Science (copyright 1974 by the AAAS).

Figure 3.2 Professor H.H. Lamb, Climatic Research Unit, University of East Anglia; and Cambridge University Press.

Figure 3.3 Professor F.K. Hare, Trinity College, University of Toronto.

Figure 3.4 Journal of Soil Water Conservation, copyrighted by Soil Conservation Society of America.

Figures 3.5 and 3.6 HMSO, London.

Figure 3.7 Methuen and Co Ltd., London.

Figures 4.5, 4.6 and 4.8 Journal of Arid Environments, copyrighted by Academic Press.

Figure 4.9 International Journal of Climatology, copyrighted by Royal Meteorological Society.

Figure 6.2 Dr. M. Kay, Silsoe College and Cranfield Press.

Figure 6.3 Longman Group UK.

Figures 10.1 to 10.6 Geographical Magazine.

Part I

INTRODUCTION

Arid regions are areas of great environmental and economic contrasts containing some of the world's most important mineral resources. The arid realm provides 82 per cent of the world's oil production, 86 per cent of iron ore, 79 per cent of copper and 67 per cent of diamonds (Heathcote, 1983). Whilst arid lands can be resource rich they are also water poor and face a number of human and ecological disasters which have focused attention upon these areas. The media frequently portray the arid realm as containing areas of economic disruption and great poverty, affected by advancing desert margins where people fall victim to the twin threats of drought and famine. In many parts of the arid realm the words drought and famine now appear almost synonymous (Agnew and O'Connor, 1991) and figure I.1 shows the widespread occurrence of drought during the 1980s.

The worst Australian drought for two centuries ended in 1983 after crop production had fallen by 31 per cent and farm income by 24 per cent; in North East Brazil an estimated 90 per cent of crops failed due to drought; Mexico's agricultural production fell by 12 per cent because of the lack of rain in 1982; and in 1988 corn production in the United States of America was reduced by 36 per cent and soyabean production by 22 per cent. Nowhere, it seems, could escape the ravages of drought in the 1980s and a comparison of figures I.1 and I.2 reveals the prevalence of drought in arid areas.

Desertification (the expansion of deserts) is often concomitant with drought and there have been many observations that desert boundaries are mobile. The Sahara Desert has received most attention with claims that it is advancing southwards at a rate of 9 km each year. Beyond the desert margins land degradation is an even more significant process leading to soil erosion and impoverishment of agricultural systems. Upon returning to the Yatenga Plateau in Burkina Faso, Harrison (1987, p. 140) observed

> vast expanses were totally bare of any ground level vegetation, covered with a fine, hard crust where silt clogged the soil's pores and kept the rain out, or with layers of rusty red gravel, all that was left of the top soil.

The removal of vegetation, associated with population increase, is held to be a

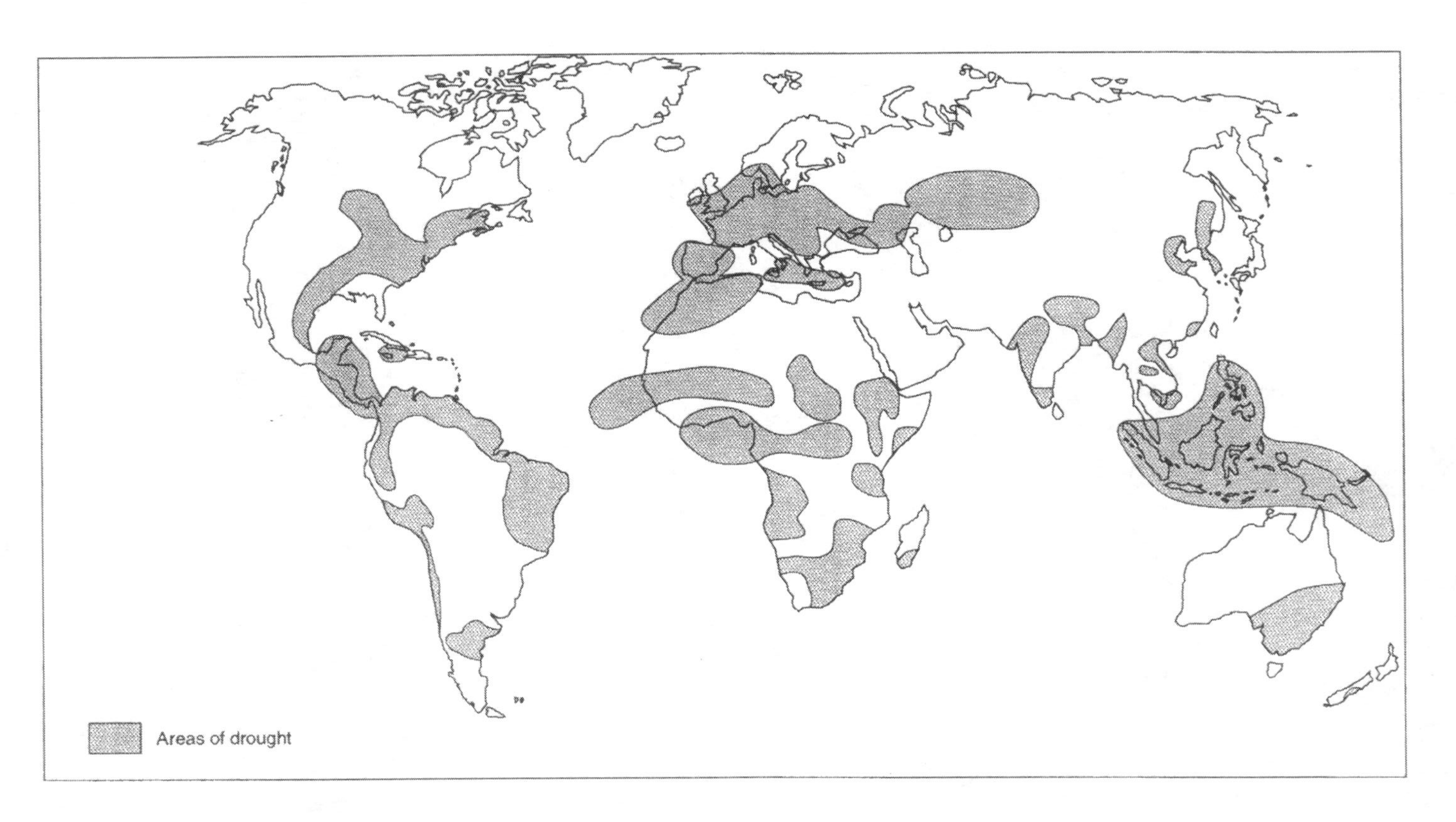

Figure I.1 Areas of drought during 1981–2 after Wilhite and Glantz (1985).

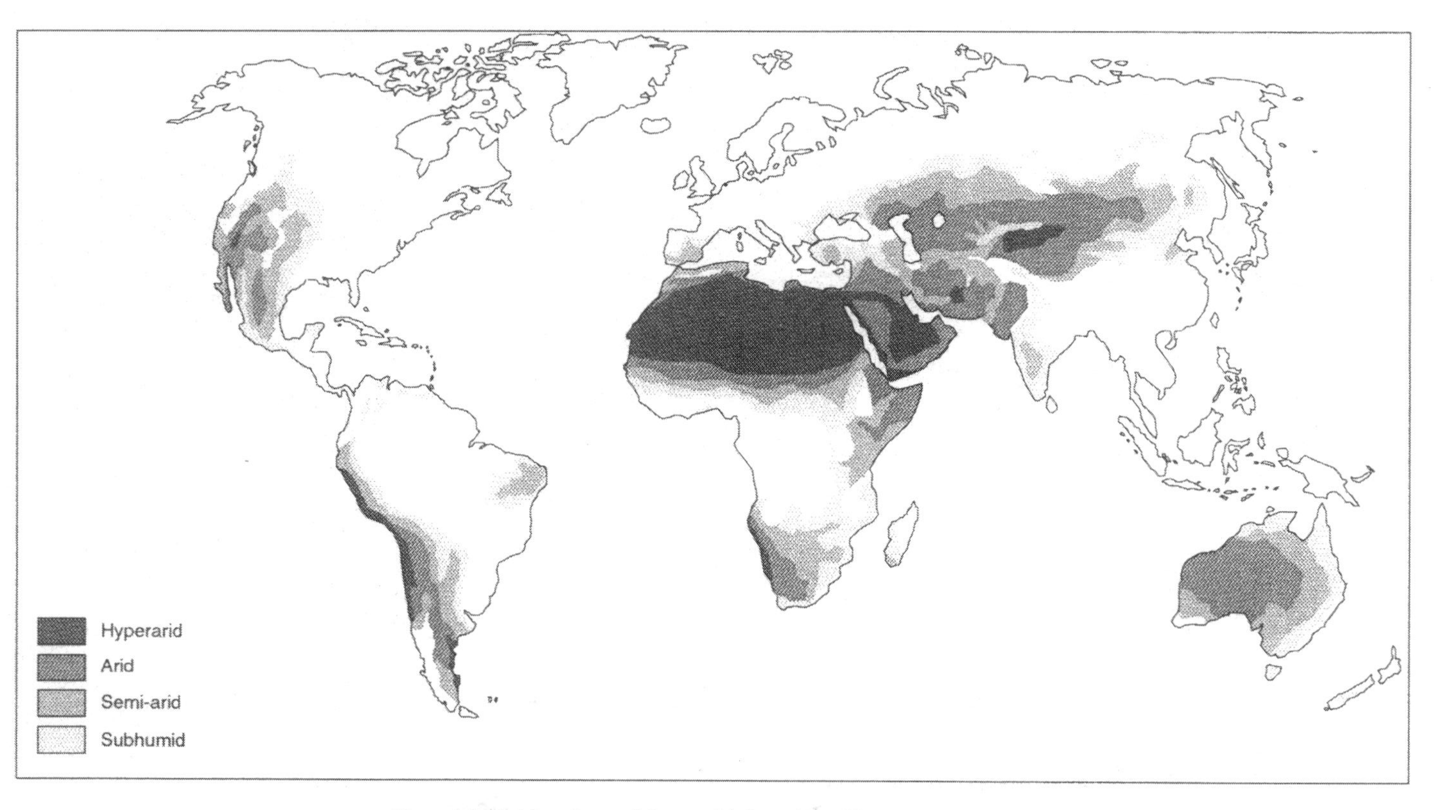

Figure I.2 Arid regions of the world, based on UNESCO (1977b).

major cause of land degradation and the World Bank (1984) reported that, in the arid region of West Africa known as the Sahel, 1 per cent of natural forest was being lost each year. Fuel wood consumption alone now exceeds the growth of new trees by a factor of 10 in Mauritania, 5 in Kenya and 2.5 in Ethiopia and Nigeria. In response the influential Club of Rome has suggested mobilising a UN military force to help combat land degradation in the Sahel while the first phase (850,000 ha) in a new forest belt 7,000 km long and 1,000 km wide has just been completed in North China to halt an advancing desert front (Haramata, 1989). It is estimated that 35 per cent (Mabbutt, 1984 and Grainger, 1990) of the Earth's surface is affected by desertification and land degradation, although this figure is disputed (Warren and Agnew, 1988).

Hare (1984a) and Macdonald (1986) note that such conditions parallel reports from the 1970s, when Copans (1983) suggested that some 100,000 people died as a result of drought in West Africa. But the fascination for the arid realm extends beyond recent concern over the impacts of drought and desertification. Heathcote (1983) believes interest in arid lands is in part due to their distinctive environments and their preservation of past civilisations; while Cooke and Warren (1973) consider exploration and exploitation have also been important. Mundlak and Singer (1977) suggest the reason for the recent upsurge in interest in these areas has more to do with population pressure in other parts of the world coupled to economic opportunity. As Batisse (1988, p. 21) states:

> To explain this (past) neglect ... the need for space, food and natural resources on this shrinking and crowded planet was not as pressing 40 years ago as it has become today.

UNESCO (1977a) state that the arid realm encompasses a third of the Earth's surface but only supports 15 per cent of the world's population while Eyre (1988) suggests only 9 per cent of the world's population are found in arid lands (table I.1).

Although the population in arid lands is relatively low on a global scale, the

Table I.1 Arid land populations (millions) (from Eyre 1988 and IUCN[1] 1989)

	1960	*1985*	*% increase*
Asia	151	271	79
Africa	50	97	95
North America	16	26	68
South America	10	18	75
Europe	1	1	–
Australia	1	1	–
Arid lands total	229 (7.6%)	414 (8.6%)	81
World[1]	3,019	4,818	56

Table I.2 Economic performance of some of Africa's dry lands (from Curtis *et al.*, 1988)

	Food production per capita		*Food aid as % of cereals imported*	*GNP per capita*	
	1974–6	*1982–4*	*1982*	*1982*	*1984*
Burkina Faso	95	92	–	116	89
Chad	83	80	43	73	–
Ethiopia	91	93	64	108	85
Mali	78	83	47	129	100
Mauritania	69	66	49	147	141
Niger	89	65	56	115	70
Senegal	114	76	16	114	88
Somalia	88	60	44	–	90
Sudan	108	92	47	119	92
	(1969–71 = 100)			(1979 = 100)	

region contains dense concentrations at a local scale. For example, 36 million people live in the five arid land cities of Cairo, Karachi, Lima, Beijing and Tehran (Eyre, 1988). Between 1969 and 1985 the population in arid lands increased by 81 per cent compared to 56 per cent for the Earth as a whole. It is therefore misleading to perceive all arid lands in the 20th century as empty, virgin territories although they do contain areas of extremely low population density and remote inaccessible places of unknown resource potential.

In addition to possible economic opportunities and environmental degradation, concern for arid lands is also growing owing to the plight of some of its inhabitants. Al-Sudeary (1988) writes that poverty is increasing with the number of rural poor in drylands growing from 500 million in 1977 to 800 million in 1983 (although these numbers are in excess of those estimated by Eyre in table I.1). From Eyre's (1988) report it is evident that 149 million (36 per cent of the total) arid land inhabitants are to be found in the poorest category by GNP and resource potential.

Table I.2 reveals the range of economic and food production performances found in a selection of Africa's arid lands with an indication of rising GNPs in some countries but all showing alarming decreases in food production per capita since the late 1960s. Norton-Griffiths (1989), however, points to the dangers of over generalising in arid lands and shows that while the countries of the Western Sahel (Senegal, Mauritania, Niger and Mali) have performed badly on agricultural and livestock production between 1961 and 1987, the Central Sahel (Burkina Faso, Chad and Sudan) has shown evidence of steady growth.

Clearly the inhabitants of arid regions face a number of environmental and economic problems. UNESCO (1977a) list three main obstacles to the development of these areas:

1 The lack of commercial development, small internal market and lack of capital.

2 The poor infrastructure hampering communications and movement.
3 The natural environment with its lack of water resources and 'hostility'.

It is the last of these upon which this text intends to focus. It will be argued below that arid lands are essentially a diverse and disparate group of countries with a variety of economic and environmental problems, the common factor being the constraints imposed upon economic activities and environmental resources by the lack of moisture at certain times. It is their lack of water that distinguishes arid lands but beyond this there is little which is common. This point was made forcibly by Gilbert White in the early 1960s and yet the habit of treating arid lands as a homogeneous region continues

> no two points of the arid zone are alike in all respects: each has its own unique combination of terrain, soil, vegetation, and moisture. For this reason, generalities about arid lands are to be approached with the caution of a hydrologist assessing a drainage basin, knowing that one steep sector may be perennially dry while a nearby valley fill is usually saturated.
>
> (White, G.F., 1986, p. 127)

The danger of treating the arid realm as a geographical region is that it is then perceived as experiencing similar problems with similar causes requiring similar solutions. The purpose of chapter 1, on arid environments, is to try to dispel some of the myths that surround the appearance and conditions of arid lands, i.e. that they are all sandy desert, that they are permanently dry, that population is sparse and that they are devoid of natural resources. These phrases do characterise some parts of the arid realm but there are other locations which support large populations, are fairly well endowed with minerals and can exploit substantial water resources.

The book is divided into four parts: part I deals with the environmental background of the arid realm; II with the water resource problems it faces through increasing demand and the variability of supplies; III with the methods available for enhancing water supplies; and finally IV with the management of this resource. We have taken the decision to focus only on the hot arid realm and to ignore cold, polar regions where precipitation is also limited. Although both areas experience similar problems over water resources there are obviously differences in the solutions to water resource problems due to their different temperature regimes. It was felt that inclusion of the cold arid lands would add unnecessary complication to the discussion of what is already a very diverse environment.

Part II considers the problems of the supply of water resources. We have decided not to discuss in detail the factors influencing **demand**, i.e. urbanisation, standards of living, health and sanitation and population growth, but to incorporate these factors into the discussion of water resources management where it is felt they can be dealt with more effectively. Part II then focuses upon the major difficulties in the effective use of water in arid lands, namely the variability of supplies, their changing nature through climatic change and the lack of an adequate data base. In addition we try to make the case that in analysing arid lands' hydrology, the techniques and instruments developed in temperate climates are often inappropriate.

Part III considers how water supplies can be enhanced. The approach is systematic but also technological with an examination of irrigation, desalination, cloud seeding and surface and subsurface storage. The last of these combines recharge dams, reservoirs, groundwater exploitation and interbasin transfers as these all face common problems and are often interdependent. Although these chapter titles may appear to stress high-technology solutions, in fact 'traditional' low-technology approaches, where appropriate (e.g. irrigation and rainfall storage), are also included to indicate the range of techniques currently employed in arid lands.

Finally, part IV addresses the question of management and emphasises the growing awareness of the strategic value of water in a chapter on the geopolitics of water. This part aims to integrate the problems of balancing supply and demand and includes a number of case studies to exemplify both the range of problems and possible solutions. Here water is not seen as a purely 'physical resource' involving hydrological and engineering investigations; but the social, institutional and political aspects of water resources development are also considered.

The book tries to draw upon examples from throughout the arid realm but given the vast area involved we are inevitably deficient. We have, however, attempted to draw a contrast between the arid lands of the oil-rich Middle East and the resource-poor nations of Africa. As World Resources Institute (1986, p. 123) state concerning freshwater supplies

> most of sub-Saharan Africa and the Middle East have chronic shortages, as do parts of large countries like the U.S., Australia, and the Soviet Union.

While Simpson (1985, p. 6), referring to Arab water resources, notes that:

> Nowhere else in the world today is so much money being spent on exploration for water and the resulting development of sophisticated methods to exploit and transport it.

We therefore feel that in placing an emphasis upon Africa and the Middle East we are reflecting areas of the arid realm where there are some of the greatest water resource problems and where much activity as a consequence is taking place.

Given the above comments on the diversity of the arid realm and the lack of homogeneity the reader may wonder why this book is not then a collection of individual case studies. We have tried to generalise to a degree over the water resources problems, techniques available and management strategies for enhancing supplies as it is only in this arena that there is 'common ground'. In this we are directing the book not only to those with a general interest in arid lands but also to those involved in water resources management, to aid their awareness of the range of issues that must be addressed and the problems specific to the arid realm. The aim has been to produce a text that is suitable for both undergraduate and postgraduate students and those actively engaged in arid land water resources. The book then lies in juxtaposition with volumes on hydrological techniques (Chow, 1964, Rodda *et al.* 1976, Shaw, 1983 and Wilson, 1984); those considering arid lands (Nir, 1974, Hills, 1966, and Heathcote, 1983); and texts examining water resources in the tropics (Balek, 1983, Barrow, 1987 and Jackson, 1977) and globally (McDonald and Kay, 1988).

The current trend in water resources development is to argue against large-scale multi-purpose projects due to their inflated costs, environmental damage and their need for government subsidy resulting in wasteful water practices. Small-scale, 'grass roots', local community projects are seen as being more sustainable and likely to succeed. Yet at the 13th International Conference on Irrigation and Drainage (1987), which focused upon water management in developing countries, it was concluded that integrated management under a single national planning structure was essential. The task for arid land water management is then to provide an overall national framework into which both medium-scale and local projects can be dovetailed while maintaining the integrity of the whole system in an environment that is highly variable and somewhat unpredictable. The key to success is flexibility, and we return throughout the book to the notion that successful adaptations in arid lands are characterised by an appreciation of the vagaries of the natural environment.

1

ARID ENVIRONMENTS

LOCATION OF ARID AREAS

Assessing resources and planning on a global scale requires the delineation of the arid realm. In the introduction we argued that arid lands contain heterogeneous environments which present a problem for mapping when it is assumed that arid boundaries demarcate a region that is in some manner homogeneous. At a global scale it is possible to generalise but the resulting maps and boundaries should be treated with caution and not used for studies at smaller scales.

Aridity is defined as a lack of moisture but this is essentially a climatic phenomenon based upon the average climatic conditions over a region. Arid lands are then most effectively identified by climatological mapping. Deserts can be considered as areas of low or absent vegetation cover with an exposed ground surface (Goudie, 1985); but McGinnies (1968) suggests that there is no universal agreement over the term desert. There is then plenty of scope for confusion between the terms **arid** (based on climate) and **desert** (based on surface characteristics). The discussion below is concerned with the arid realm, encompassing arid and semi-arid environments from 'desert' through to 'steppe' landscapes. The term desert will be used to convey hyper-arid conditions where rainfalls are particularly low and vegetation is sparse.

The effect of the lack of moisture is manifest through the soils, vegetation and topography of arid lands. Identifying arid lands can therefore be undertaken through a variety of non-climatic criteria and Oliver (1973) and Nir (1974) note that arid lands can be defined in several ways ranging from pedology through to botany, but Wallen (1967) argues that climate is their fundamental characteristic. Although climate is the most important environmental characteristic it is also the most variable. Climatic classification has the advantage of being a sensitive indicator of change and there is a reasonable coverage of climatic stations in many parts of the world. Unfortunately the arid realm is not well endowed with climatic observations and the changeable nature of the world's climate means that a large number of observations through time and space are required to distinguish between change and fluctuation, and to establish climatic 'norms'.

Plant, soil and geomorphological characteristics are less variable in turn, but then these criteria have the disadvantage of possibly no longer reflecting current environmental conditions, i.e. of being a relic of a drier past. Nevertheless all of these criteria have at some time been used to identify arid lands. The consideration of the demarcation of arid areas will then commence with geomorphological characteristics followed by soils, plants and finally climate.

Geomorphological definitions

Scientific interest in desert landforms stretches back at least as far as the end of the 19th century with North American scientists particularly active in a debate over landform evolution and cycles of erosion (Lustig, 1968 and Small, 1972). The 20th century has witnessed great controversy over the relative importance of geomorphological agents in deserts, in particular the roles of wind and water. Graf (1988, p. 3) wrote quite forcibly:

> One of the most startling paradoxes of the world's drylands is that although they are lands of little rain, the details of their surfaces are mostly products of the action of water.

There exist several texts such as those by Cooke and Warren (1973), Mabbutt (1977) and Thomas (1989) dealing with arid lands' geomorphological processes and features, which suggest that no single process dominates arid environments. It has long been mooted, however, that (in parts) the stable geological environment, lack of vegetation cover and particular climatic conditions can produce distinctive landforms. One might expect arid lands to be vast expanses of dunes and badlands with the occasional salt flat. In fact arid lands vary from the tectonically active mountainous regions in North and South America to the geologically stable shield areas in Africa and Australia which contain, in places, a deep regolith and flat peneplain expanses. For instance, Lustig (1968) lists plains, flats and depressions as the major features of arid lands (e.g. the Sahara, Australia and Arabia) but then contrasts these to the basin and range topography of North America. Heathcote (1983) records that arid lands in the USA are characterised by 38 per cent desert mountains, 31.4 per cent alluvial fans and only 21 per cent desert flats, while Australia's arid interior encompasses only 16 per cent mountains, 38 per cent dunes and sand, 32 per cent bedrock plains and 13 per cent clay plains. Arid zone landscapes then comprise a number of features of which the following are most typical (based upon Goudie, 1985, Heathcote, 1983 and Thomas, 1989):

Alluvial fan a fan-shaped deposit found at the foot of the slope, grading from gravels and boulders at the apex to sand and silt at the foot of the fan. Called a bajada when coalesced.

Dunes an aeolian deposit of sand grains (unconsolidated mineral particles) forming various shapes and sizes depending upon the supply and characteristics of the material and the wind system.

Bedrock fields including pediment, a plano-concave erosion surface sloping from the foot of an upland area; and hamada, a bare rock surface with little or no vegetation or surficial material.
Desert flats with slight slopes possibly containing sand dunes, termed playa when the surface is flat and periodically inundated by surface runoff.
Desert mountains the most common feature of arid lands.
Badlands well dissected, unconsolidated or poorly cemented deposits with sparse vegetation.

Dregne (1976) comments upon Martonne and Aufrere's (1927) mapping of interior drainage basins (with no surface runoff or runoff not reaching the oceans) to delineate arid regions with a resulting estimate of 29 per cent of the Earth's surface being arid or 35 per cent of the non-polar land area. Petrov (1969, 1976) examined the occurrence of playas, sebkha's, dunes, badlands, etc. and found that 36.2 per cent of the world's surface is arid. The diversity of arid landforms, however, should not be underestimated. For example, a terrain classification of the Wahiba sands, Oman, identified 24 different terrain units from north–south mega ridges up to 100 m high and 1–3 km apart, barchans up to 50 m high with north-north-east-facing slip faces, transverse sand dunes and outcrops of aeolianite (Jones *et al.*, 1988).

Although the lack of vegetation cover in the hyper-arid region can enhance the effectiveness of geomorphological agents, arid areas are not particularly dynamic. This is in part due to the lack of moisture although aeolian processes, salt weathering and surface runoff can be dramatic and destructive under certain conditions. Geomorphological classifications of arid lands will therefore tend to map areas where arid conditions have prevailed for some time but are not necessarily current. For this reason they can also fail to reveal changes in the location of arid areas. The geomorphological approach to mapping the arid realm presupposes that it has a recognisable landscape that is markedly different than those elsewhere. Yet many authors (e.g. Beaumont, 1989 and Smith, 1968) have noted that the geomorphology of desert areas is broadly similar to the features in other lands with the same geology although arid landforms may be more angular and abrupt. Cooke and Warren (1973, p. 6) state:

> In our view the formulation of simple and comprehensive generalisations about the nature of desert geomorphology, especially if they are based on the recognition of relations between desert landforms and desert climates, will be difficult, and in the present state of knowledge is impossible.

That is, similar landforms can be found in different climate areas, produced by different processes. McGinnies *et al.* (1968, p. 4) concur and state:

> While there are certain geomorphological characteristics associated with arid regions, these are not sufficiently sensitive to biological and physical conditions to serve as boundary markers for the desert.

In the new edition of their book, Cooke and Warren demonstrate the difficulty of using geomorphological features to map arid lands by showing the vast differences obtained in the position of the Sahara Desert depending upon whether one uses active or stable sand dunes as the criteria. While there continues to be great interest in the processes shaping arid landscapes and their relative rates, it must be concluded that geomorphological mapping does not lend itself to the reliable delineation of arid environments.

Pedological definitions

It follows from work on zonal soil groupings (Brady, 1974) that relationships existing between climate, climax vegetation and soil type provide the basis for an alternative approach to arid zone mapping. Many areas in arid lands are too steep for soils to develop or bare rock is exposed, and hence arid land soils are limited in extent (McGinnies, 1979). Their character is primarily influenced by low rainfalls, high evaporation rates and low amounts of vegetation cover. These result in soils with low organic matter, an accumulation of salts at the surface, little development of clay minerals, a low cation exchange capacity, a dark or reddish colour due to desert varnish and little horizon development because of the lack of percolating water (for a more detailed account see Fuller, 1974).

It would appear that vast areas are covered by thin, infertile soils nearly useless for agriculture. Dregne (1976) does record that:

> **Entisols** cover 41.5% of arid lands (immature soils ranging from barren sands to very productive alluvial deposits),
> **aridisols** cover 35.9% (red–brown desert soils, dry and generally only suitable for grazing without irrigation) and
> **vertisols** cover 4.1% (moderately deep swelling clay which is difficult to cultivate).

However, **mollisols** (one of the world's most important agricultural soils) cover 11.9 per cent of arid lands, **alfisols** (high base saturations, reasonably high clay contents, agriculturally productive) cover 6.6 per cent and **vertisols** are used widely for irrigation. There are then arid lands where soils are highly productive providing tremendous potential for agriculture.

Shantz (1956) mapped arid soils (pedocals) as covering over 43 per cent of the Earth's surface, seemingly an overestimate compared to the climatic calculations shown in Table 1.1. Dregne (1976) maps arid land soils covering 46 million k m^2 or 31.5 per cent of the land area excluding polar regions. Although the total area is of the same order as for the climatic delineation of arid lands, Dregne notes several discrepancies, including the mapping of vertisols in India as depicting an arid land environment when they are also found in humid areas; the same is observed for entisols in Australia. Given the time taken for soils to develop and the range of processes which influence soil characteristics, the same conclusion as for geomorphological mapping of arid lands must be reached. That is, soils

Table 1.1 Climate classifications of arid lands (% of world land area) (from Heathcote, 1983, p. 16)

	Koppen	*Thornthwaite*	*Meigs (1953)*	*UNESCO (1977b)*
Hyper-arid	12.0	15.33	20.5	19.5
Semi-arid	14.3	15.24	15.8	13.3
Total	26.3	30.57	36.3	32.8

having the characteristics associated with arid climates can be fossil relics or may have been formed by non-aridity controls. Hence soil mapping of arid areas should only be undertaken at global or regional scales.

Botanical definitions

Plant adaptation to water availability has been used as a system of botanical classification distinguishing between hydrophytes (aquatic species) through to the xerophytes (tolerating arid conditions). Whittaker (1975) provides a list of the various mechanisms by which plants have adapted to aridity:

- deep or extensive root systems,
- water storage in tissues,
- protective wax coating of hairs,
- reduced leaf area and shedding leaves,
- use of green stems for photosynthesis,
- reversal of stomatal functions, i.e. C4 plants,
- tolerance of desiccation,
- high osmotic pressures,
- rapid germination and growth.

For example, some plants such as sagebrush or the creosote bush can tolerate significant loss of water and still survive (Varley, 1974), while Laurie (1988) found that vegetation in Oman achieved water potentials of an extraordinary −30 bars. Not all mechanisms are fully understood. For instance, at the ICRISAT centre, Niamey, it was found that tree growth (*Faidherbia albida*) was associated in some way with termite activity. In Oman, Prosopis trees exuded an oil which in contact with the calcareous sands acted to reduce infiltration rates around the base of the tree, possibly to impair plant growth although this also reduces water entry and possibly increases erosion. The same tree also appears to be able to absorb water through the leaves (reverse osmosis) as a sapling (Brown, 1988) but this has not been demonstrated in mature trees.

Evenari *et al.* (1986) have produced a comprehensive description of the flora found in arid lands on a region by region basis. It has proved difficult, however, to produce a distinctive classification of such vegetation due to the diversity of arid environments. Beaumont (1989) notes that hydrophytes, mesophytes and xerophytes can all be found in the arid realm. Nevertheless, McClearly (1968) attempted to distinguish between the flora of:

- **temperate arid areas** (Patagonia, Gobi, Thar, Mojave)
- **warm arid areas** (Sonara, Chihuahuan, Sahara, Arabia)
- **coastal deserts** (Atacama, Namib)
- **Australia.**

Within the classification, Australia is dominated by grasses and small shrubs while temperate arid lands are characterised by sagebrush (*Artemisia tridentata*), salt desert shrubs and trees (Prosopis and Acacia). Coastal deserts include quantities of woodland including tree cacti (Cereus sp.) found in the Atacama. The warm arid areas are perhaps the most difficult to characterise because of their greater diversity. McClearly suggests that the following families represent up to 40 per cent of the flora: Leguminosae, Graminae and Compositae. It should be noted, however, that the above classification does not extend into the trees and grasses of the semi-arid savanna regions.

Odum (1971) distinguishes between desert biotic communities receiving an annual rainfall of 250 mm or less with savanna (semi-arid) receiving 250–750 mm a year, emphasising the importance of moisture supply. A linear relationship between rainfall and dry matter production in arid areas is plotted by Odum with 100 mm of rainfall producing 1,000 kg of dry matter per hectare and

Mean Annual Rainfall mm	Region	Physical landscape	Natural vegetation	Land use (irrigation excepted)	
				Animal rearing	Cultivation
50	SAHARIEN		None except scattered plants in depressions	Large scale nomadism (~300km)	
100		Stoney sandy desert, dunes		Pastoralism with camels and sheep	
200			Widely scattered		
300			Open steppe	Nomadic transhumance (~150km) - Pastoralism with camels and cattle	
400	SAHELIEN		Wooded steppe		
500		Patches of dunes in eroded soils with depressions (bas-fonds)		Semi-sedentary pastoralism with camels and cattle	No diversification (Niébé, millet, short cycle ground-nuts)
600					
700			Wooded savannah	Sedentary pastoralism	Little diversification (Millet, sorghum, ground-nuts)
900	SOUDANIEN	Eroded slopes			Wide diversification Long-cycle millets and sorghum, groundnuts cotton
1100					Rice (in heavy soils)
1300	GUINEEN		Dense forest		
1500					

Figure 1.1 Rainfall and associated landscape components, after Davy *et al.* (1976).

600 mm producing 6,000 kg. Schmida *et al.* (1968) report that the relationship between rainfall and biomass is in fact curvilinear and further suggest that although water availability is the prime control over vegetation distribution, the variability of water supply is also of great significance. That is, arid environments are characterised by pulses of water with corresponding biological pulses leading to both spatial and temporal variation in vegetation. Schmida *et al.* then distinguish between two systems: one extensive and enduring where vegetation has adapted to arid conditions (through drought resistance, stomatal mechanisms, structural adaptations and rooting systems); and an episodic pattern that responds rapidly to changing moisture conditions (desert winter annuals). McGinnies (1968) also distinguished between perennial plants such as succulents or dwarf or woody vegetation and annuals or ephemerals which appear after rain.

Until the advent of satellite imagery the mapping of vegetation over large areas was an arduous and time-consuming task. It proved simpler to relate vegetation and landscape components to climate as shown in figure 1.1. Hence there were attempts to demarcate arid boundaries based upon a plant–rainfall classification. Dregne (1976) lists the following results from work by Shantz (1956):

- **semi-arid vegetation 7 million km^2** (scherophyll brushland, thorn forest and short grass), 5 per cent world land area
- **arid vegetation 33.4 million km^2** (desert grass savanna, desert grassland, desert shrub), 23 per cent world land area
- **hyper-arid areas 6.3 million km^2** (little or no vegetation), 4 per cent world land area.

Good (1970) argues that the distribution of plants is affected not only by climate, but also by edaphic factors such as soil and topography, and that temperature is perhaps more important than rainfall. While it could be argued that this observation is only true on a global scale and that within arid lands rainfall is dominant, McGinnies (1988) compared climatological and vegetation classifications of arid lands and, perhaps surprisingly, found little correlation concluding that more work should be done in this field. Plant communities are then affected by more than the prevailing climate yet this relationship is often assumed. The southern margin of the Sahara Desert in Sudan was mapped by this approach (using rainfall to map vegetation) in 1958. In 1975 the desert margin was again mapped using vegetation cover from aerial reconnaissance. A comparison of the results suggested that the desert had advanced southwards by 90 km in 17 years. However, plant communities are affected by localised drought and anthropogenic interference (overgrazing, fuel wood collection, etc.). The assumption that there is a simple relationship between vegetation and climate is no longer tenable and recent studies of the desert edge in the Sudan by the University of Lund, Sweden, have found little support for the notion of a southward migration of the desert edge.

The recent interest in desertification and availability of satellite imagery has renewed attempts to map arid boundaries based upon vegetation, but care should be exercised over this approach given the above comments and the difficulty of using remote sensing when vegetation cover is sparse. Commenting upon the mapping of arid areas using a definition of less than continuous vegetation cover, Good (1970, p. 18) cautions:

> It is not easy, however, to know exactly where the line is to be drawn, and this definition breaks down if applied too narrowly.

Climatic definitions

Wallen (1967) listed three approaches to climatic classification: classical, index

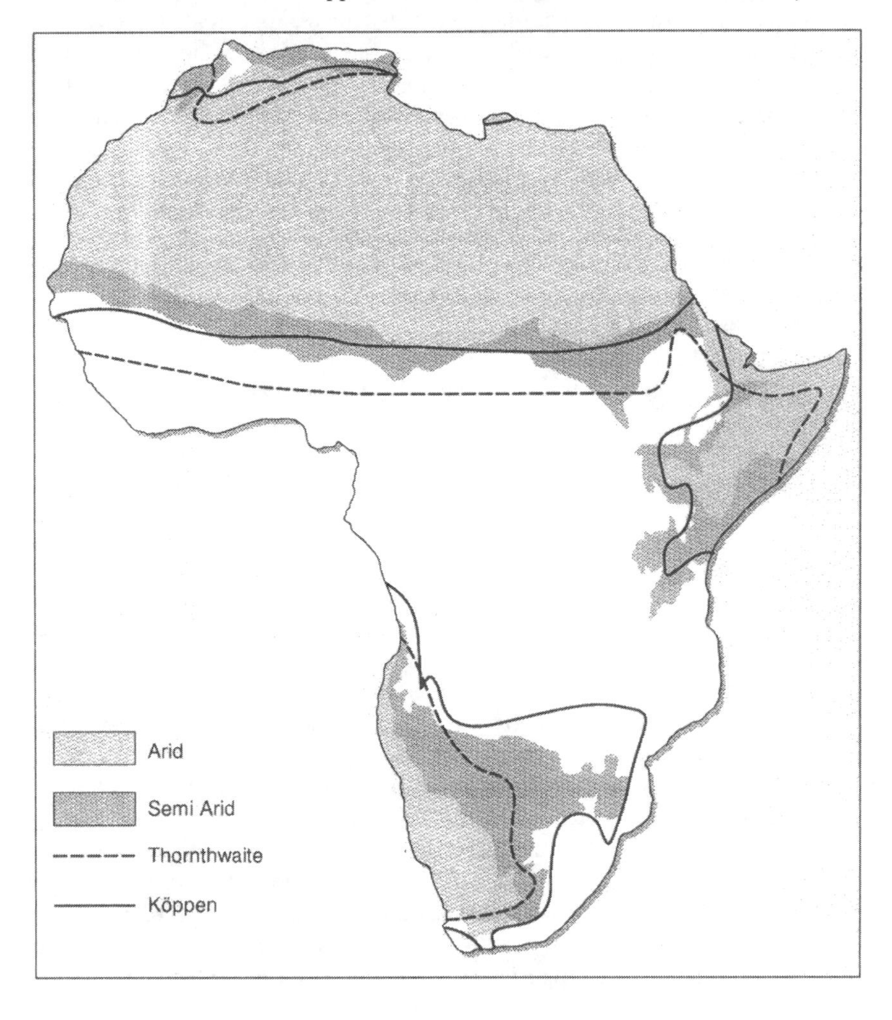

Figure 1.2 Climatological boundaries of arid lands in Africa based upon Köppen (1931), Thornthwaite, (1948, 1955) and UNESCO (1977b).

and water balance. **Classical** approaches involve the delineation of arid boundaries through the threshold relationship between, say, rainfall and plant growth, e.g. Le Houreou and Popov (1981) used 100 mm as the boundary for deserts and 600 mm for semi-arid regions; while UNESCO (1977a) employed 600 mm to delimit arid lands with a summer rainfall regime and 400 mm for winter rainfall regimes. This type of approach has been used to ill effect when determining the expansion of deserts due to poor data and questionable assumptions (Oliver, 1981, Warren and Agnew, 1988). **Climatic indexes** such as those employed by Köppen (1931) relate temperature and rainfall to vegetation with graduations from arid through semi-arid to humid climates, with some account of seasonal variations. The more 'rational' **water balance** approach employed by Thornthwaite (1948) considered the moisture available to plants by an index of aridity determined by a water balance calculation of rainfall and potential evaporation.

Although the Köppen classification has low data demands and is useful educationally, Mather (1974) criticises this approach for failing to take adequate account of water supply. Furthermore, because temperature and rainfall are only generally related to major vegetation changes, Mather suggests that any correct results obtained are fortuitous. Thompson (1975) disagrees, and notes that arid lands are reasonably accurately delineated. Wallen (1967) found Köppen's classification could only be used to establish boundaries over large areas and is not applicable for the comparison of one place with another, while Thornthwaite's calculations tended to overestimate moisture supplies. Given the lack of a satisfactory map of arid areas, UNESCO requested Meigs (1953) to develop Thornthwaite's method in order to assess the potential for global food supplies. Cold areas were excluded and a subdivision of 'extremely arid' was added. This map was subsequently reinterpreted by J. Mallet and R. Ghirardi (UNESCO, 1977b) using Penman's (1948) formula for evaporation in the light of improved data and understanding and this has become the accepted 'standard' for the distribution of the world's arid lands (figure I.2).

Figure 1.2 compares boundaries for arid lands based upon the Köppen (1931), Thornthwaite (1948) and UNESCO (1977b) approaches where:

Köppen (1931)
Arid boundary $= P/T < 1$
Semi-arid boundary $= 1 < P/T < 2$
P = mean annual precipitation (cm)
T = mean annual temperature (°C)
(after Yair and Berkowicz, 1989).

Thornthwaite (1948, 1954)
Arid boundary $I_m = < -66.7$
Semi-arid boundary $-33.3 > I_m > -66.7$
$I_m = 100[(P/P_e) - 1]$
P = mean annual precipitation (mm)

P_e = mean annual potential evapotranspiration (mm)
(after Mather, 1974).

UNESCO (1977b)

Arid and hyper-arid boundary P/ET_p = less than 0.20
Semi-arid boundary P/ET_p = 0.50 to 0.20
P = mean annual precipitation (mm)
ET_p = mean annual potential evapotranspiration
(after UNESCO, 1977b).

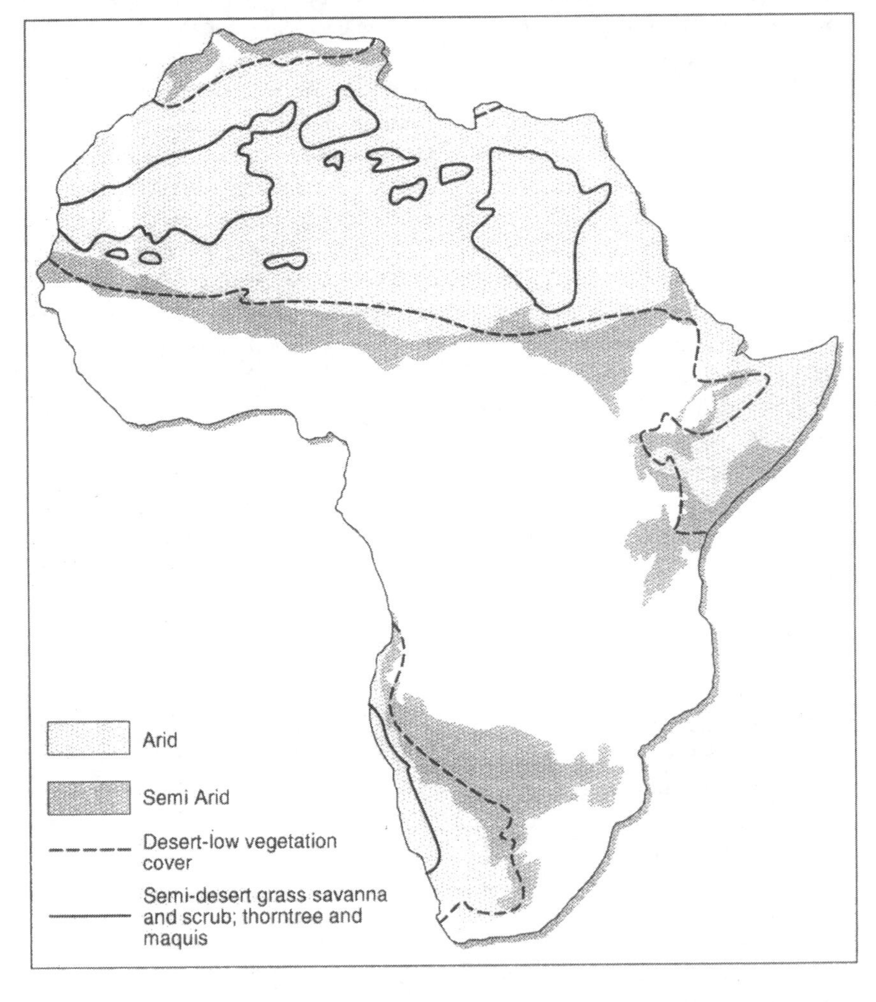

Figure 1.3 Boundaries of arid lands in Africa based upon UNESCO (1977b) and vegetation distributions, after Griffiths (1972).

In figure 1.2, while there is broad agreement on the general pattern of arid lands, there are noticeable differences in their exact location. The UNESCO map is based upon the atmospheric supply of water to vegetation and there is again a general correspondence between climatic classification and vegetation mapping of arid lands in Africa evident in figure 1.3 (vegetation details from Griffiths, 1972). There is some confusion, however, over which categories of vegetation should be included, e.g. UNESCO (1977b) typify maquis and chaparral as subhumid climates while they have been mapped in figure 1.3 as being found in semi-arid environments.

Table 1.1 (from Heathcote, 1983) reveals that estimates of the world's arid areas based upon climatic classifications vary from 26 to 36 per cent with the greatest differences occurring in the location of hyper-arid areas. A substantially larger figure of 43 per cent was suggested by Nir (1974) and McGinnies (1988) using shortage of moisture as the criterion (within the root zone for the latter). Frankenberg (1986) employed a climatological aridity line defined as the difference between rainfall and potential evaporation to distinguish between humid and arid areas and also found that nearly half of the world's land surface could be classified as arid. Yair and Berkowicz (1989) have suggested it is time that aridity is redefined to take account of the sensitivity of an area to low rainfalls by consideration of the intensity and duration of rainfalls, soil salinities and the ratio of soil cover to bare rock.

There is then quite a range to the areas defined climatically as being arid and Gilbert White (1960) quoted by McGinnies (1988) concluded:

> There are no clear boundaries of arid lands; there are as many lines on maps as there are measures of aridity in climate, plants, landforms and

The purpose of such classifications must then be to identify the features of arid lands in order that their global significance may be recognised and any major changes taking place identified. Thus development decisions may be taken with some consideration of resources available and environmental impacts. However, there is also a grave danger with such an approach in that arid lands are treated as a homogeneous entity with similar environments and similar problems having similar causes.

It would appear that roughly a third of the Earth's surface is covered by the arid realm with over three-quarters found in the continents of Africa (37 per cent), Asia (33 per cent) and Australia (14 per cent) (Grainger, 1990), but clearly it is dangerous to generalise too much over the physical features found in this region. The only characteristic over which there is a degree of unity is the lack of moisture for at least part of the year. In other words the only common physical feature of arid lands is that they are arid. The delineation used in this text for arid areas conforms to the UNESCO (1977b) climatic map shown in figure I.2 which excludes the cold, polar regions. We now turn to the climatic processes involved in the creation of the world's arid regions before discussing their water resource problems in chapter 2.

THE CLIMATE OF ARID AREAS

Causes of aridity

Most explanations of aridity concern processes acting at global or regional scales. Thompson (1975) lists four main processes:

1 **High pressure** Air that is heated at the equator rises, moves polewards and descends at the tropical latitudes around 20° to 30° latitude. This descending air is compressed and warmed, thus leading to dry and stable atmospheric conditions covering large areas such as the Sahara Desert (see figures 1.4 and 1.5).

2 **Wind direction** Winds blowing over continental interiors have a reduced opportunity to absorb moisture and will be fairly stable with lower humidities. These typically dry, north-easterly winds (in the northern hemisphere) are seasonally constant and contribute to the aridity of South West Asia and the Middle East.

3 **Topography** When air is forced aloft by a mountain range (figure 1.6) it will cool adiabatically (A to B) at the saturated adiabatic rate once the dew point is reached (B to C) with possible precipitation. On the leeward side of the mountain the same air descends (C to D) warming at the dry adiabatic rate and hence the descending air is warmer at corresponding altitudes compared to the ascending air. Hence a warmer, drier wind blows over the lands to the leeward side, providing the ascent is sufficient to reach the dew point temperature.

4 **Cold ocean currents** Onshore winds blowing across a cold ocean current close to the shore will be rapidly cooled in the lower layers (up to 500 m). Mist and fog may result as found along the coasts of Oman, Peru and Namibia, but the warm air aloft creates an inversion preventing the ascent of air and hence there is little or no precipitation. As this air moves inland it is warmed and hence its humidity reduces.

More than one of these processes can lead to aridity at a particular location but Hills (1966) believes that the major cause of aridity is explained through global atmospheric circulation patterns, although topography and cold ocean currents have local significance. Logan (1968) uses the above four processes to distinguish between subtropical, continental interior, rainshadow, and cool coastal arid lands as follows:

Subtropical areas (e.g. Sahara, Arabia, Sonora, Australia and Kalahari) are characterised by anticyclonic weather producing clear skies with high ground temperatures and a marked nocturnal cooling. Seasonal contrasts are evident but rarely will winter temperatures reduce to freezing and the climate is one of hot summers and mild winters. Despite high temperatures, convectional rainfalls will only develop when moist air invades the region.

Continental interior areas (e.g. arid areas of Asia and Western USA)

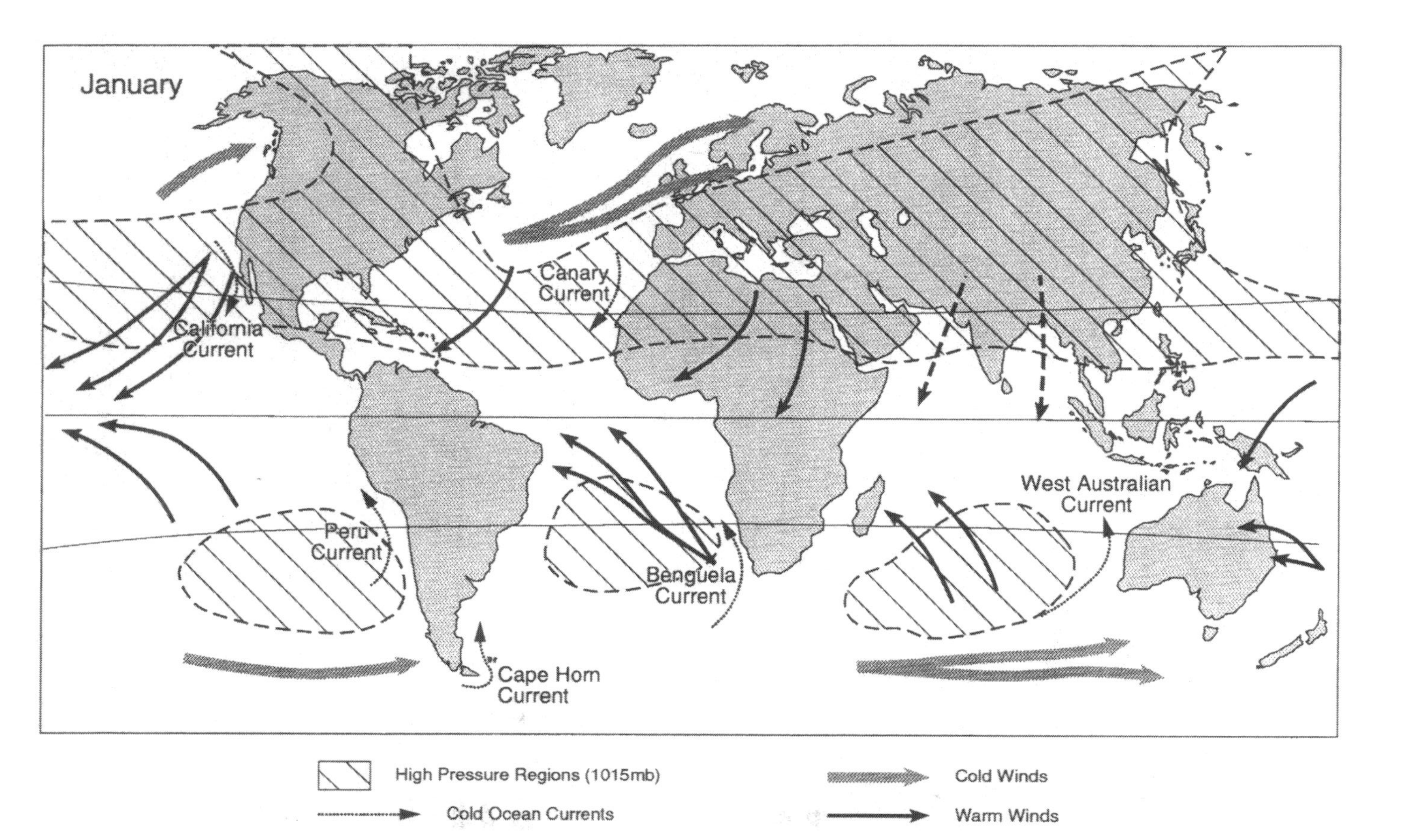

Figure 1.4 General atmospheric circulation during January.

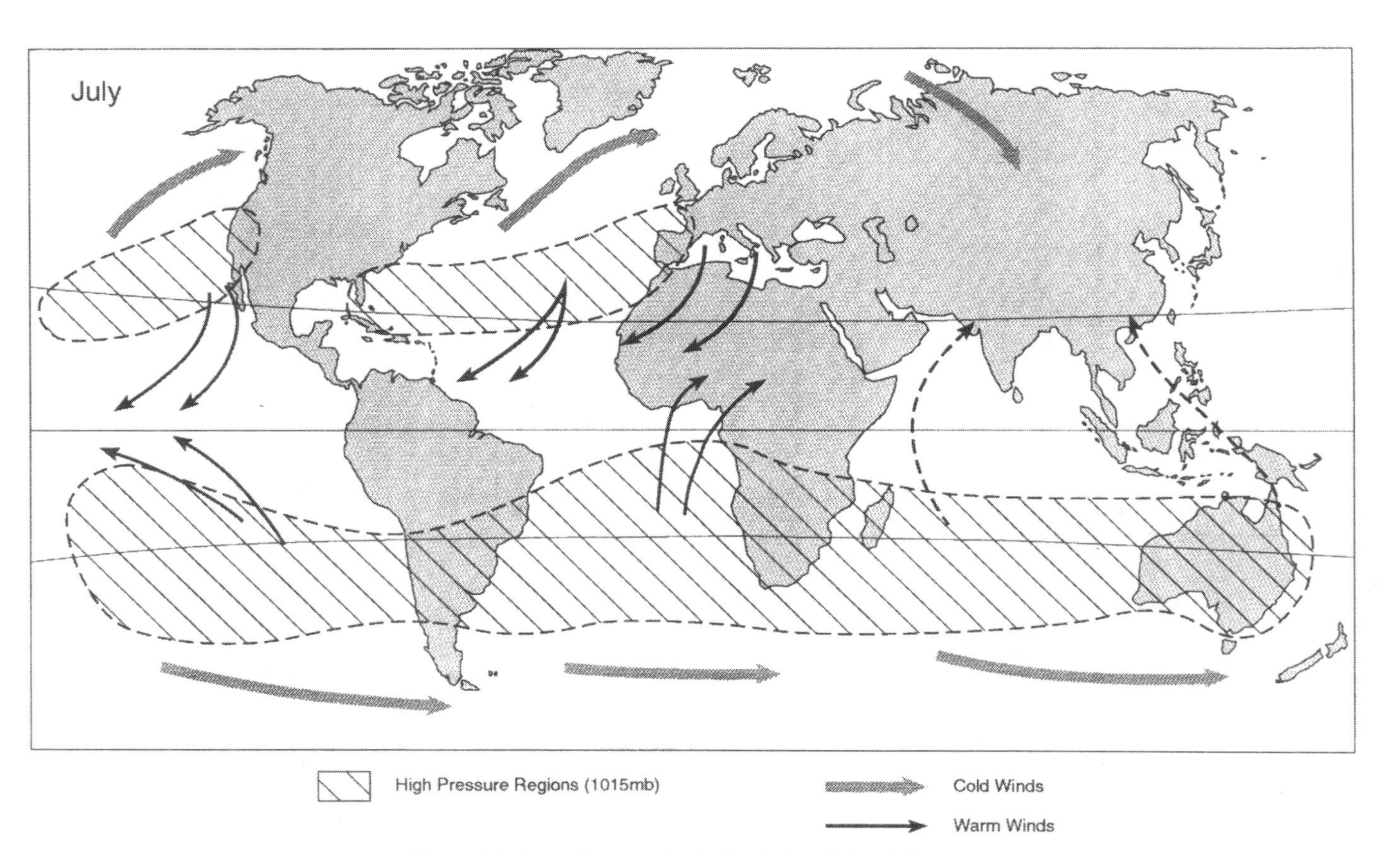

Figure 1.5 General atmospheric circulation during July.

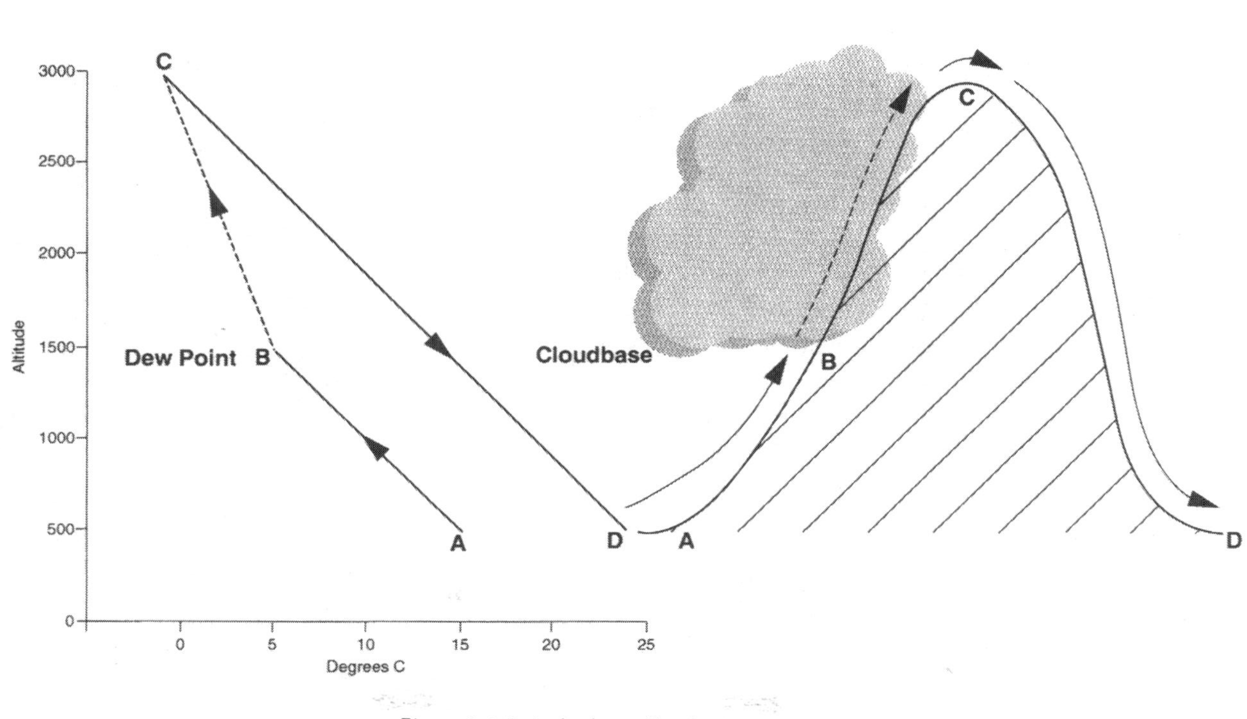

Figure 1.6 Rainshadow effect leading to aridity.

demonstrate large seasonal temperature ranges from very cold winters to very hot summers. Although snow can fall its effectiveness may be reduced by ablation as it lies on the ground through winter. Rainfall in the summer months is unreliable and can occur as violent downpours.

Rainshadow areas (leeward sides of mountain ranges such as the Sierra Nevada, the Great Dividing Range in Australia and the Andes in South America) occur where conditions are diverse given the range of latitudes in which these areas are found. The extreme conditions of the continental interior arid lands are unlikely to be found, however.

Cool coastal areas (e.g. Namib, Atacama, Pacific coast of Mexico) occur when conditions are reasonably constant with a cool humid environment. Thunderstorms can develop when the temperature inversion is weakened by moist air aloft.

Subtropical arid climates are characterised by a large diurnal range. High-pressure systems with an inversion aloft and stable air beneath produce clear skies which will enhance daytime solar radiation receipts and consequently nocturnal radiation cooling. Temperatures, humidities and possibly windspeeds can demonstrate significant differences over 24 hour periods as shown by figures 1.7 and 1.8. Note the differences between conditions at ground level and those at a height of 1 metre. The greatest contrast occurs around midday due to surface heating, but then conditions gradually become homogeneous in the lower layer of the atmosphere following nocturnal cooling. This is not always the case; for

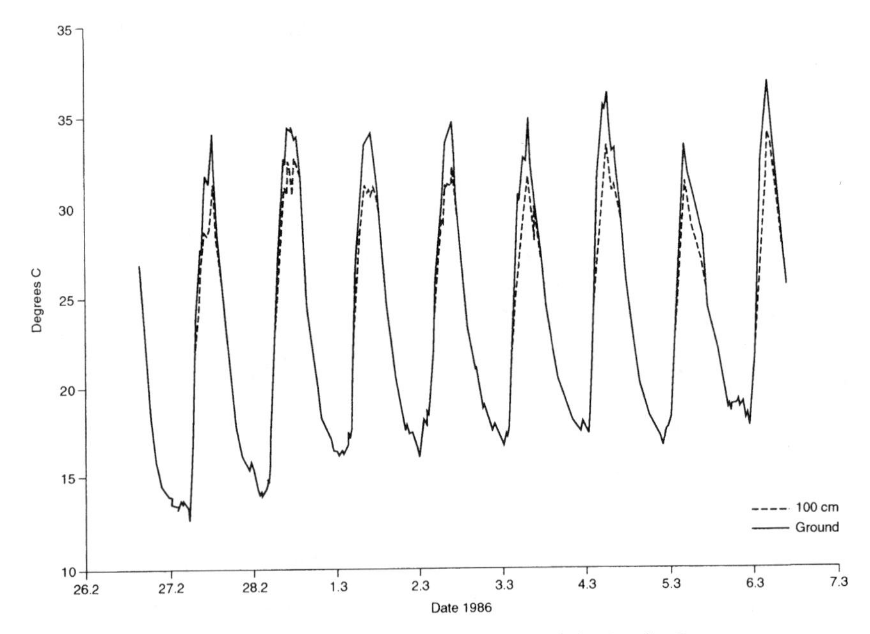

Figure 1.7 Diurnal temperature changes, Wahiba Sands, Oman.

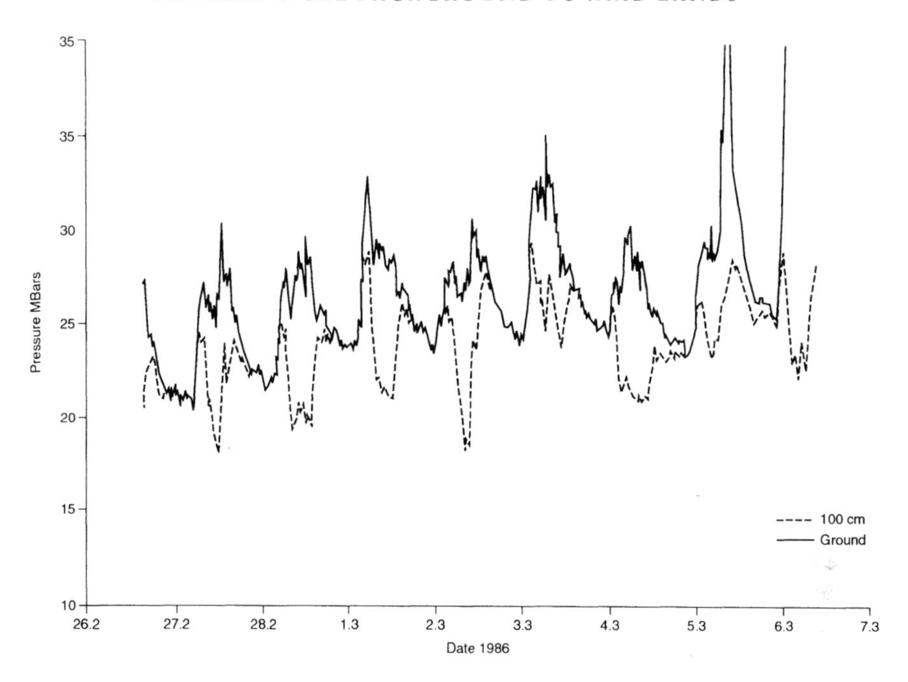

Figure 1.8 Diurnal changes in atmospheric moisture contents, Wahiba Sands, Oman.

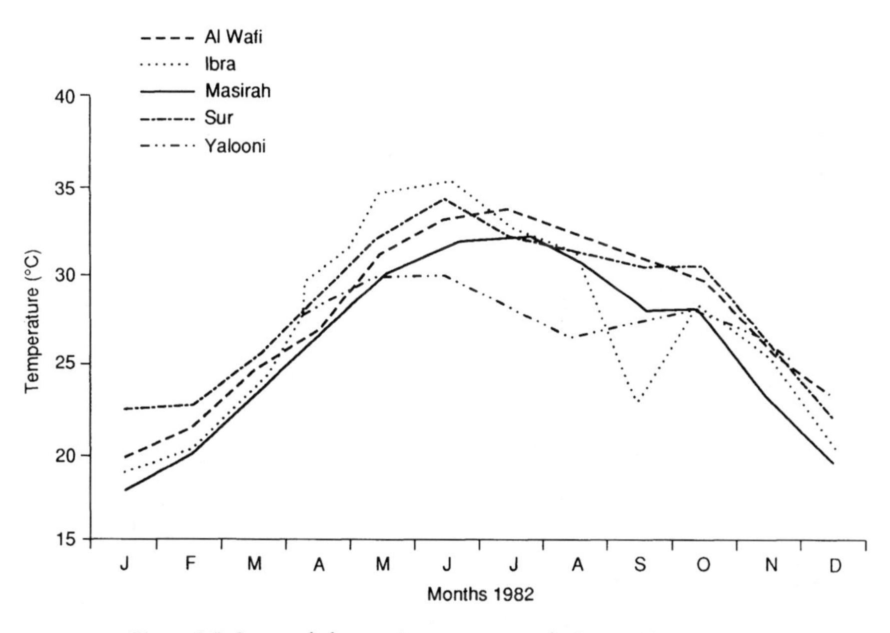

Figure 1.9 Seasonal changes in temperature during 1982, Oman.

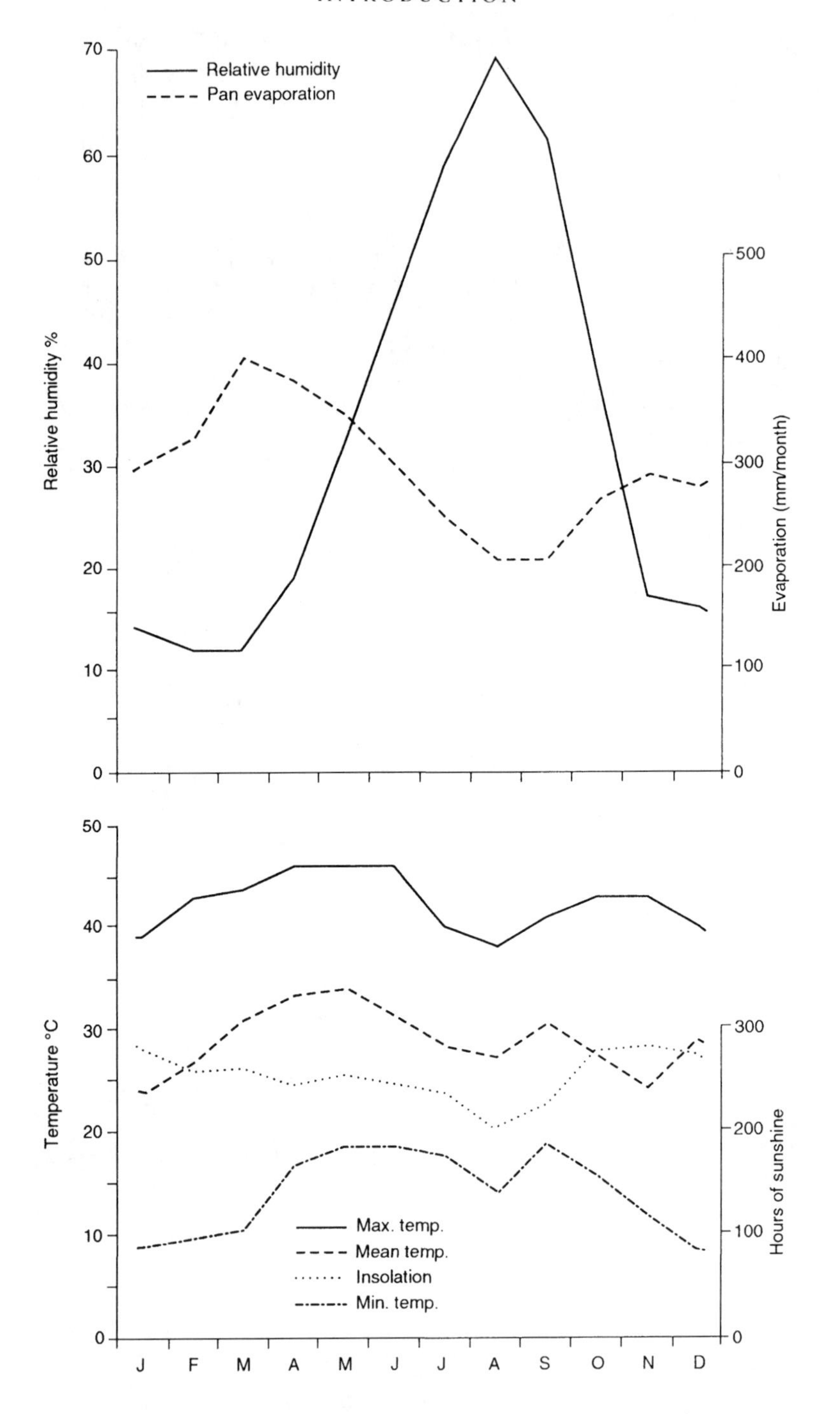

Figure 1.10 Average seasonal climatic conditions, Niamey, Niger.

example, with heavy cloud cover there may be little change.

As air masses compete for dominance with the passage of the Sun overhead so seasonal changes are also evident. Figures 1.9 and 1.10 show a simple seasonal pattern for Oman and the Sahel although, in the case of the latter, cloud cover and precipitation during the summer months reduce energy losses and increase humidities while the seasonal mean temperature range is greater in Oman. Both examples display quite high temperatures even during winter months, yet in some arid areas frost and even snow can be expected. Clearly arid lands are not hot and dry all the time but most lie in the tropics and can therefore receive substantial inputs of solar radiation. Nir (1974) for instance commented that Amman in Jordan received four times the amount of solar radiation compared to Manchester in England and that many parts of the Middle East receive near to their maximum possible.

This energy can be expended on heating the environment and on wind generation, go into storage or be utilised in evaporation. The actual energy balance depends upon surface characteristics and the moisture available but with such high energy levels and often low humidities the potential evaporation can be quite high. For example, values of up to 8 mm a day were found in Northern Tunisia; 14 mm per day were found at Niamey in the Sahel; with similar values for the Wahiba Sands area of Oman (Agnew, 1980 and 1986). However, the actual amount evaporated will depend upon moisture supply and a mean value of only 0.182 mm per day was obtained for bare sand in Oman (Agnew, 1988a), although Greb (1983) reports that 50 to 70 per cent of precipitation is lost through evaporation in the Central Great Plains of North America. (Problems concerning the estimation of evaporation in arid lands are discussed in chapter 5.)

The image of arid lands as a dune-covered, windswept, parched earth, devoid of vegetation, with the occasional sandstorm has some merit but is clearly an overgeneralisation. Both Cooke and Warren (1973) and Heathcote (1983) comment that windspeeds, for instance, are no greater in the arid realm and may be less than in other areas (although their geomorphological impact may be more significant). The shortage of atmospheric moisture at some times is then the only common feature in regions that can register nearly all extremes of climate.

Arid lands can alternatively be subdivided by their rainfall and thermal regimes. For example, McGinnies (1979) lists:

- two rainy seasons (e.g. Venezuela)
- winter rains (e.g. North African coast)
- summer rains (e.g. Sahel)
- seldom having rain (e.g. Sahara Desert)
- fog and mist (e.g. Namib).

Or based upon temperature:

- tropics; little change in monthly temperatures (e.g. Venezuela, Somalia)

- subtropics; considerable temperature changes (e.g. Thar and Australian Deserts)
- temperate; characterised by cold winters (e.g. the arid lands of Iran, Syria, Gobi and North American Plains)
- cold highlands (e.g. Andes Mountains, Tibet).

Thomas (1989) reports the following based on Meigs (1952):

- hot arid lands cover 43 per cent (Central Sahara, Australia)
- mild winters 18 per cent (Kalahari, Southern Sahara)
- cool winters 15 per cent (Atacama, Mojave)
- cold winters 24 per cent (Gobi, Turkistan).

Precipitation

Rainfall occurs when moist air is cooled sufficiently through ascent, mixing, radiation cooling or contact cooling. The processes leading to aridity tend to prevent such cooling through maintaining air stability, creation of inversions or through the warming of the atmosphere, thus lowering humidities. In any arid area several of these processes may be operating but their influence depends upon atmospheric conditions. When rainfall does occur it can be a dramatic, intense and localised downpour as moist air 'breaks through'. This point is perhaps best demonstrated by referring to specific examples.

The Sahel region of West Africa is dominated at the surface from November until April by a high-pressure cell, an offshore cold current and dry winds from the north-east. From March to May, (Griffiths, 1972) moist air gradually penetrates from the south as the Inter Tropical Convergence Zone (ITCZ) migrates northwards following the Sun's path with a lag of some 4 to 6 weeks (Cocheme and Franquin, 1967, Farmer, 1989 and Sumner, 1988). Trewartha, (1961) identified a zone immediately south of the ITCZ where rainfall is suppressed by an inversion caused by hot dry air from the north rising over the denser moister air from the south (see figure 1.11). As the height of this inversion increases southwards, so the opportunity for rain improves through convection (Kidson, 1977). Garnier (1967) identifies local thunderstorms providing variable and sporadic rainfall in addition to disturbance lines (a line of thunderstorms travelling east–west). These disturbance lines (Rasmusson, 1987) occur when the stability of the overlying easterlies is disturbed by moving waveforms leading to roughly north–south belts of convergence in their lower layers (Gregory, 1965).

These surface conditions are determined by upper air flows, in particular the position and intensity of the subtropical jet streams (Kidson, 1977 and Hastenrath, 1985). During winter (dry season) months when anticyclonic conditions prevail over the Sahel, the jet stream becomes convergent towards the equator (Hayward and Oguntoyinbo, 1987) producing a downward shift of air to feed high pressure at the surface. In summer months divergence aloft results in convergence of the south-westerly and north-easterly winds at the surface

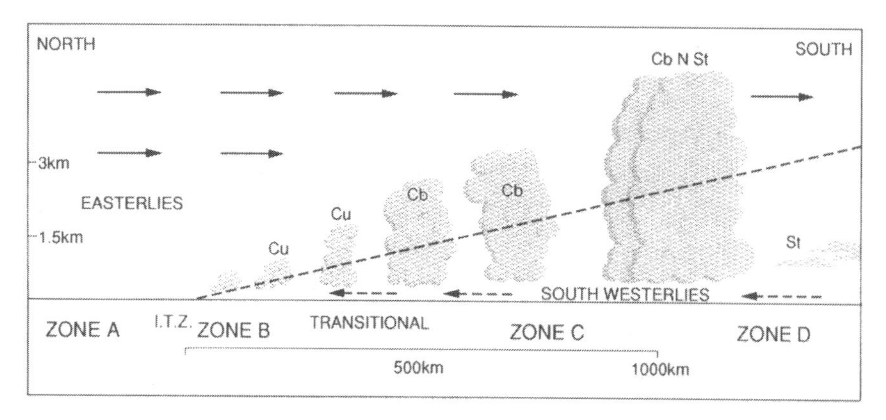

Figure 1.11 Cross section of atmosphere during the rainy season in West Africa, after Trewartha (1981).

drawing in moist air with increased opportunities for precipitation. Low rainfalls correspond with a weak easterly jet associated with weaker circulation in the middle and upper troposphere. Changes in the jet stream then have consequent impacts upon precipitation resulting in variable amounts. There have been several attempts to explain changes in Sahelian rainfalls by general circulation patterns but Kidson (1977) comments that these explanations have not proved very convincing, although this is the most likely cause. Figure 1.12 shows that the resulting precipitation varies from season to season with a varying onset and quite different seasonal distributions, although the most significant changes in rainfall amounts appear to take place during the middle of the rainy season, i.e. around August.

Like the Sahara, arid conditions in the Sultanate of Oman in the Middle East are due to a combination of factors including the direction of prevailing winds and a cold offshore ocean current. During winter (Shamal) months, gentle easterlies blow in from the Indian Ocean coupled with depressions tracking down the Gulf bringing rain to the northern coast which is often prevented from passing into the dry interior by a mountain range. The region's climate is dominated by the south-westerly monsoon (Kharif) shown in figure 1.13. During the summer months the Arabian high-pressure cell migrates northwards drawing in air from the southern hemisphere which is deflected eastwards by the Coriolis force after crossing the equator. These moist, often strong south-westerly winds only produce reliable precipitation in the south of the country as they are stabilised by the cold offshore current to the north. Humidities do increase, however, with significant amounts of mist and dew possible. Both Oman and the Sahel therefore undergo seasonal changes when moist air penetrates into the area but from which rain will only fall providing other meteorological conditions are favourable. Precipitation is therefore difficult to predict and somewhat unreliable (see the list section of chapter 2).

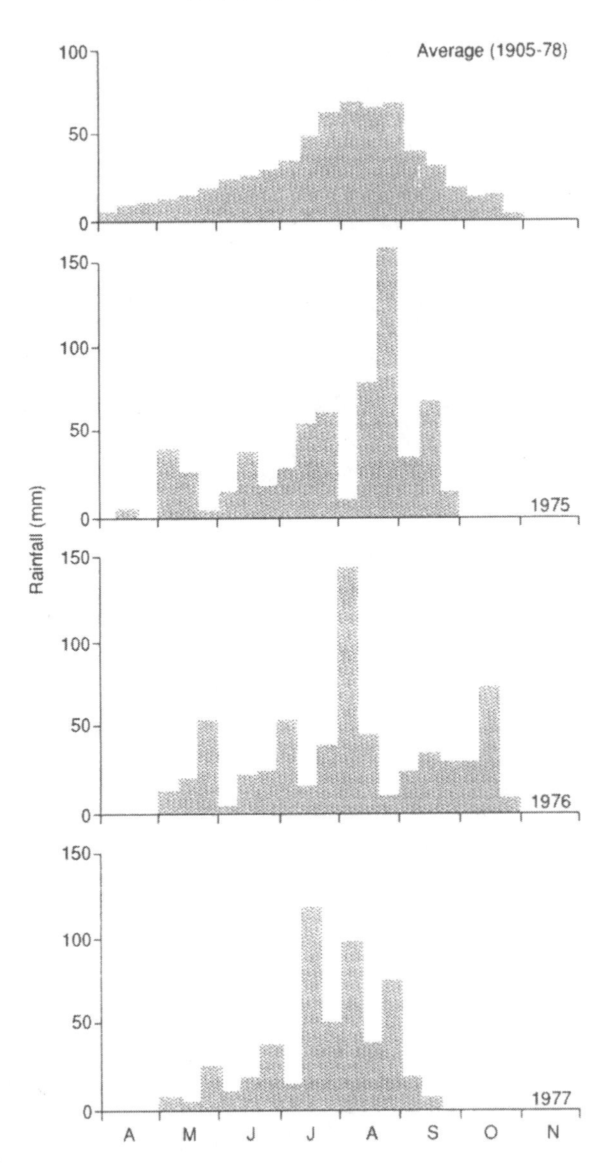

Figure 1.12 Seasonal rainfalls at Niamey, Niger.

It is estimated that 68 per cent of Australia receives less than 500 mm each year while 37 per cent receives less than 250 mm (Linacre and Hobbs, 1977) leaving a vast, dry interior as shown by figure 1.16. Aridity in Australia is mainly due to air subsidence over the continental interior, the distance from the ocean and topographical barriers. The arid conditions reach the north-west coast of Western Australia but are not as extensive along this western coast as may be

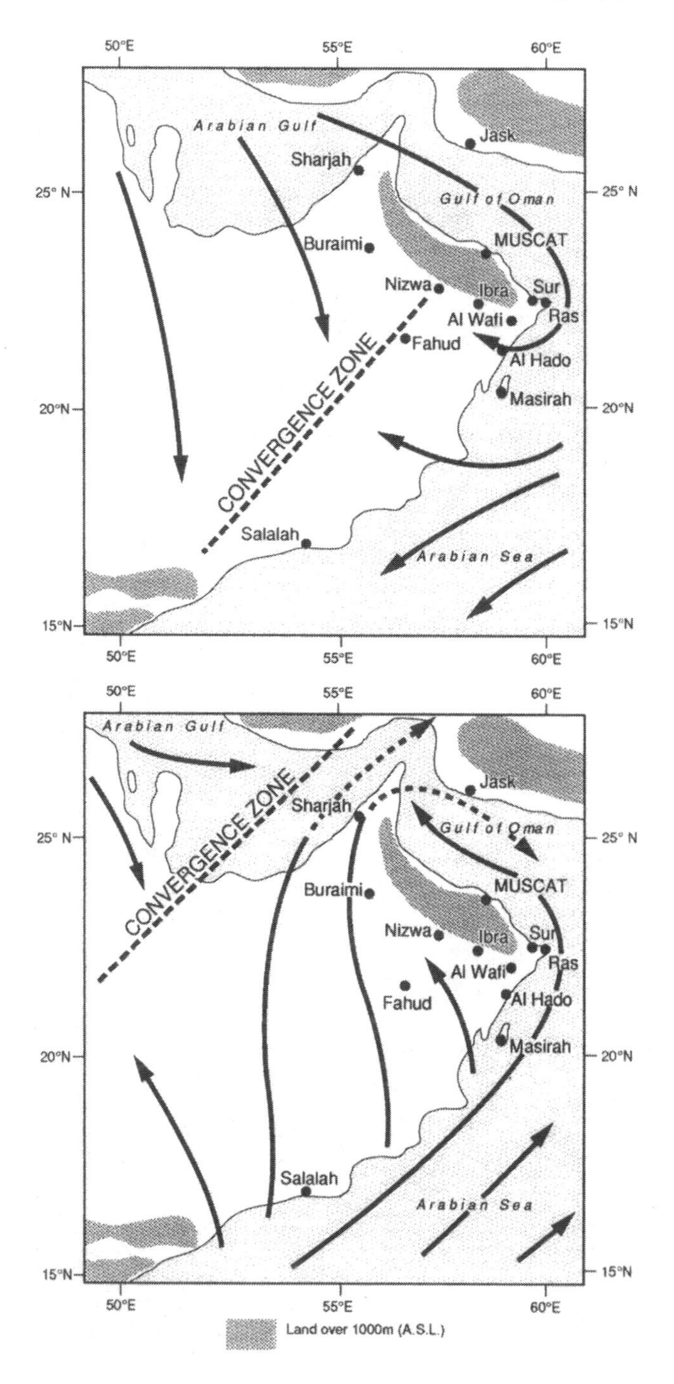

Figure 1.13 General circulation over Sultanate of Oman, after Brunsden and Cooke (1986).

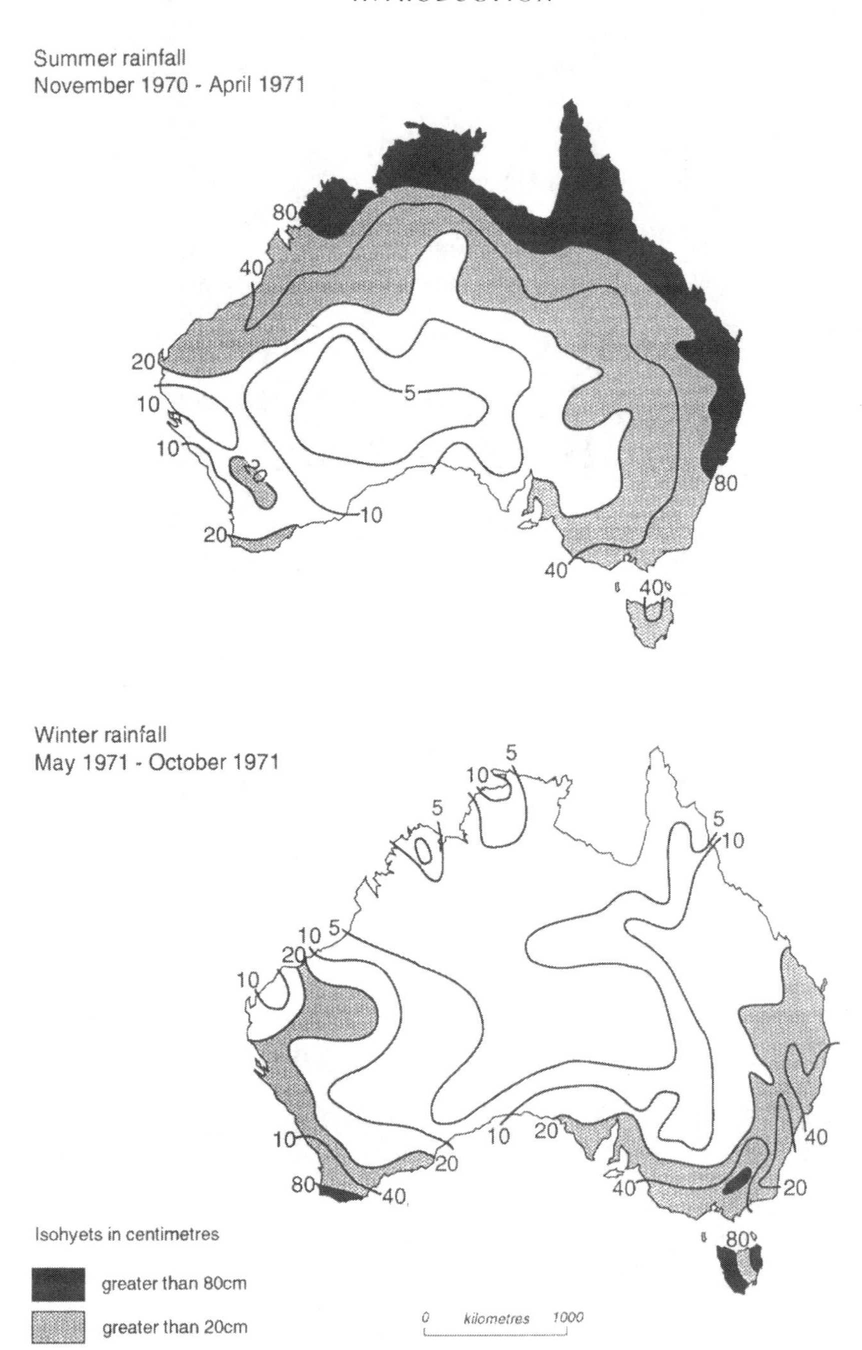

Figure 1.14 Distribution of rainfalls in Australia, after Linacre and Hobbs (1977).

expected in comparison with other desert areas. This is due to the effect of the cold offshore ocean current not being as pronounced because of the influence of monsoonal circulations reducing the oceanic temperature differences. Linacre and Hobbs note that the main form of precipitation is convectional, caused by warm, moist air incursions into the interior being heated by the Sun and producing locally unstable conditions. This gives rise to local, intense storms which are difficult to predict as the anticyclones are well established over the continent and restrict penetrations of moist air associated with the southward migration of the ITCZ during summer months. Tropical cyclones are also prevalent at this time but are deflected by the Great Barrier Range to the east while they can penetrate the continental interior via the north-west coast, bringing occasionally high rainfalls. Thus the rainfall in the arid interior of Australia is variable and infrequent as the various air masses battle for supremacy with the march of the sun. This is, however, a very generalised description and Trewartha (1981, p. 88) notes that work is still progressing towards a better understanding of the association between synoptic conditions and rainfalls.

AGRICULTURE IN ARID LANDS

It follows from the diversity of physical environments in arid lands that patterns of human occupation are equally if not more diverse. Beaumont (1989) has reviewed the evolution of occupation of drylands by humans and the development of settlements and permanent cultivation. The most enduring patterns of habitation are to be found in agricultural systems that have adapted to the variability of water supplies. The focus below will be upon these food production systems which involve the bulk of the population; ignoring commercial systems, in particular irrigation and ranching. Similarly industrialisation and mining have been omitted in this discussion on life in arid lands. The emphasis here is to demonstrate the dynamic array of strategies developed by arid land inhabitants to combat water resource shortages and other environmental constraints, while texts such as Beaumont (1989), Heathcote (1983) and Hills (1966) already provide a wide perspective of technological and commercial enterprises.

As with vegetation, agricultural activities can be characterised as:

- **endurers** (pastoralists who inhabit the most arid parts)
- **avoiders** (dryland farmers who are only active in wet seasons)
- **escapers** (irrigation farmers who create an ‘artificial’ environment).

Traditional irrigation systems will be discussed in chapter 6, but table 1.2 provides some idea of the relationship between rainfall and other agricultural activities.

Table 1.2 Agricultural landuse in arid lands (based on Ruthenberg, 1976, p. 298)

Rainfall	*Production system*	*Livestock*
under 50 mm	Pastoralism using pastures only infrequently	Camel
50–200 mm	Pastoralism with long migrations	Camel
200–600 mm	Pastoralism supplemented by dryland farming and sheep	Cattle, goats
over 600 mm	Dryland farming with some herding	Cattle, sheep

Dryland farmers

Cannell and Dregne (1983) suggest that dryland farming takes place in areas of scanty precipitation but that attempts to delineate such areas on the basis of annual rainfalls will be misleading due to the variability of the effectiveness of rainfall. Nevertheless Willis (1983) notes that at least 180–250 mm of soil water are required to produce grain for wheat and similar crops.

The dryland farmers have to tackle problems ranging from poor soils to attacks by pests and disease but perhaps the most intractable problem of all is the variable rainfall. Soil fertility is maintained by growing leguminous crops, manuring fields with animal and human wastes and crop rotations (UNESCO, 1978). The use of fallow land to replenish soil fertility, control weeds and maintain soil moisture is perhaps one of the key practices of arid land farmers. For instance, Gibbon and Harvey (1977) working in Western Sudan note that on the poorer soils land is cropped for 3 to 4 years with millet and then left fallow for up to 20 years although groundnuts may be used to reduce the fallow period down to 3 years. In the Great Plains of North America, the fallow period is much reduced, and Willis (1983) describes an 11 month winter wheat fallow lasting from September to August, and a 19 month spring wheat fallow lasting from September to March a year and a half later. The fallow technique has been changing over this century and Greb (1983) lists the following in North America:

- Maximum tillage (1910s, 1920): ploughing and shallow harrowing kills weeds and reduces evaporation losses but also destroys the stubble and leaves the soil prone to wind erosion. Water efficiencies (percentage of soil water stored compared to precipitation during the 14 months fallow) below 20 per cent.
- Conventional bare fallow (1930s, 1940s): commences in spring with reduced soil disruption through shallow land preparation with water efficiencies 20 to 25 per cent.
- Stubble mulch (1950s): the mulch absorbs raindrop impact and lowers surface heating thus reducing erosion and evaporation with water efficiencies of around 30 per cent.
- Minimum tillage (1960s, 1970s): using herbicides instead of mechanical means; capable of water efficiencies above 30 per cent.

Fallow land can, however, have negative effects and these have been noted in the Canadian Prairies where it can lead to accelerated loss of organic carbon possibly affecting soil structure, erosion may be enhanced and increased drainage can produce saline seeps (de Jong and Steppuhn, 1983). Despite being an important dryland farming practice, fallow is not universally adopted and Unger (1983) reports that in the Southern Great Plains of the United States, wheat was planted after fallow on only 4–16 per cent of land in the east and 29–41 per cent in the west. The distribution was affected by land set-aside schemes and the prevailing economic conditions.

In West Africa, seeds are not sown until there is sufficient soil moisture to allow germination and 10–15 days growth (INRAN, 1977, Irvine, 1969 and Valet, 1978). If the initial rains are followed by a dry period it may be necessary to resow several times with shorter growing season varieties. This has the additional benefit of producing a field with variable harvest dates which reduces the risk of losing the whole crop through pest or disease attack. However, this is a very labour-intensive strategy which places limitations on the amount that can be produced (Kowal and Kassam, 1978). Gibbon and Harvey (1977) found that during the rainy season the control of weeds required 40 per cent of all time spent. Finally the harvested crop is placed into a granary designed to keep pests out; invariably the seed is left on the ear as this ensures better preservation and it may be covered with ashes or other substances. A variety of arid land crops are grown (Anochili, 1978, Arnon, 1972 and Purseglove, 1972) but millet and sorghum are the most common in semi-arid environments of Africa and millet is considered one of the most nutritious of cereal grains. These plants have evolved a number of mechanisms to combat aridity, (Dancette, 1980, ICRISAT, 1984 and Maiti and Bidinger, 1981) including the apparent ability to remain dormant during periods of drought. The food production system is further diversified with, for example, over 50 per cent of households keeping livestock within the Niamey Département, Niger: 29 per cent have poultry, 27 per cent have cattle, 22 per cent have sheep and goats, 11 per cent have horses, 11 per cent have asses and 4 per cent have camels. Thus by using crops adapted to arid conditions, and a number of techniques to reduce environmental degradation, farmers are able to minimise risks and ensure a harvest as often as possible.

Pastoralism

Pastoralists inhabit a more hostile and drier environment and in order to understand better their various strategies we shall focus upon a particular group, the Tuareg (see Keenan, 1977), who inhabit the country of Niger in the middle of the Sahel region. Niger lies between the grassy savanna of Nigeria and the Sahara Desert with almost half the country receiving rainfall of less than 100 mm. Although the country is roughly five times the size of the UK its population is only a tenth, i.e. 5 million, reflecting its arid climate (Stonehouse, 1985). To the north the Tuareg pastoralists keep camels and goats, to the south Hausa farmers cultivate millet and sorghum, while in between the Fulani herd cattle and sheep with some farming.

Table 1.3 Herd sizes for selected pastoral groups (based on Monod, 1975)

Group	*Camels*	*Cattle*	*Goats and sheep*
Ahaggar Mountains	10	–	15
Ahaggar Mountains	–	–	40 to 50
Western Niger	–	40	75
Western Niger	–	25	180
Tahoua region	10	40	200
Mali	1	100	100

Pastoralists herd animals in the most inhospitable areas of the world where cultivation of crops is not feasible. They may be nomadic by not having a permanent place of residence, e.g. Tuareg, or semi-nomadic like the Fulani who may settle around water points and even harvest crops depending upon conditions. Animals are important because they provide a number of products and services:

- Sustenance – milk, meat and blood.
- Shelter – wool and skins.
- Fuel – dung.
- Transport – camels and donkeys.
- Exchange – for crops, salt or tools.
- Status – bridewealth.

Thus animals enable the nomads such as the Tuareg to inhabit the arid lands where camel milk may be the only means to avoid dehydration for periods of time. Pastoralists have been called 'economically perverse' because they tend to release animals when demand is low (during 'good' years) while retaining them when prices are high (Baker, 1974). One of the keys to understanding the pastoral economy is that the animals are the resource not the environment. Just as a farmer will try to acquire more land in order to increase production and to be more secure, so a nomad will try to maximise herd size. How then is this achieved? Table 1.3 shows that a number of permutations of different herd sizes can be found.

It appears that herds are larger to the wetter south where cattle numbers are higher while camels are more common to the north. Goats and sheep can be found in most pastoral herds throughout the region. The reasons for this diversity lie in the means by which the pastoralist combats aridity. There are six basic strategies by which the herders may reduce risks, as follows.

Movement

The essence of nomadic pastoralism is 'environmental opportunism' in that the animals and herders travel together in search of good grazing following rainfall. However, these movements are not aimless nor totally random but as figure 1.15

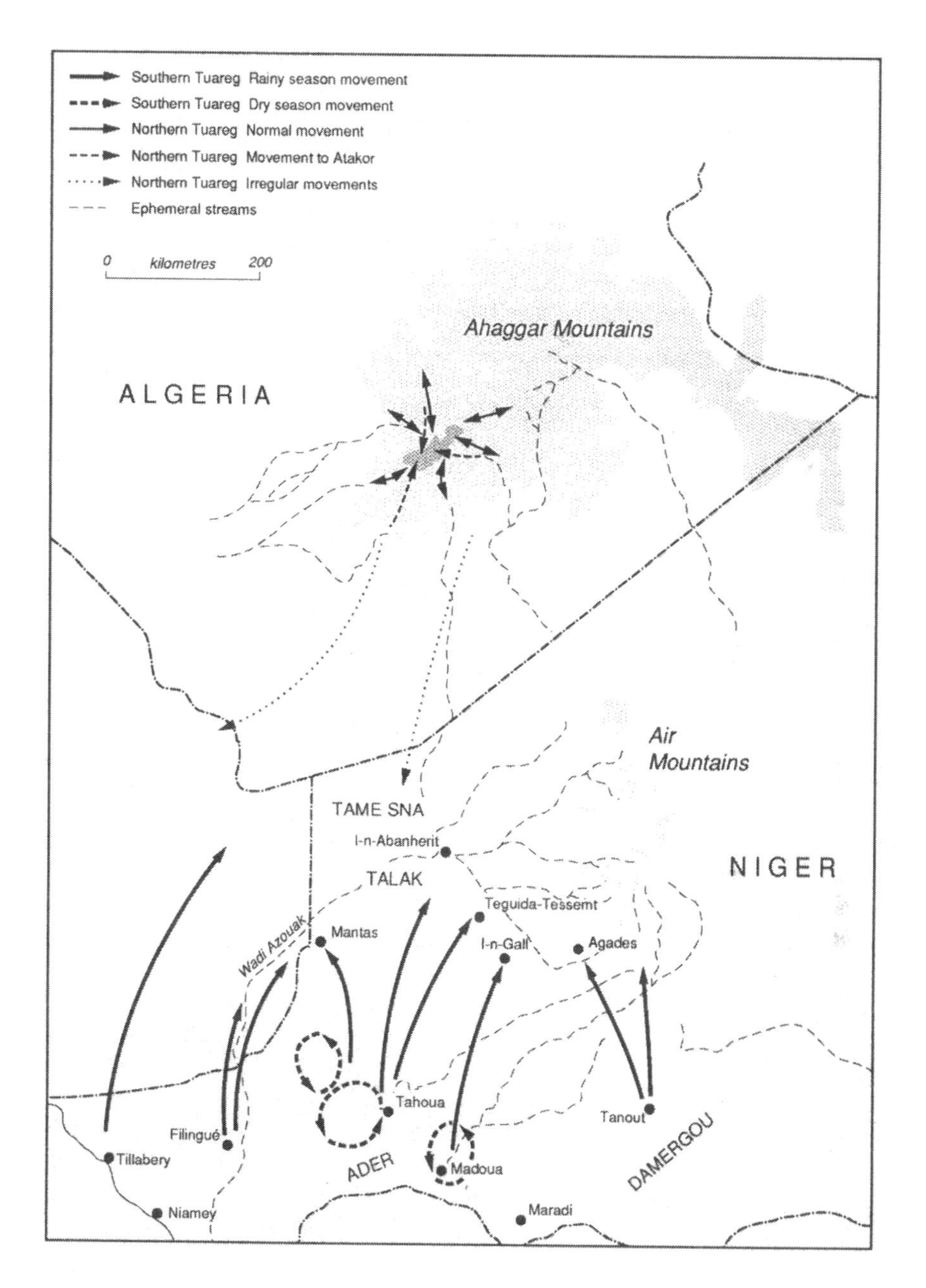

Figure 1.15 Pastoral migration in Niger, after Johnson (1969).

shows follow a regular pattern with some identification of grazing rights between different groups. Goat herders in the Ahaggar Mountains exploit the seasonal and altitudinal changes in temperature and grazing conditions. Livestock herds also spend the cold season and dry season to the south of the Sahel around wells and water courses with a daily movement in search of fodder of up to 10 km for cattle but tens of kilometres for camels (Nir, 1974). The stubble left over from the millet and sorghum harvest may be grazed, providing manure for the next season's crop. As the dry season nears its end water may become scarce and this is the most dangerous period for pastoralists, with rainfall highly variable at the start of the rainy season. Fodder, however, is the major constraint and animals are more likely to die from starvation than thirst. With the first downpours, fresh succulent grasses emerge and the herds migrate northwards following the rains (Johnson, 1969). This movement reduces grazing pressure on the dry season pastures while avoiding some of the insects and diseases that accompany more humid conditions. Although figure 1.14 shows the general case, in practice when pressure on the land increases there may be disputes over access to pastures and watering points.

Migration is, however, one of the Tuareg's chief strategies to combat the vagaries of the environment and is assisted by the resilience of the camel and goat in these circumstances. A camel can store 25–30 per cent of its body weight as water, it can withstand large body temperature fluctuations thus reducing the need to perspire, its fur insulates against temperature extremes while its urine can be concentrated to reduce moisture losses further. Camels can travel 25 to 30 km every day and in the wet season can last without water for up to 3 weeks by relying upon the moisture in vegetation. Louw and Seely (1982) report that a camel was observed to drink 186 litres of water at one time. Goats are also hardy animals by being able to graze on a large variety of plants but do require more attention from herders. Louw and Seely (1982) note that the black Bedouin goat can withstand a 30 per cent loss of body water, comparable to a camel, while a 12 per cent loss is critical for most mammals. Goats are browsers as well as grazers and can therefore exploit more of the range than sheep which are essentially grazers. Both goats and sheep lactate when fodder is fresh and in good supply whereas camels lactate for much longer emphasising their importance for desert dwellers (Goudie and Wilkinson, 1977). Cattle cannot survive longer than 2 to 3 days without water and tend to be found in the wetter areas to the south. Each animal therefore exploits a particular ecological niche with different abilities to migrate and different labour requirements. Hence the variety in table 1.3 is in part explained by the changing rangeland conditions throughout the year.

Herd diversification

By diversifying both the types of livestock and their ages and sex ratios the risks of total loss of the herd are minimised, i.e. 'don't put all your eggs in one basket'.

As each animal exploits different environments, a herd with a variety of livestock is better able to respond to changes and provides a greater range of products and services. To explain this last point let us take an 'idealised' pastoral family of a man aged 30, woman 25, a girl 15, boy 18 and two young children. Dahl and Hjort (1976) suggest this is reasonable given that older children are often 'loaned' out to relatives to provide additional labour and to spread out risks. Such a family would require 13,800 kcal each day (Dahl and Hjort, 1976) and 318 g of protein. A 'typical' camel herd (few bulls, mostly adult dams and heifers) of 28 animals would be required to satisfy these nutritional requirements. In practice camel herds are much smaller than this because they are mainly used for milk and transport. They are seldom slaughtered as they reproduce more slowly than cattle or goats. Cattle are important milk producers as their lactation period lasts 7 to 9 months which is longer than a goat's although shorter than a camel's. The above family would require 17 kg of milk a day. In the rainy season about 2 kg per cow each day can be expected but as only some of the animals would be lactating a herd up to 60 would be required. During the dry season milk yields decline and then some 400 to 600 animals would be required. If, however, the animals were periodically slaughtered and the meat consumed then a herd of only 64 would be necessary (in some cultures cattle blood is also drunk to supplement the diet). Pastoralists may be reluctant to slaughter camels and cattle but this does not apply to goats and sheep which have been called the 'small change' in the pastoral economy. Their milk yields are not very high but their meat can be consumed rapidly because of their size and consequently there are no problems over preservation of this food supply following slaughter. In order to satisfy the calorific demands of the above family, they would need from 100 to 200 sheep and goats in total or around 70 to 80 to cover just the protein requirements. Clearly large permutations are possible depending upon the products consumed from the herd, the grazing conditions, the herd composition (bulls:cows:calves), and the variety of animals in the herd. Cattle then are kept in wetter areas for their milk, camels are found in drier regions for their milk and transport while sheep and goats are fairly ubiquitous providing milk and meat.

Animal sharing

There are a variety of ways in which livestock are shared. Animals may be given as an outright gift to a destitute household through kinship ties or loaned for, say, the period of lactation to avert hardship. The sharing of animals also distributes a household's herd amongst relations thus reducing risks. Thus the animals in any given herd may belong to several different owners.

Storing food

Storage is difficult because of the heat and need for movement. Camel milk lacks rennet and therefore cannot be turned into cheese. Milk from goats and cattle can be used to produce yoghurt, butter and cheese, lasting from days, weeks and months respectively. Hence the importance of these livestock. The storage of meat is particularly a problem for large animals which tend only to be slaughtered for special occasions. Goats may have four kids a year which are fully mature 12 months later. Slaughtering small livestock does not therefore impair the reproduction of the herd while the meat can be rapidly consumed by a household.

Hunting and gathering

Hunting of game and gathering wild fruit are relied upon in times of crisis. Camels are important in the former but with poaching banned in many areas and game reserves established, plus an increasing pressure upon the land, this strategy is becoming less of a viable option, although the Bushmen of the Kalahari supply 80 per cent of their diet by plant gathering (Nir, 1974) and the Australian Aborigines have also adapted in this way to their arid environment.

Raiding and trading

In the past if herd losses could not be recouped by loans or rapid regeneration by small livestock then pastoralists might have resorted to raiding their neighbours. This could provide crops from dryland farmers, slaves to help with the herding and even additional animals. In all cases the camel was important for mobility. This practice has steadily died out during the 20th century with trading being of more importance. If we again examine the nutritional needs of a pastoral household, the protein demands could be met by 43 live animals but they would supply too few calories. This could be supplemented by 720 kg of cereals (millet, sorghum and maize) obtained from farmers in exchange for animals, salt and tools.

Pastoralists have then evolved a series of strategies involving the types of animals they keep, where they graze them and the products they use from them. Thus they have a dynamic set of responses to combat both the aridity of their environment and its variable nature. Roughly one camel provides as much food as 2.3 cattle or 3.5 sheep or 7 goats. If these figures are compared with table 1.3 it is evident that the Tuareg in the Ahaggar Mountains appear ill-found while those in the west are just adequate and those to the south have a reasonable safety margin. Although this calculation is crude and over simplistic, it does reflect the fact that not all food production systems are successful. There is a grave danger in assuming that some 'golden age' existed when all was in ecological balance and sustainable – what Hopkins (1975, p. 10) called the 'myth of merrie Africa'.

During the pre-colonial period, arid lands in Africa faced problems of famine, warfare, disease and slavery. Within the Tuareg, a feudal system operated with the camel herders operating as 'warriors', subjugating and protecting the goat-herding 'vassals' (Agnew, 1985). Oases were cultivated under a sharecropping arrangement with four-fifths of the harvest going to the Tuareg 'landlord' who supplied land, seeds, water and tools as opposed to the farmers' labour. In taking the above, somewhat pre-colonial view of pastoralism and dryland farming we may in fact be describing a rather peculiar landuse system that was in operation after decimation of both human populations and livestock herds through slavery and disease. Nevertheless it is evident that a number of strategies have evolved to combat physical constraints in arid lands. Yet there is growing awareness that the system is failing with widespread reports of livestock death, falling yields and persistent drought. Some have blamed the environment but attention is also being given to mismanagement of the food production systems through ill-conceived development attempts (Garcia, 1981, Glantz, 1976 and 1987 and Wijkman and Timberlake, 1985). For example, the sinking of boreholes has encouraged animal numbers to increase resulting in overgrazing, (Goudie, 1986) although IUCN (1989) cast some doubt on the extent of this particular problem. The intensification and commercialisation of agriculture, sometimes using herbicides and fertilisers, can lead to a reduction or even abolition of the fallow period possibly causing falling yields and soil erosion (Franke and Chasin, 1980 and 1981). The opening up of these once isolated areas to world economic forces has clearly had unforeseen and tragic consequences but it does highlight the plight and precarious nature of arid land dwellers.

The inhabitants of the arid realm have evolved a way of life that minimises risks rather than maximises opportunities. Any reduction in the flexibility of pastoralists and dryland farmers to respond to the vagaries of their environment may produce human and environmental problems, e.g sedentarisation policies, sinking of boreholes, closing of national boundaries and promotion of cash cropping. Changes involving water resources therefore have to be sensitive to the realisation that any change entails a degree of risk that may be perceived by those affected as too great. We shall return to this point when considering modern technologies for enhancing water supplies; but first we turn to discuss in more detail the variable nature of water resource supplies in arid lands.

Part II

WATER RESOURCE PROBLEMS

An often quoted statistic is that 97 per cent of the world's water is contained in the oceans (Speidel and Agnew, 1988), leaving only 2.59 per cent as freshwater. The vast proportion (76.6 per cent) of all the freshwater is frozen (ice caps and glaciers), and 22.9 per cent is contained by aquifers (groundwater) leaving only 0.54 per cent of freshwater in lakes, rivers, soil and the atmosphere. Fortunately these terrestrial supplies are periodically renewed through the operation of the hydrological cycle and substantial surface resources can be found.

Arid lands contain some of the world's largest river systems (the Murray–Darling extends 3,717 km, Rio Grande 3033 km, and the Indus 3,186 km) but the distribution of these freshwater surface resources is uneven. Africa's four major natural lakes cover an area of 160,000 km^2 while the Nile extends for 6,580 km, the Niger for 4,180 km and the Congo for 3,500 km, yet vast areas are unprovided for and the flows that do occur may not be exploitable. L'Vovich (1979) estimates that only 31 per cent of these freshwater supplies are stable (counted as base flow in rivers and surface storage) with 24.4 per cent of precipitation lost as flood surface runoff. Tolba (1980) reports that 30 per cent of the world population lack reasonable access to potable supplies. The need to regulate and develop water supplies is therefore of prime importance but table II.1 suggests a much wider perspective is necessary than just focusing on water supply.

The predominance of arid land countries is evident in table II.1 The inclusion of the Netherlands, Poland and Belgium is perhaps surprising but reflects the significance of demand as well as water supply. World Resources Institute (1986) report that the Ancient Romans had access to approximately the same amount of water as Romans today but that the population is now four times as great. As Hagan (1988, p. 9) said of arid lands, 'actually a water problem is often really a people problem'.

Figure II.1 shows projected global population growth with the increases in developing countries fuelling the growing demand for more water. Eyre (1988) reports that the rate of growth of arid land's population between 1960 and 1985 was 81 per cent compared to 56 per cent for the Earth as a whole. Although global population statistics are always of questionable reliability it is clear that con-

Table II.1 Countries with the lowest per capita water supplies (1,000 m^3 per year) (from World Resources Institute, 1986 and 1989)

	1987	*1985*
Malta	–	0.07
Egypt	0.02	1.20
Saudi Arabia	0.18	–
Libya	–	0.19
Barbados	0.21	0.20
Singapore	0.23	–
Oman	–	0.54
Kenya	0.66	0.72
Netherlands	0.68	–
Belgium	–	1.27
Poland	1.31	1.57
South Africa	1.47	1.54
Haiti	1.59	1.67
Peru	1.93	2.03

tinued global population growth is forecast and that much of this growth will take place in arid lands, particularly in cities which will accelerate the demand for potable supplies. By the 1970s there were 350 arid realm cities with populations greater than 100,000 and 30 had more than 1 million inhabitants (Cooke *et al.*, 1982). Eyre (1988) estimates that 36 million people already live in just five arid realm cities: Cairo, Karachi, Lima, Beijing and Teheran. In 1950 less than 30 per cent of the population in the Middle East lived in towns but by the 1980s this number had increased to an average of 45 per cent with 70 per cent in some places (Beaumont, 1988).

Balek (1983) notes that the per capita drinking requirements in hot, dry environments is between 10 and 20 litres a day whilst in industrial cities consumption increases to 800 litres a day with 1,000 expected in the future. UNESCO (1978) report water demands of 600 litres per capita per day in New York, 500 in Paris, 260 in London, 100–200 from rural areas and 20–30 in developing countries. The average water consumption in the United States rose from between 100 and 150 litres per capita per day around 1900 to between 400 and 500 litres per capita per day by the 1970s. In Jordan, water consumption is presently 77 litres per

Table II.2 Water requirements of selected manufacturing processes (based upon Biswas, 1980, p. 18)

1 litre of petrol	requires	7–34 l	of water
1 can of vegetables	requires	40 l	of water
1 ton of coal	requires	1,400 l	of water
1 ton of steel	requires	8,000–61,000 l	of water
1 ton of paperboard	requires	62,000–376,000 l	of water

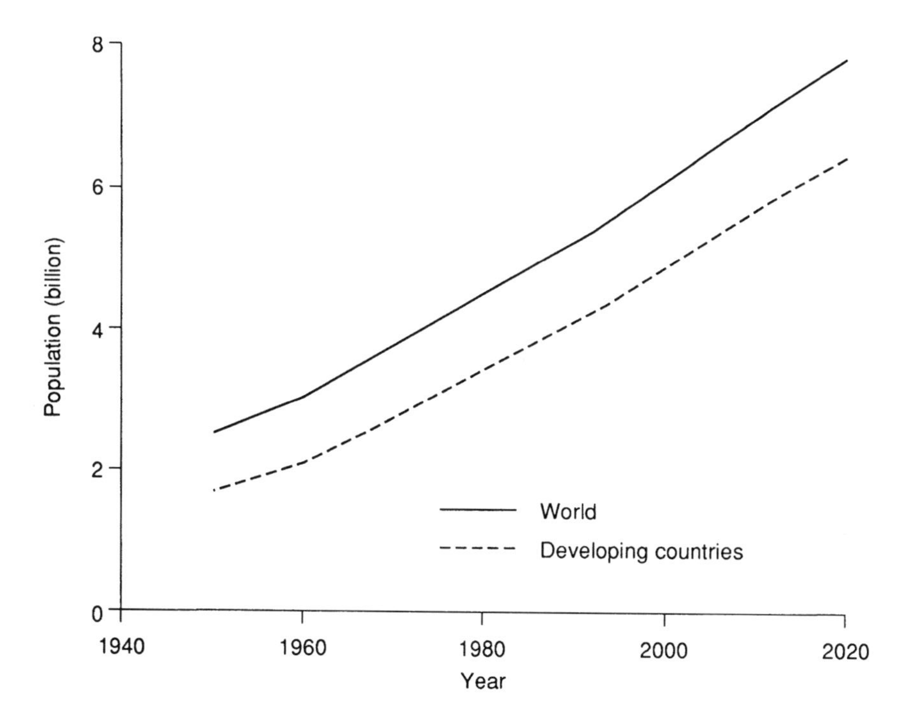

Figure II.1 Projected world populations, data from Eyre (1988).

capita per day in urban areas and 60 litres in rural areas (Sahouri, 1988). Water demands are rising in arid lands and will continue to increase following population growth, urbanisation and industrialisation. The significance of the last factor can be seen from table II.2.

Improving health and sanitation also adds to water consumption. Guinea worm infects 10–48 million people each year with people reinfected 50 to 100 times in a lifetime in places like Andhra Pradesh, India. Yet Biswas (1981) reports that piped potable water in Nigeria reduced guinea worm infections from 60 per cent to zero in some villages while sieving contaminated water can also have dramatic results. In the Yemen, where water consumption can be as low as 5 to 6 litres per capita per day, less than 25 per cent of the rural population have access to clean water and 30 to 40 per cent of the deaths of children less than 5 years of age are related to water contamination (Vincent, 1987). Lean *et al.*, (1990) report that 25,000 people worldwide die from water-related deaths each day; diarrhoea alone kills 4.6 million children every year. In India, water-borne diseases cost the country 73 million working days each year amounting to US$ 600 million in lost production and health care. These horrific statistics of death and disease could be drastically reduced by the provision of potable water with additional economic benefits.

The demand for large amounts of water as health and sanitation improve is illustrated by the following figures for the UK (from Open University 1974): 37 per cent of domestic water use is expended on sanitation; 33 per cent on washing and bathing; 21 per cent on laundry with only 3 per cent for drinking and cooking. Cairncross (1987) cites an example of two Mozambique villages: one with a standpipe used 12.3 litres per capita each day; while the inhabitants of the other village who had to walk for 5 hours to collect water used 3.2 litres. The demand for water will inevitably increase beyond population growth and concern is becoming acute as shortages become ever more apparent. World Resources Institute (1986, p. 128) state:

> Increasing water withdrawals for municipal purposes, shown as a ratio of freshwater renewal, indicate the extent to which some areas, particularly North and Sub Saharan Africa and the Middle East, were already coming close to their limits a decade ago.

Arid realm countries which consume annually more water than is naturally replenished are (percentages refer to ratio of consumption to naturally renewable supplies): Libya (374 per cent); Qatar (174 per cent); Yemen AR (147 per cent); UAE (140 per cent); Yemen PDR (129 per cent); and Saudi Arabia (106 per cent). In addition high ratios occur in Egypt (97 per cent); Malta (92 per cent); Israel (88 per cent); Tunisia (53 per cent); and Afghanistan (52 per cent) (from

Table II.3 Urban and rural populations served with safe water supplies in selected Middle East countries (from Arab–British Chamber of Commerce, 1985)

Country	*Urban pop. (1,000s)*			*Rural pop. (1,000s)*		
	total	*safe supply*		*total*	*safe supply*	
Jordan						
1980	1,550	1,550	(100%)	683	444	(65%)
1985	1,982	1,858	(94%)	819	696	(85%)
1990	2,227	2,227	(100%)	982	933	(95%)
Libya						
1980	2,596	2,596	(100%)	649	584	(90%)
1985	3,564	3,564	(100%)	396	316	(80%)
1990	3,838	–		307	–	
Saudi Arabia						
1980	6,358	5,832	(92%)	1,150	1,000	(87%)
1985	7,852	7,488	(95%)	940	970	(100%)
1990	9,346	9,346	(100%)	1,045	940	(90%)
Syria						
1980	3,436	3,367	(98%)	5,543	2,982	(54%)
1985	4,123	4,123	(100%)	6,658	4,437	(67%)
1990	4,866	4,866	(100%)	7,908	6,352	(80%)
Yemen (PDR)						
1980	637	539	(85%)	1,288	316	(25%)
1985	717	597	(83%)	1,220	394	(32%)
1990	831	790	(95%)	1,394	552	(40%)

(1990 figures were predictions)

World Resources Institute, 1990). The concentration of these water-deficient countries in the Middle East prompted the Centre for Strategic and International Studies, Washington, DC, to announce (Alam, 1990):

> By the year 2,000 water not oil will be the dominant resource issue in the Middle East.

However, recent events in this region suggest this underestimates the equal significance of oil.

The concern over the growing demand for water resources has precipitated a number of international initiatives to combat the problem. In 1950 UNESCO created an Advisory Committee on Arid Zone Research. Thirty titles were published between 1953 and 1969, reflecting concern over water resource issues (Batisse, 1988). In the 1960s the Arid Zone Research Program ceased due to the launch of the International Hydrological Decade and a new programme on Man and the Biosphere. Part of this initiative addressed the need to provide potable water to all urban and 25 per cent or rural dwellers by the end of the decade, i.e. 1980. The 1976 Conference on Human Settlements set new goals of potable water for all which were endorsed by the UN Conference of 1977. The World Health Organisation estimated that by 1980, only 29 per cent of rural communities had access to clean water but that there were huge disparities with 93 per cent achieved in Egypt as against 12 per cent in Kenya (Biswas, 1981). In 1980 the UN declared the 1980s as the International Drinking Water Supply and Sanitation Decade.

There are then no lack of initiatives at the international level and some success with Snowden (1990) reporting an increase from 35 per cent of Tanzania's villagers having access to water in 1982 compared to 45 per cent by 1988. There have also been considerable achievements in the Middle East. Table II.3 shows that the supply of drinking water to urban dwellers is now high with an equally high percentage of rural inhabitants served or indications of significant improvements during the 1980s. Lowes (1987) suggests that, globally, water provision is keeping pace with urban water demands (34 per cent of urban dwellers not served in 1970, 23 per cent not served in 1985) but is falling behind in rural sanitation and rural water supplies. By 1985, 64 per cent of rural inhabitants lacked reasonable access to potable water. Lean *et al.* (1990) report that there has been a 'tailing off' of water and sanitation provision during the 1980s such that, out of a global population of more than 5 billion today, 1.4 million lack safe sanitation and 1.2 billion lack potable drinking water. Tolba (1990) suggests even higher estimates of 1.7 billion lacking access to potable water and 3 billion lacking proper sanitation.

Despite demand being an essential component of water resource management there is often comparatively little that can be done to reduce rates of population increase, of urbanisation and industrialisation; or to curb improvements in health and sanitation. Indeed some of these changes are promoted as part of development. Arid land nations are therefore likely to face growing demands for water and need to address urgently the supply of water resources. Lowes (1987)

Plate 1 This bath house dating back to 1632 is found in Ethiopia near to Lake Tana revealing that the history of water management in Africa need not be considered only over the last 100 years.

reviews the progress of water provision during the last decade and suggests the major constraints in order of importance are:

- lack of human resources,
- operational and maintenance inadequacies,
- institutional constraints,
- lack of finance,
- management.

The management of water resources and the role of governments are dealt with in chapters 10 and 11 but the above list ignores a fundamental problem of water resources development in arid lands; that is, the variability of supplies. One cannot consider the management of water resources in arid lands without taking account of the inherent difficulties introduced by the physical environment. Chapter 2 deals with the variable nature of the hydrological cycle while chapters 3 and 4 consider the specific impacts of climatic change and drought. This part then ends with chapter 5 which examines the particular problems of assessing the magnitudes of water resources in arid lands.

2

VARIABLE WATER SUPPLIES

RAINFALL

The discussion in the previous chapter on the climate of arid areas points to rainfall variability in both space and time coupled with high potential evaporation rates as the essential characteristic of these lands. Jain (1968) characterises Rajasthan (India) as one lean year in every three and one famine in every eight. These phenomena need to be understood in order to explain the environmental constraints imposed upon human activities and in order to investigate common arid lands' problems. Figure 2.1 shows the global distribution of patterns of rainfall variability. Comparison with figure I.2 reveals that arid land regions suffer from the highest variability with a coefficient of variation of 40 per cent common, and in extremely arid areas values in excess of 100 per cent. The reasons are threefold: atmospheric processes, inadequate data base and statistical analysis.

Atmospheric processes and temporal variations

The previous chapter illustrated the ephemeral and often localised nature of precipitation which may only fall every few years or so. Nir (1974) mentions a frequency of once every 8 years in the Sahara and every 18 years in Peru while figure 2.2 shows the infrequent occurrence of rainfall in Oman. Most arid lands are located beneath semi-permanent anticyclones into which moist air, rain-bearing frontal systems or tropical cyclones can only occasionally penetrate (Graf, 1988). There is progress in the understanding of changing circulation systems but it is still proving difficult to explain and predict the changes that do occur. The UK Meteorological Office has been working for some time on arid land rainfall predictions (Owen and Folland, 1988) along with other scientists (Druyan, 1989). Some success using sea surface temperatures raised expectations but the failure to predict correctly the heavy downpours in the Sahel during 1988 (Toulmin, 1988b) suggest that there is still much work yet to be done. Whether the conditions that promote rainfall or aridity dominate depends upon a variety of factors including surface albedos, energy budgets, sea surface temperatures, air

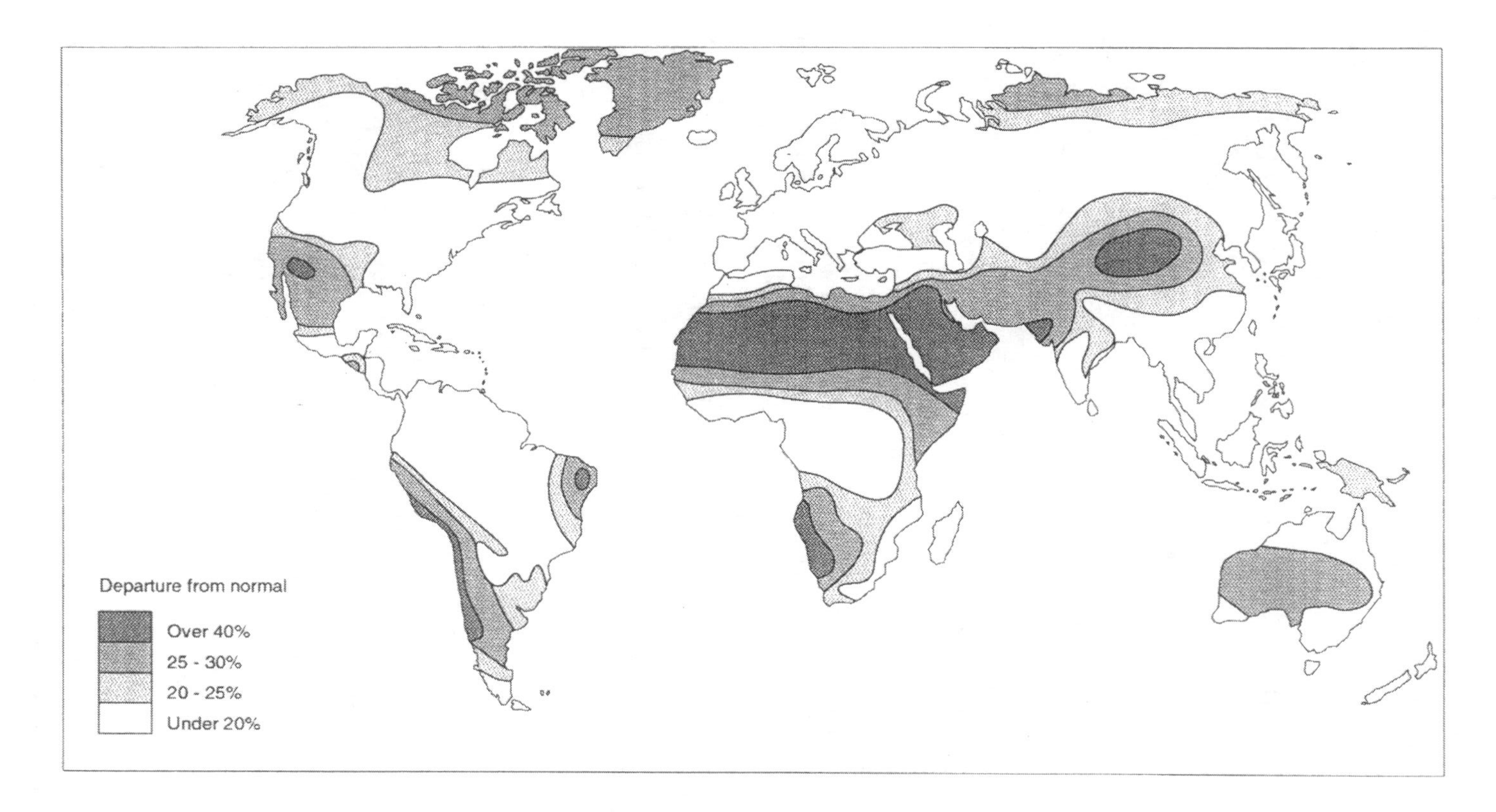

Figure 2.1 Variations in annual rainfalls, after Rumney (1968).

turbidity, soil moisture and even the preceding year's conditions. Atmospheric conditions over arid regions are then prone to marked changes that are proving difficult to explain and predict fully. These result in variable rainfalls as shown in figures 2.3 and 2.4 taken from the same Sahelian counry of Niger. It is noticeable that while some stations display marked trends, others appear merely variable around the mean.

Spatial variation and point measurements

Although there have been vast improvements in areal measurements of rainfall through radar (Collinge and Kirby, 1987) and satellite imagery (Barrett and Martin, 1981, Dugdale *et al.*, 1991) most rainfall data is still collected by individual observers (see chapter 5). Such rainfall stations occur where there are permanent settlements, and many are therefore isolated with vast areas lacking even simple non-recording raingauges. This problem is aggravated by the highly localised nature of convective rainfall in arid lands. UNESCO (1977a) for example report that both Bakel and Dakar (Senegal) lie on the same 500 mm isohyet yet a difference of 270 mm was recorded in 1972. Sharon (1972) found in a hyper-arid area of Israel that, for three observing stations all within 15 km of each other, on 70 per cent of rain days the two other stations received less than half of the amount of rain falling at the third. Examples are given of the following observations: 22.1 mm and 0.0 mm; 2.8 mm and 23.0 mm; 1.1 mm and 1.0 mm. A denser network of raingauges revealed that only 20 per cent of the area received intense rainfalls which were highly localised, falling as a cell with a diameter of around 5 km. Studies of rainfall in Central Namib showed that there was little correlation between stations 25 to 35 km apart (Sharon, 1981). Given

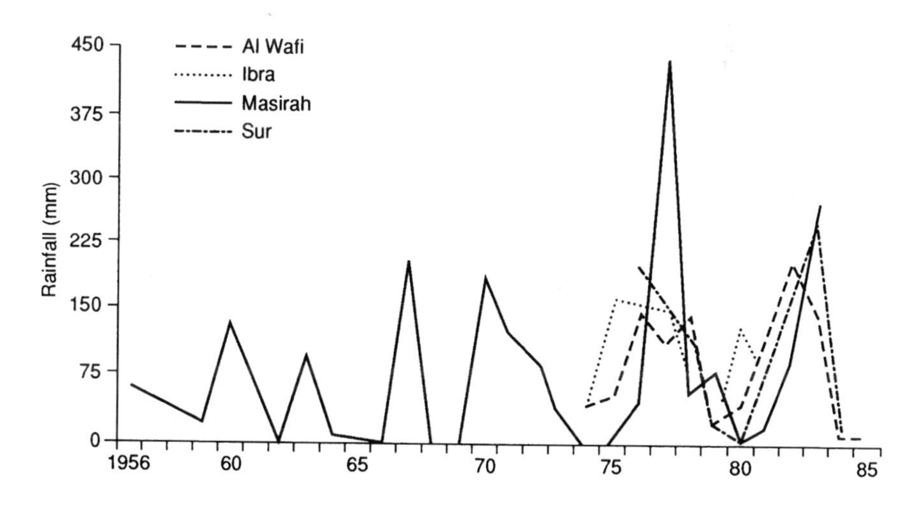

Figure 2.2 Annual rainfalls, Oman.

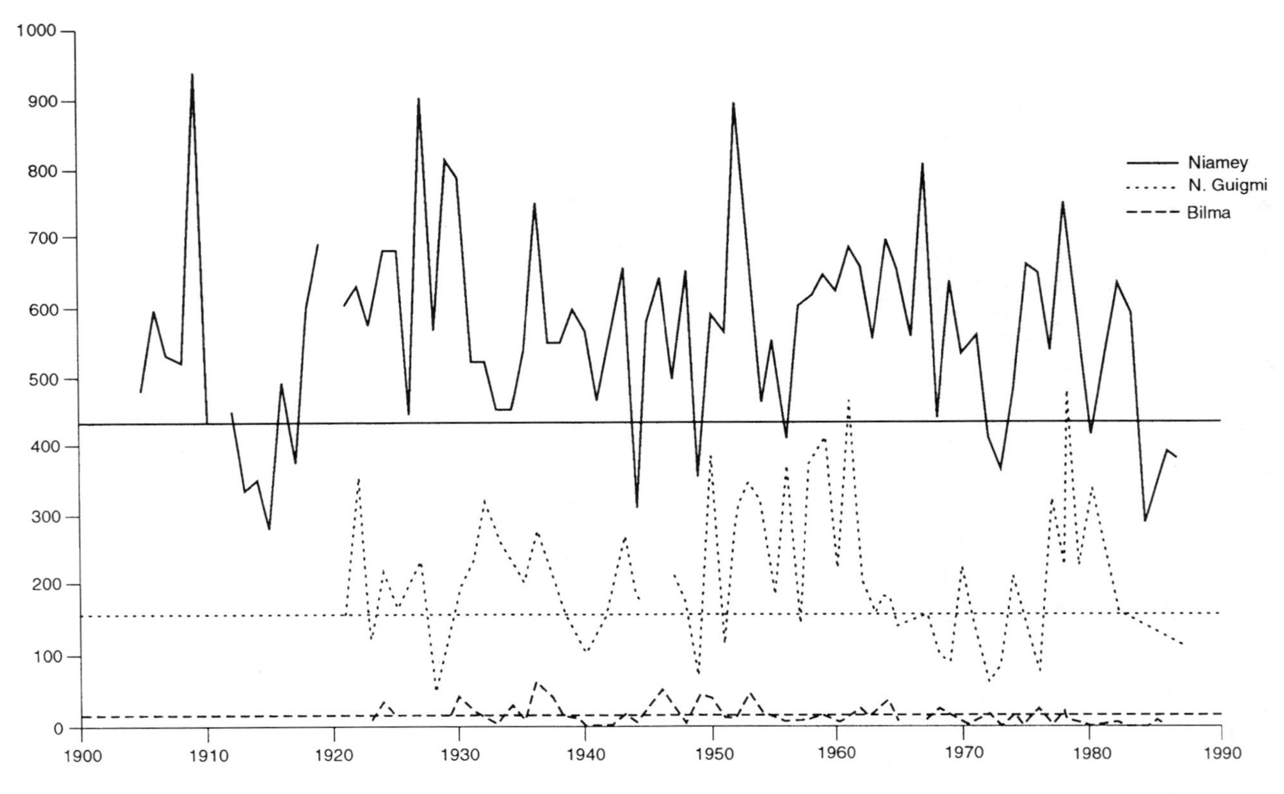

Figure 2.3 Annual rainfalls for Naimey, N'Guigmi and Bilma, Niger (horizontal lines are 80% of mean annual rainfall – see chapter 4).

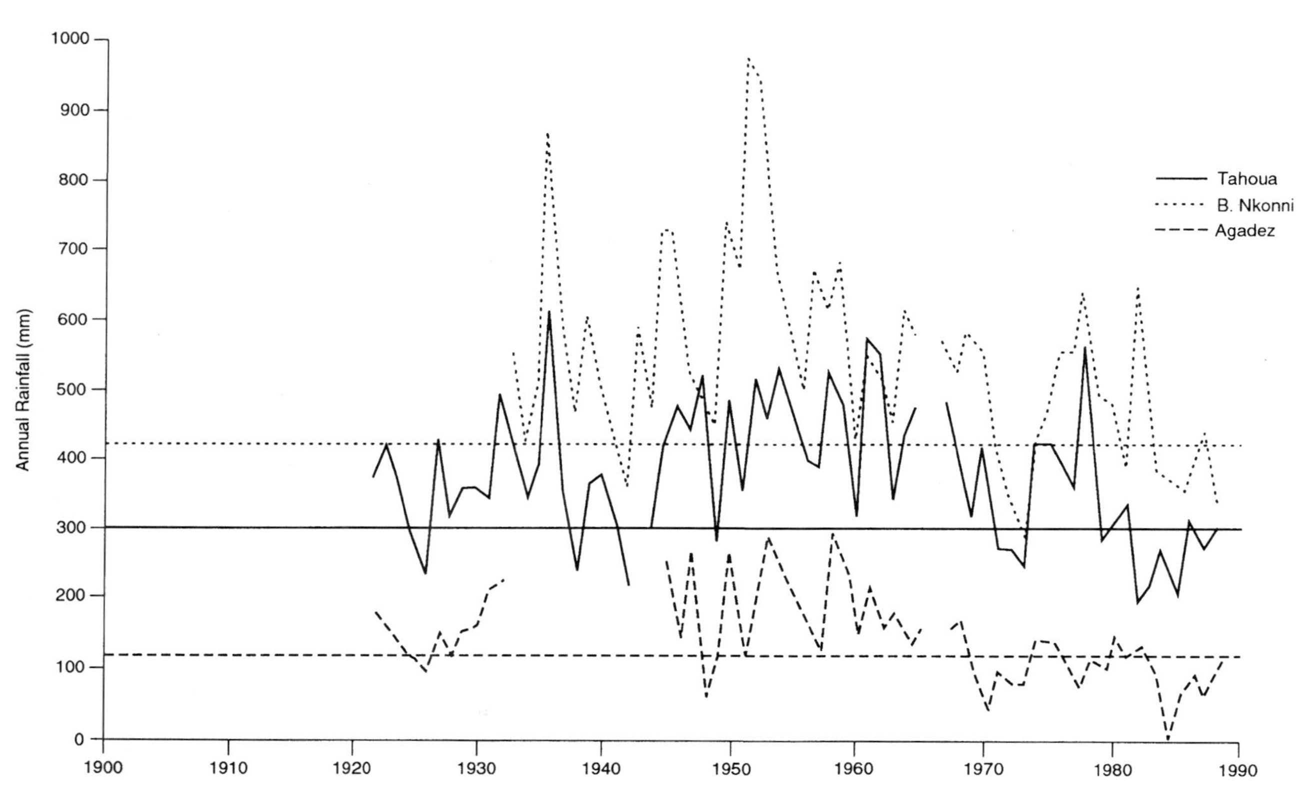

Figure 2.4 Annual rainfalls for Tahoua, B'Nkonni and Agadez, Niger (horizontal lines are 80% of mean annual rainfall – see chapter 4).

Table 2.1 Pairwise dependency of decadal rainfall at Niamey, Niger

Decad	*Chi-square calculated*	*Chi-square statistic*	*Remarks*
Apr 1 – Apr 2	0.03	3.84	independent
Apr 2 –	0.57	3.84	independent
Apr 3 –	1.19	5.99	independent
May 1 –	7.89	9.49	independent
May 2 –	4.64	9.49	independent
May 3 –	3.55	9.49	independent
Jun 1 –	1.36	9.49	independent
Jun 2 –	1.09	9.49	independent
Jun 3 –	4.64	9.49	independent
Jul 1 –	2.73	9.49	independent
Jul 2 –	7.09	9.49	independent
Jul 3 –	4.57	9.49	independent
Aug 1 –	2.62	9.49	independent
Aug 2 –	9.26	9.49	independent
Aug 3 –	4.04	9.49	independent
Sep 1 –	6.00	9.49	independent
Sep 2 –	5.26	9.49	independent
Sep 3 –	1.03	9.49	independent
Oct 1 –	1.15	5.99	independent
Oct 2 – Oct 3	11.44	3.84	dependent

the localised nature of some rainfall it is possible for rainfalls a few kilometres away to remain undetected by a raingauge, producing not only unreliable areal estimates but also considerable variations at specific points of measurement (see Goudie and Wilkinson, 1977).

The variability of atmospheric conditions over an area can be illustrated by their independence from one another (i.e. the rain falling during a specific period is not affected by the rain that fell during preceding periods). The dependence of daily rainfall amounts has been widely proven or assumed (Gabriel and Neuman, 1962, Smith and Schreiber, 1973 and Haan, 1977). Mooley (1971) commented upon the reduction in persistence with increasing time period such that monthly rainfalls are often independent from each other. The dependency of decadal (10 day totals) rainfall at Niamey, Niger, was investigated by constructing a contingency table using pairs of decadal rainfalls from adjacent time periods t and $t+1$ and the association was tested against the chi-square statistic at the 0.05 level of significance (Walpole and Myers, 1972).

The results listed in table 2.1 which show decadal rainfalls (10 day totals) in the Sahelian region of West Africa were found to be independent from each other, i.e. the amount falling in one period had no apparent influence upon the amount falling in the following period. This would not be the case where atmospheric conditions were persistent. This points to the erratic movement of the ITCZ over the region and the changeable nature of atmospheric conditions producing variable rainfalls.

Table 2.2 Variation of decadal rainfall, Niamey, Niger (1906–78)

Decad	*Mean (mm)*	*Std dev.*	*Coeff. variation (%)*
Apr 1	4.1	5.15	125
Apr 2	9.0	11.27	125
Apr 3	10.4	13.87	135
May 1	12.6	18.19	144
May 2	14.6	19.00	130
May 3	20.4	20.82	102
Jun 1	24.1	24.00	99
Jun 2	26.0	19.44	75
Jun 3	31.3	22.35	71
Jul 1	34.0	22.78	67
Jul 2	50.9	29.21	57
Jul 3	64.6	40.14	62
Aug 1	70.5	40.56	58
Aug 2	67.0	36.86	55
Aug 3	70.7	48.76	69
Sep 1	42.8	27.87	65
Sep 2	33.4	25.49	76
Sep 3	19.0	17.95	95
Oct 1	12.3	13.26	108
Oct 2	16.4	22.07	135
Oct 3	6.4	7.60	120

Statistical analysis

As the mean declines so the coefficient of variation increases because small deviations become more significant:

Coefficient of variation = (Standard deviation/Mean) × 100 per cent

Hence areas with low rainfalls almost inevitably record the highest variation (see figure 2.1 and Table 2.2) even though the magnitude of the variability away from the mean is smaller. Cooke and Warren (1973) comment that (in absolute terms) the variability of rainfall in arid areas may not be much greater than that in more temperate regions. However, for areas receiving around 500 mm of rainfall annually, Gregory (1972) reports coefficients of variation in the USA of between 10 and 20 per cent while for the same annual rainfall in West Africa values in excess of 50 per cent are found (Hayward and Oguntoyinbo, 1987). But discussions over the interpretation of 'variability' too often ignore the fact that for areas with low rainfalls, even small variations are much more significant. Jackson (1977) pursues this point and notes that the variability of rainfall can only be considered within the context of a given landuse, i.e. the significance of a fluctuation in rainfall can only be examined in relation to what was required. Jackson (1977, p. 46) states:

Table 2.3 Chi-square test for normality using non-zero rainfall (decadal rainfall, Niamey, Niger)

Decad	*Chi-square calculated*	*Chi-square statistic*	*Accept H_0 of normality*	N
Apr 3 –	4.36	3.84	NO	25
May 1 –	32.49	5.99	NO	37
May 2 –	24.48	5.99	NO	45
May 3 –	14.20	7.82	NO	57
Jun 1 –	17.88	9.49	NO	62
Jun 2 –	14.64	11.07	NO	64
Jun 3 –	12.60	12.59	?	65
Jul 1 –	18.58	11.07	NO	65
Jul 2 –	15.88	11.07	NO	66
Jul 3 –	11.65	14.07	YES	65
Aug 1 –	3.77	11.07	YES	66
Aug 2 –	10.36	12.59	YES	66
Aug 3 –	13.42	11.07	NO	65
Sep 1 –	20.51	11.07	NO	65
Sep 2 –	11.77	11.07	NO	60
Sep 3 –	9.12	11.07	NO	60
Oct 1 –	9.30	5.99	NO	42
Oct 2 –	1.59	3.84	NO	27

(N = number of non-zero rainfalls, from a possible total of 66. H_0 = null hypothesis.)
(This analysis was based upon a separation of the data into zero and non-zero proportions because of the highly skewed distributions of early and late season rainfalls; after Jennings and Benson, 1969 and Haan, 1977.)

> Rainfall variability has been the subject of much comment, but in reality comparatively little quantitative analysis of this factor has been undertaken.

Whilst the variation of annual totals presents problems for identifying climatic change and making predictions, changes over a shorter time scale present much more immediate problems for the inhabitants of arid lands. Figure 1.12 shows the variation in the onset and cessation of the rainy season in West Africa. Average values coincide with the gradual march of the sun across the region but for individual years there is considerable variation. This variability is so marked that it produces a very skewed rainfall distribution with a large number of low observations and infrequent but very significant extreme events. An analysis of decadal (10 day totals) rainfall (from the Sahel region) shown in table 2.3 reveals that they could not be considered to follow a normal distribution (i.e. where the mean, mode and median values coincide) except during August. Instead decadal rainfall follows a gamma distribution (Thom, 1958 and Thorn, 1972) (see table 2.4) with a zero lower boundary and an infinite upper boundary. Figure 2.5 shows the normal and gamma probability distributions for the Niamey decadal rainfall from June 21 to June 30 which closely approximates to the normal distribution but finds a more reliable estimate of probability with a gamma

Table 2.4 Chi-square test for gamma distribution using non-zero rainfall from Niamey, Niger

Decad	*Chi-square calculated*	*Chi-square statistic*	*Accept gamma distribution?*
Apr 2 –	0.20	3.84	YES
Apr 3 –	0.92	3.84	YES
May 1 –	21.72	7.82	NO
May 2 –	3.16	9.49	YES
May 3 –	6.96	11.07	YES
Jun 1 –	4.24	12.59	YES
Jun 2 –	14.18	12.59	NO
Jun 3 –	9.75	14.07	YES
Jul 1 –	7.11	14.07	YES
Jul 2 –	9.30	11.07	YES
Jul 3 –	4.51	12.59	YES
Aug 1 –	8.81	12.59	YES
Aug 2 –	4.48	14.07	YES
Aug 3 –	2.58	14.07	YES
Sep 1 –	3.13	12.59	YES
Sep 2 –	3.31	9.49	YES
Sep 3 –	5.85	11.07	YES
Oct 1 –	0.85	7.82	YES
Oct 2 –	1.18	3.84	YES

(This analysis was based upon a separation of the data into zero and non-zero proportions because of the highly skewed distributions of early and late season rainfalls; after Jennings and Benson, 1969 and Haan, 1977.)

distribution. These results question the use of means to indicate the 'central tendency' of rainfall in arid lands and all associated calculations such as the coefficient of variation.

Thus the variability of daily, decadal and annual rainfalls poses problems for scientists examining change, for the inhabitants relying upon rainfalls and for planners who need prediction of future conditions. Farmer (1989) admirably demonstrated this problem by calculating deviations of Sahelian rainfalls away from a mean value calculated for two different time periods shown in figures 2.6 and 2.7. Rainfall either has been markedly below average since the 1960s or it is now 'normal' following a particularly wet period. Figure 2.8 shows annual rainfalls from stations around Lake Tana in Ethiopia. There appears to have been a marked decline since the 1970s but does this indicate below average conditions in the 1980s or abnormally wet conditions in the 1970s?

SURFACE RUNOFF

Arid lands contain some of the world's largest rivers (e.g. Nile, Indus and Murray) which are classified as **exogenous** because they originate outside of the arid zone. Perennial flows that are **endogenous**, i.e. originate within arid lands,

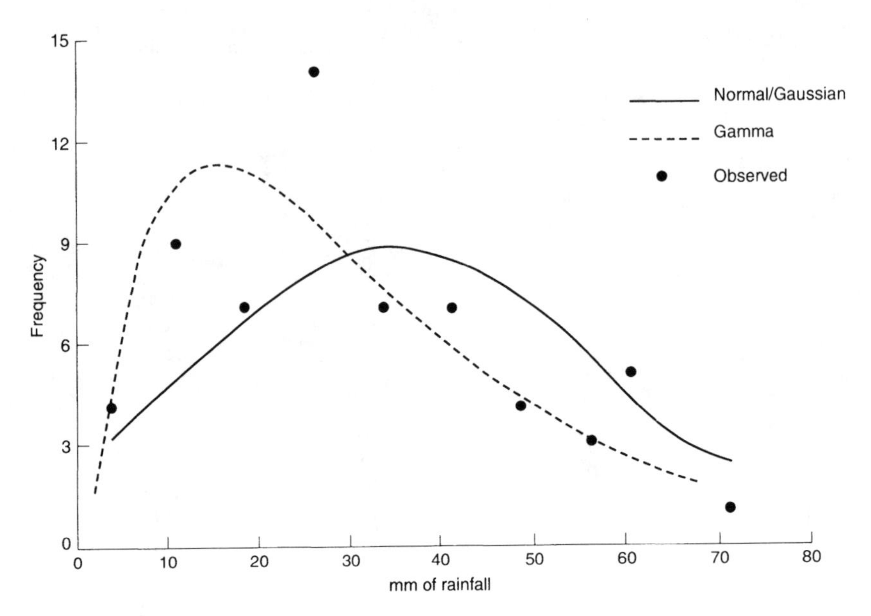

Figure 2.5 Rainfall probability distributions for the decadal rainfall June 21–30 at Niamey, Niger.

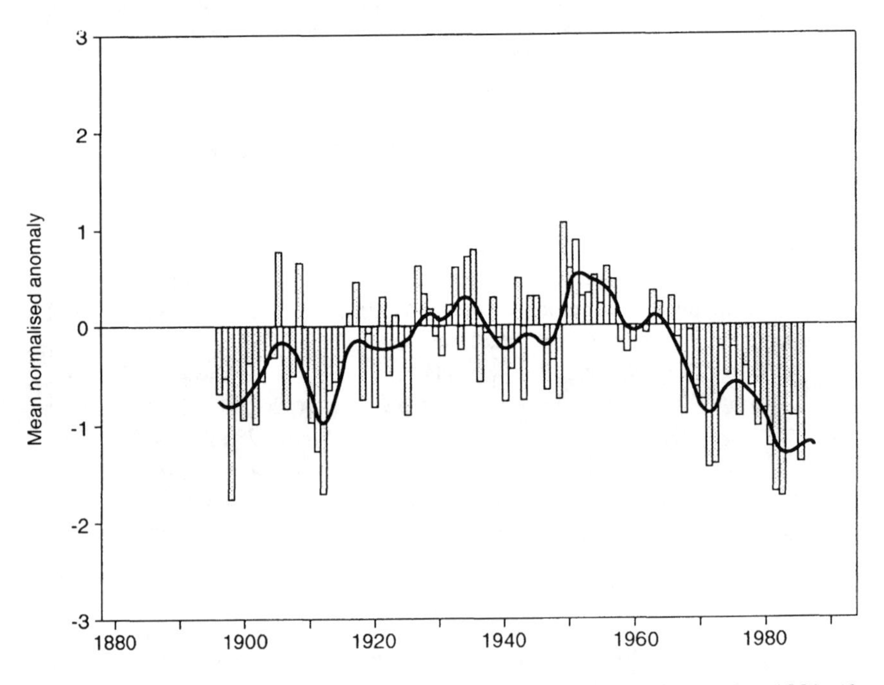

Figure 2.6 Sahelian rainfall departures from a mean calculated on observations 1931–60, after Farmer (1989).

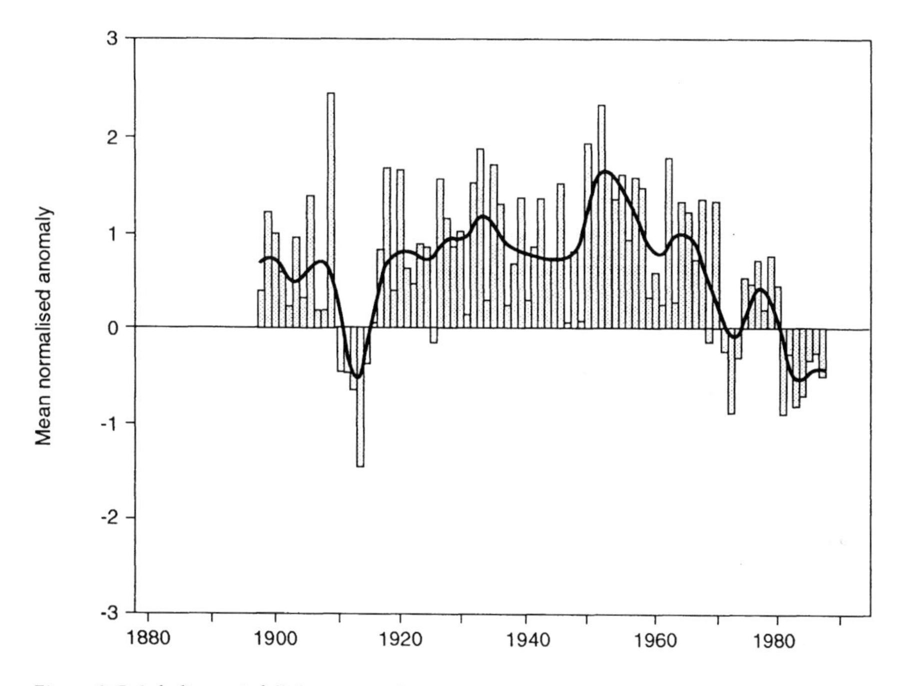

Figure 2.7 Sahelian rainfall departures from a mean calculated on observations 1960–80, after Farmer (1989).

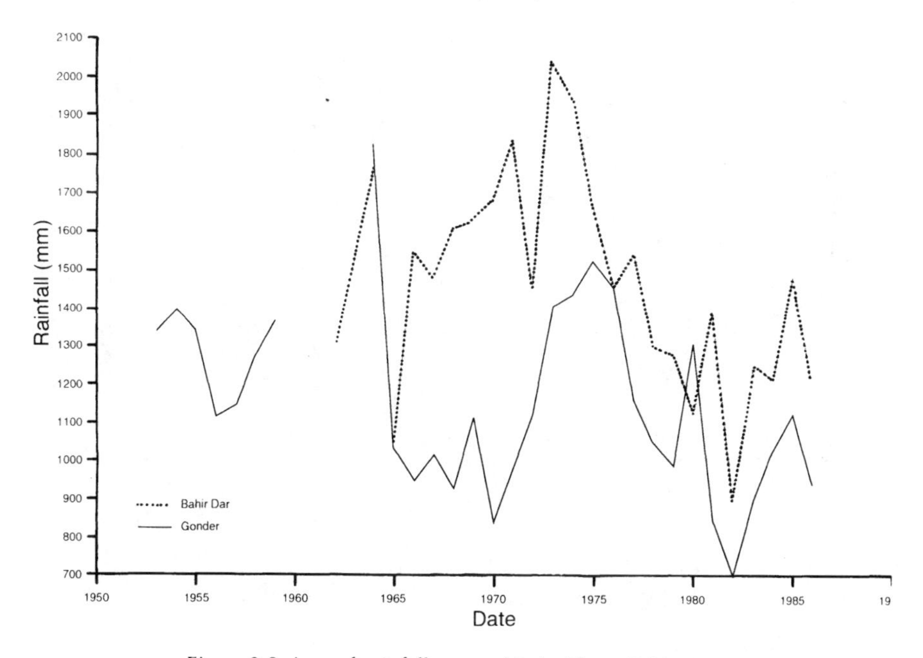

Figure 2.8 Annual rainfalls around Lake Tana, Ethiopia.

are inevitably springs maintained by groundwater. Simpson (1985) notes that significant arid land springs tend to issue from limestone massifs, for example the Atlas Mountains and the Jebels Akhdar of Libya and Oman, although the basaltic rocks also support springs at the Jebel Al Arab on the Jordan–Syria border. These surface flows originating within arid lands often fail to reach the sea and drain into inland drainage basins, a characteristic which has been used to classify arid areas. Another distinguishing feature of surface runoff is the wadi which Goudie and Wilkinson (1977, p. 40) describe as

> a stream course which is normally dry but sometimes subjected to large flows of water and sediment.

The term 'stream course' is more appropriate than 'dry valley' as wadis can be found in relatively flat arid areas while dry valleys can also be found in humid areas. Although annual discharges are lower in arid areas compared to temperate regions, the peak discharges can be significantly higher and are often much less predictable, Nir (1974, p. 25) states,

> more people have drowned in desert rivers than perished from thirst.

The variability of precipitation in space and time inevitably has consequences for the generation and prediction of surface runoff, particularly those lands with mountainous areas. When rainfall is so localised it is possible for surface runoff

Plate 2 Perennial flow is maintained by water stored in Wadi gravels, providing a vital supply sustaining communities in otherwise desolate areas.

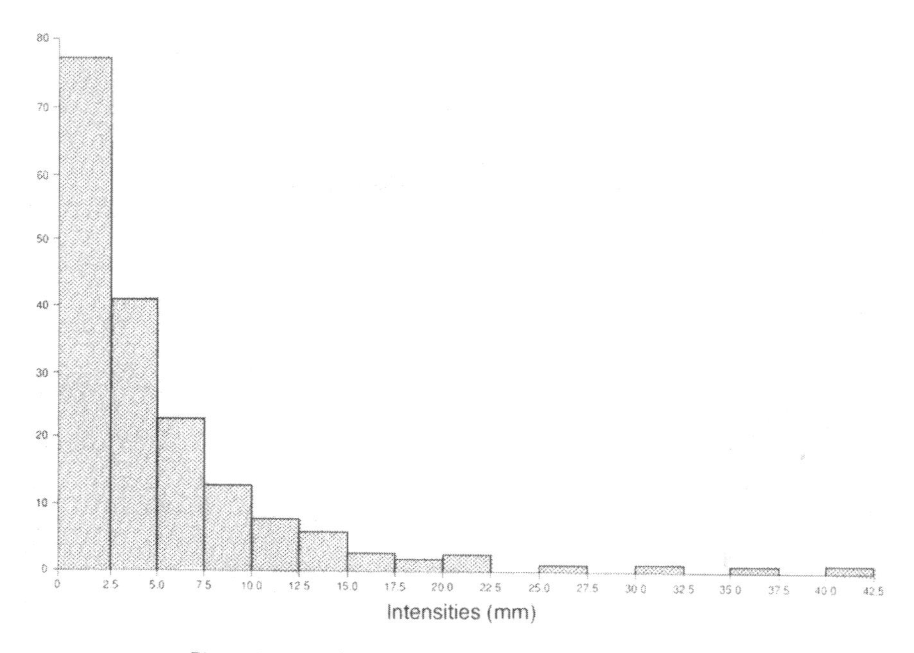

Figure 2.9 Rainfall intensities, Niamey, Niger (1973–8).

to occur in only some wadis while adjacent ones remain dry. Rainfalls in tropical latitudes can be very intense as well as being localised. Jackson (1977) notes that in the tropics 34.2 per cent of raingall intensities exceed 25 mm an hour and Nir (1974) mentions that intensities of 40 to 50 mm an hour are not uncommon. Figure 2.9, however, indicates that high-intensity rainfalls were not in fact as frequent in Niger as Nir suggests for arid lands in general, and figure 2.10 shows that when such events did occur they were of short duration and rarely exceeded 40 mm an hour at Niamey. Delwaulle (1973) noted that rainfalls in West Africa only reached 150 to 172 mm an hour twice during the period 1966 to 1971. However, a recent study by Hoogmoed (1986) in Niger reported the following peak rainfalls:

Niamey Ville	386 mm per hour, total rainfall of 22 mm (1977)
Zinder	325 mm per hour, total rainfall of 14 mm (1982)
Gaya	436 mm per hour, total rainfall of 26 mm (1975)

A major reason for such seemingly contradictory evidence of maximum rainfall intensities is the lack of operational rainfall rate recorder devices and hence the need to estimate rainfall rates from daily rainfall charts. Such interpolations can be prone to error and the results depend upon the smallest time scale used. The introduction of tipping bucket rainfall gauges linked to data loggers is a useful development although care needs to be given to the time interval over which observations are taken and the need to maintain such instruments.

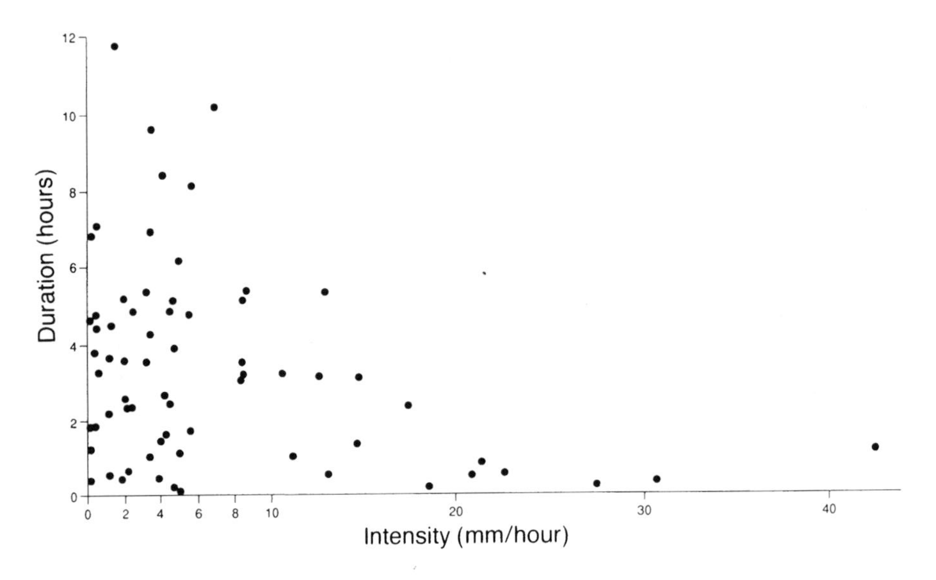

Figure 2.10 Rainfall intensity: duration at Niamey, Niger (1977–8).

Figures 2.11 and 2.12 show infiltration rates for sands in Oman and West Africa which suggest that rainfall intensities in excess of 100 mm an hour would be necessary to initiate overland flow in these coarse sands, hence producing rapid runoff in channels following the ideas of Horton (1933 and 1945). Given that in many arid lands, soils are not deep and often there are surface crusts and vast areas of bare rock, these infiltration rates appear excessively high. Hoogmoed (1986) suggested an average surface storage of only 5 mm and infiltration rates of 30 mm per hour were more appropriate. Taking these parameters he estimated that 5 to 15 per cent of annual rainfall would run off (see table 2.5).
Overland flow is generally assumed to be the mechanism operating in arid environments although Scoging (1989) notes that this is a gross oversimplification and reviews a number of attempts to model this process by including surface characteristics and profile controls. Graf (1988) suggests that numerical models of infiltration, storage and surface runoff are, however, not widely proven for

Table 2.5 Surface runoff for selected dryland sites (after Hoogmoed, 1986)

Site	*RO(%)*	*RO(mm)*	*P(mm)*
Patancheru, India	4.5	33	738
Gaya, Niger	8.7	67	774
Niamey, Niger	14.4	74	514
Maradi, Niger	15.6	56	356
Zinder, Niger	10.4	38	365

(Where RO = surface runoff, P = annual rainfall.)

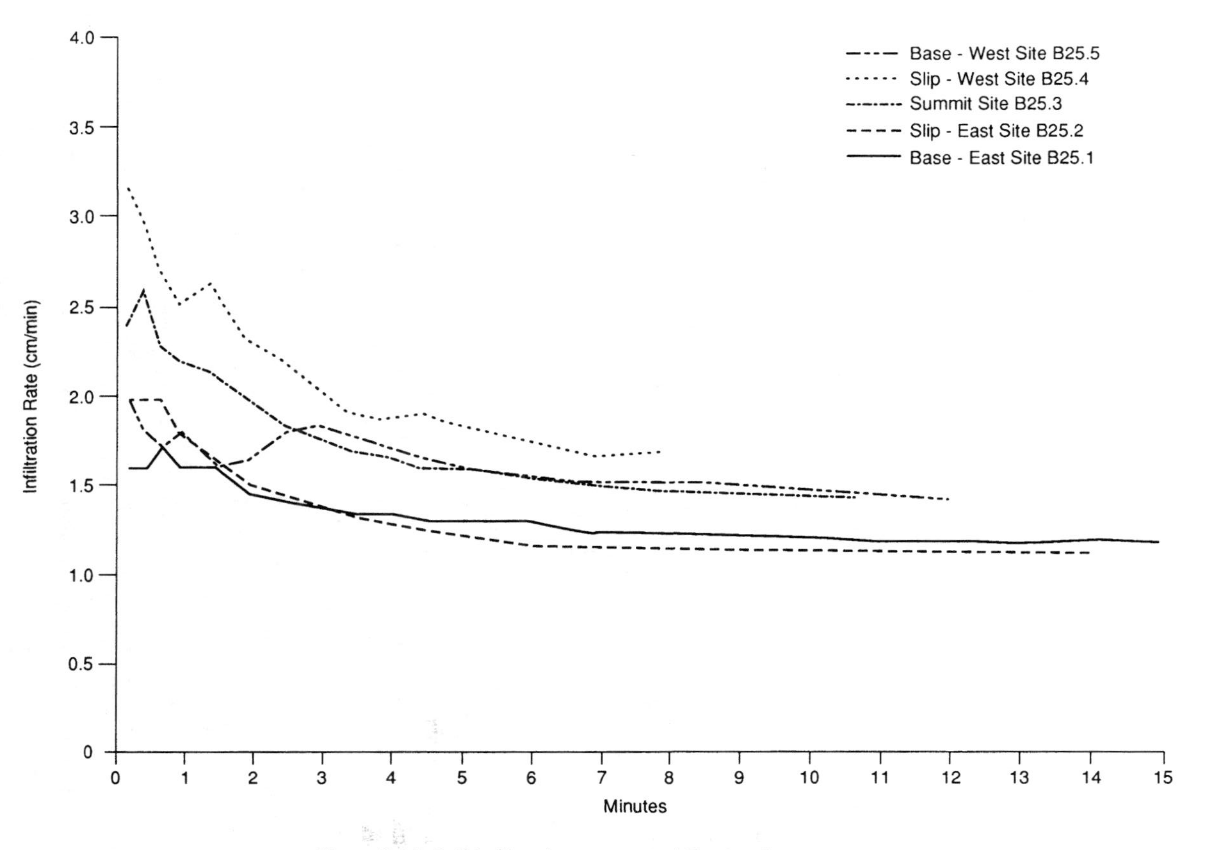

Figure 2.11 Soil infiltration rates, Wahiba Sands, Oman.

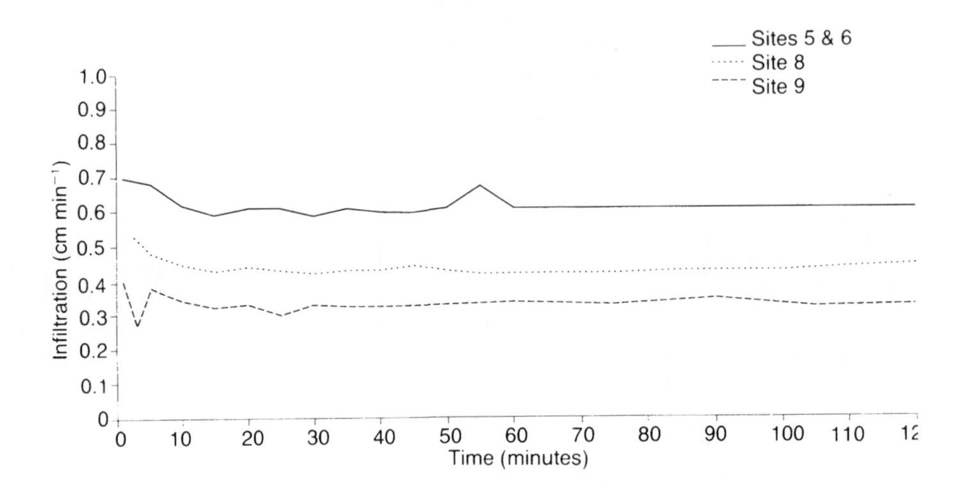

Figure 2.12 Soil infiltration rates, ferruginous sands, Agrhymet Centre, Niger.

arid lands and require too much data. Instead Graf advocates empirical approaches, suggesting that of the US Soil Conservation Service (1972) which takes account of soil infiltration groups (gravels to clays); land cover and condition (range, meadow, woodland, roads); and rainfall amount.

Rodda *et al.* (1976) review criticisms of Horton's ideas of overland flow and discuss an alternative theory that in vegetated areas with deep soils throughflow is more likely to account for storm runoff (Hewlett and Hibbert, 1963 and Kikby and Chorley, 1967). However, in arid lands with intense downpours and often only a thin covering of soil, the surface layers can quickly become saturated, with both overland and throughflow distinct possibilities. Such high-intensity rainfall events, while not very common, can be of great hydrological significance especially for those areas with steep slopes and only a thin surface of weathered mantle. In such conditions the surface may be quickly saturated leading to the generation of very rapid runoff producing dangerous and destructive flash floods, although not all rainfalls produce channel flows.

Graf (1988) distinguishes between:

Flash floods, limited to basins of 100 km^2 or less, produced by convectional rainfall and characterised by a wall of water with maximum discharge reached in a matter of minutes.

Single peak floods, of a longer duration from a few hours to a few days, as the product of tropical storms over basins of thousands of square kilometres.

Multiple peak floods, when rainfall remains persistent for days or even weeks over a region.

Seasonal floods, found in exogeneous flows, e.g. the rivers Nile and Senegal.

The flood hydrograph associated with flash floods is characterised (Reid and

Plate 3 Intense and sudden downpours can release large amounts of water and sediment in arid areas, causing flooding and cutting off communications.

Plate 4 Flooding at the Sultan Qaboos University, Oman.

Frostik, 1989) by an incredibly rapidly rising limb to a very sharp peak and an equally rapidly falling limb. Ward (1978) cites the example of Colorado Springs (USA) where a flood of 12 million m^3 passed through the city in just 2 hours causing immense damage while the stream bed, normally dry for most of the year, could have accommodated this amount of water if it had taken 20 hours or longer to flow. It was the speed and intensity of the flood rather than the total amount of water that created the problem. Flooding can also be a problem with perennial rivers with, for example, the Nile breaching its banks in 1988 resulting in damage and destruction in the Sudan, while in 1990 it was reported from Mogadishu that tens of thousands had to be evacuated in Southern Somalia because villages in a normally arid area were submerged by flood waters (*Independent*, 30 April 1990). Inevitably such events are difficult to monitor and predict, whilst much of this water may run off into oceans and therefore be lost to agriculture and other activities. Associated with such flash flooding is a high sediment load. The debris carried by arid realm streams can be quite high and Nir (1974, p. 24) quotes values of

20 kg/m^3 Chaco R., Argentina,
78 kg/m^3 Colorado R., United States,
144 kg/m^3 Rio Grande, United States,
250 kg/m^3 Hwang-Ho, China,

compared to the load of the Missouri of 2.7 kg/m^3. The reasons for such a high sediment transport are the infrequent occurrence of flood events; the high flows that do occur; and during low flow periods the build-up through weathering of detritus material which is readily available for transport due to the lack of vegetation and organic matter. Runoff events of a recurrence interval of 100 years or more are quite capable of removing all the available sediment in small catchments (Schick, 1985). Schick cites the example of a 48 hour February rainstorm with a recurrence interval of 1 in 50 years or more over the Sinai Peninsula that produced a deltaic deposit 300 m long and 500 m wide at the outlet of the Wadi El Arish, while in 1973 a 1 in 200 year rainfall event at Oued Medjerdah, Tunisia, produced a deposit over a 138 km^2 area.

Where there are perennial (and probably exogenous) rivers a different problem may be presented by their marked seasonal regime shown in figure 2.13 (see Rodier, 1985). The river responds with a time delay of several months to inputs during the rainy season (June–September) so that peak flows occur well into the dry season. This can have advantages for irrigation but the high temperatures and low humidities plus different day lengths compared to the rainy season can present difficulties for some crops. There is little base flow at the end of the dry season to maintain the regime so flows decline dramatically by the start of the rainy season and discharges vary from 2,000 to 1 cumecs. A comparatively new and worrying phenomenon, occurring at this location, is the alteration of hydrological regimes by anthropogenic activities, e.g. irrigation in the headwaters in Mali (Rodier and Chaparon, 1970). The historical data and derived past

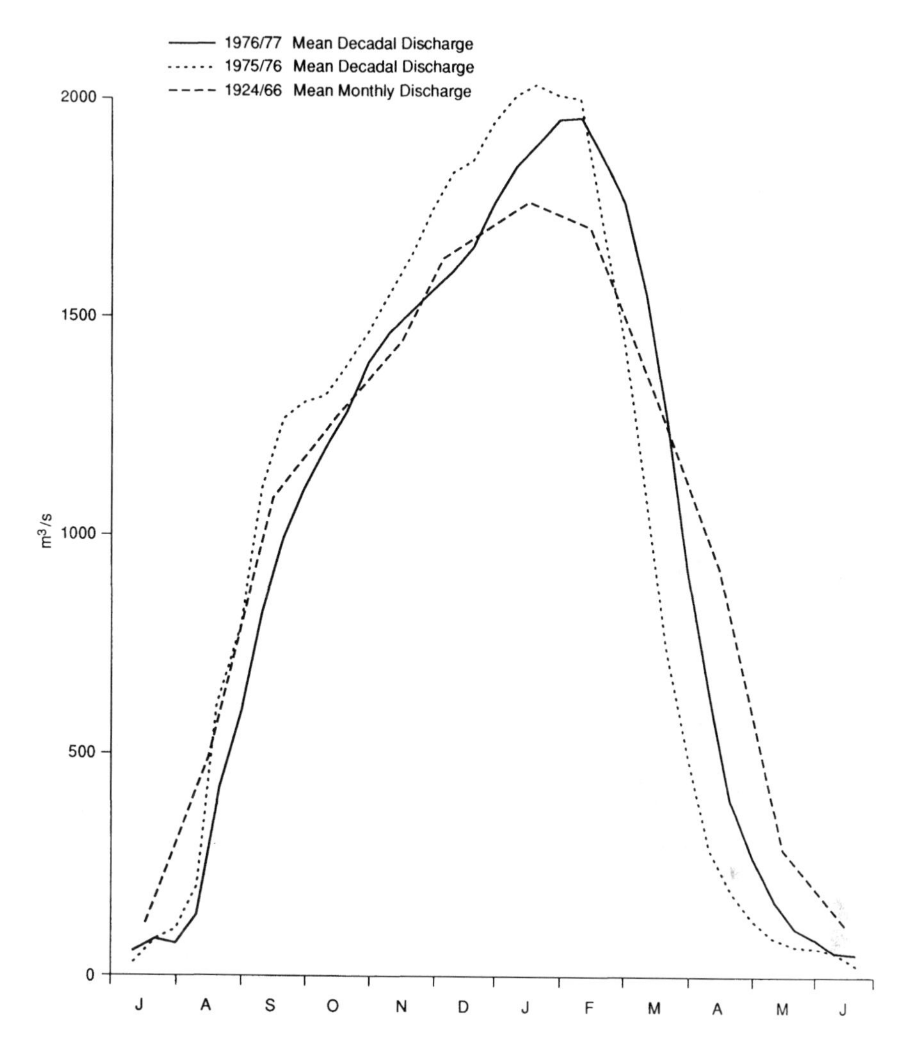

Figure 2.13 Seasonal flows of River Niger, Niamey, from Bureau des Mines et Hydraulic, Niger.

relationships of flows may no longer be valid for predictions of discharge which hampers flood warnings, navigation control and efficient planning of extraction rates. Other authors (see chapter 3) have commented upon climatic change affecting flows in the rivers Nile, Senegal and Niger.

GROUNDWATER

From the introduction to this section it is evident that on a global scale there are vast amounts of groundwater resources compared to surface stores in lakes and rivers and the atmosphere. But only a small fraction of groundwater takes part in the hydrological cycle, feeding base flows of streams, maintaining lake levels and being recharged through surface flows and percolating rainwater. Groundwater has become an increasingly important source of freshwater (see chapter 9) with UNEP and the World Bank funding attempts to survey the arid land groundwater resources and to develop appropriate extraction techniques. Such programmes are much needed as, unfortunately in many arid lands, there is only a crude idea of the extent of groundwater. UNESCO (1978) estimate that 24 per cent of the world's groundwater resources can be found in Africa, yet World Resources Institute (1987, p. 120) state:

> Most sub-Saharan African countries do not have enough information about their groundwater locations, recharge areas and available volumes to develop plans.

Partly this is due to the costs of identifying and quantifying groundwater resources through borehole drilling, mapping springs and monitoring wells; but it is also a function of the discontinuous occurrence of groundwater coupled to the great depths at which it can sometimes be found. Table 2.6 shows the details of a number of large groundwater systems in arid lands varying in extent from 136 to $1,800 \times 10^3$ km^2.

Groundwater is separated into the saturated region or **phreatic zone** where all voids and pore spaces are filled with water, the top of which is demarcated by the **water table**, above which lies the **vadase zone** where unsaturated, aerated conditions prevail. The water table is free standing where atmospheric and fluid pressures are in equilibrium, but **confined** conditions can occur when an aquifer is overlain by a less permeable bed. This condition is shown in figure 2.15 with a

Table 2.6 Selected arid land groundwater sources (based on Margat and Saad, 1984)

	Australian great basin	*Algeria + Tunisia*	*Libya*	*Niger Mali*	*China*
Geology[1]	Ss,S	Ss,S	Ss,S	Ss,s	a
Area (10^3 km^2)	1,700	600	1,800	500	136
Storage (m^3)	2×10^{13}	6×10^{13}	6×10^{12}	$4-8\times10^{10}$	$0.5-1\times10^{12}$
Renewable resources (10^6 m^3/year)	1,100	270	1,000	800	35,000
Total dissolved solids (g/l)	0.5–1	0.5–6	–	<0.5	–
Withdrawal[2] (10^6 m^3/year)	600	200	160	–	10,000
Fall in water level (m/year)	1.45	2.07	3.75–5.0	–	3.75–5.0

Notes: [1] S = sand, Ss = sandstone, a = alluvions
[2] rate based on 1970s observations

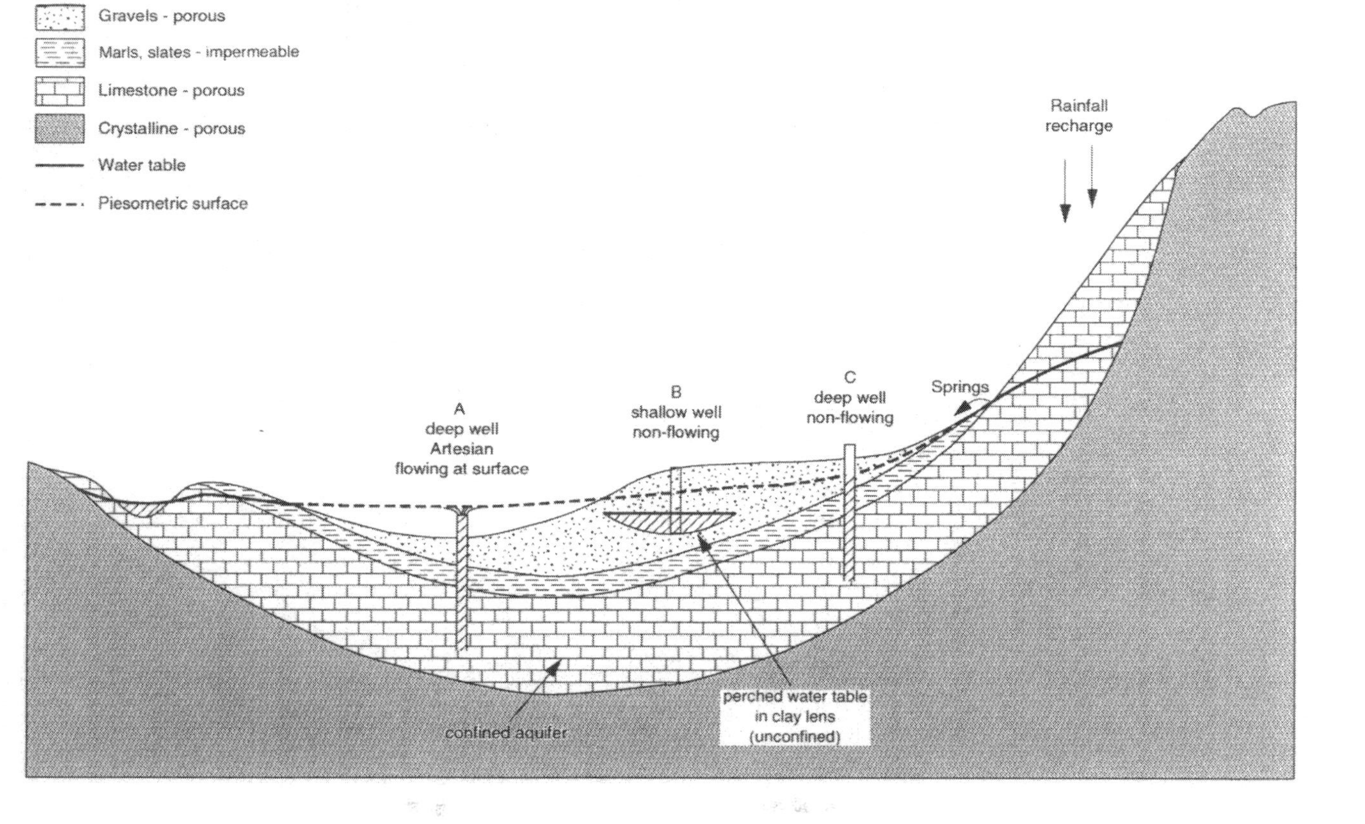

Figure 2.14 Occurrence of groundwater.

synclinal basin containing porous and impermeable rocks. Rainfall in the mountains recharges groundwater in the limestone aquifer. This water is channelled down into the limestone aquifer but the overlying impermeable layer of marls and slates prevents upward movement and creates confined conditions. A hydraulic pressure can be created such that the piesometric surface (height to which water will rise if free to do so) is above ground level, which will force water up any well tapping into the aquifer with the possibility of water flowing naturally to the surface (Artesian conditions) as in the case of well 'A'. Water at well 'C' rises up to the piesometric surface but does not reach the ground level. Confining beds can be discontinuous and interbedded with porous material, i.e. clay lenses in alluvial gravels associated with a fan deposit. It is then possible for **perched groundwater** to occur which is prevented from draining into a lower saturated zone, producing a multi-layered water table as shown in figure 2.14. Thus water can be found in the shallow well 'B' but is only available locally, and the supply is more limited than for wells tapping into the deeper, more extensive aquifer (the water table may not follow the surface topography when recharge is localised).

These water resources are stored in the pore spaces and joints of rocks and unconsolidated sediments or in the openings widened through fractures and weathering. The term **aquifer** is given to a rock or sediment of which all or part is saturated with water and is able to yield significant quantities. Lloyd (1981) suggests that the term should also include the qualifier that the water supply is sufficient for a particular economic use. That is, rocks that merely contain water do not necessarily warrant the term aquifer. Rocks which are porous and can therefore contain water but which are unable to transmit it rapidly enough to supply a well produce an **aquiclude** (Goudie, 1985). This definition is obviously dependent upon the yields required from the well. Shaw (1983) provides a more distinct term of **aquitard** for a semi-porous bed which only allows some slow seepage of water and hence acts as a barrier to groundwater movement, as distinct from **aquifuge** which applies to a bed that neither absorbs nor transmits water (Bear, 1979). The Australian Water Resources Council (1975) prefer to avoid possible confusion over the terms aquiclude, aquitard and aquifuge by employing **confining bed** as a 'less permeable material overlying or underlying an aquifer'. That is, it is the relative permeability of the rock or deposit compared to the aquifer characteristics rather than some absolute measure (Lloyd, 1981).

Table 2.7 shows the different materials in which groundwater commonly occurs with sand, gravel and sandstone clearly important. Limestones can equally contain significant groundwater resources when joints and cracks have been enlarged to form voids in which large quantities of water may be stored. Nir (1974) provides examples of Palaeozoic dolomites in Zambia, Palaeozoic limestones of the Anti-Atlas in Morocco and limestones under the Sahara. Jones *et al.* (1981) make the point that there is less difference between arid and temperate regions concerning groundwater occurrence than for surface water resources, although carbonate rocks tend not to be as highly karstified, and extensive cave

Table 2.7 Hydrological properties of rock types (based on Australian Water Resources Council, 1975)

Unconsolidated sediments	*Porosity (%)*	*Hydraulic cond. (m/day)*	*Specific yield (%)*	*Comments*
Clay	50	$5 \times 10^{-1.5}$ to 10^{-7}	5	water not easily yielded
Silt	45–50	5×10^{-4} to 10^{-1}	15	water not easily yielded
Sand	35–45	5×10^{-2} to 10^{-1}	30	good yields
Gravel	35	5×10^{3} to 10^{2}	20	good yields
Sedimentary rocks	*Porosity (%)*	*Hydraulic cond. (m/day)*	*Specific yield (%)*	*Comments*
Sandstone	5–30	5×10^{-2} to 2.5	5–15	variable but yields good
Shale	1–10	5×10^{-7} to 10^{-2}	0.5–5	water in fractures
Fractured rocks	*Porosity (%)*	*Hydraulic cond. (m/day)*	*Specific yield (%)*	*Comments*
Limestone	1–10	high in fractured	0.5–5	good yields are possible
Basalt	up to	and	high	good yields are possible
Granite	35% where	weathered	where	good yields are possible
Schist	fractured	rocks	fractured	good yields are possible

systems are relatively rare, although they may be of great local significance.

The low hydraulic conductivities of clay and silt reveal their impermeable nature with very slow movement of water through such a medium. Hence such deposits have a low specific yield (the volume of water that a water-bearing rock or soil releases from storage under the influence of gravity, Goudie, 1985, p. 401) of only around 5 per cent. While fractured rocks can contain substantial amounts of groundwater, it is only available locally, whereas sedimentary rocks and unconsolidated deposits can provide extensive groundwater resources with good specific yields. Beaumont (1988) reports that in the Middle East there are two main types of aquifer:

1 Alluvial aquifers which are the commonest and smallest, formed from sand and gravel unconsolidated deposits and locally very important.
2 Solid rock aquifers (of limestone and sandstone) which tend to be deeper and much more extensive but may contain fossil waters that are not being recharged.

Fractured rocks are found in Saudi Arabia but do not yield water as rainfall is too limited (Jones *et al.*, 1981). Clearly then the occurrence of groundwater varies spatially both in terms of depth and location depending upon geological conditions and the presence of suitable materials for containing and yielding water.

Goudie and Wilkinson (1977) point out that fortunately arid lands contain large sedimentary deposits created by the infilling of synclinal basins and large aquifers can be found in the valleys of the Nile, Senegal River and San Joaquin Valley, California. However, the occurrence of groundwater is also a function of the input of water, i.e. recharge. Shallow aquifers may be recharged by contemporary hydrological processes via percolating rainfall or seepage from lakes and surface runoff. The Salt River, Arizona, lost between 17 and 29 per cent of flow through transmission losses to groundwater storage (Graf, 1988). S.N. Davis (1974) believes that apart from transmission losses, groundwater recharge is a rare event with most rainfall lose through evaporation. Exogenous surface flows, percolation of irrigation water and rainfall where aquifers outcrop outside of arid regions are then seen as the most important mechanisms. Beaumont (1988), however, argues that contemporary groundwater recharge is greater than previously thought and Agnew (1988a) suggests that rainwater percolating through sands in Oman may in fact reach the water table some 30 m below the surface. Sand dunes and talus deposits at the base of slopes are then important areas for recharge and Beaumont (1989) reports that in Saudi Arabia some 25 per cent of rainfall can reach the groundwater table in dune fields.

Some deep aquifers are regarded as 'fossil' resources because they were saturated some time ago and are no longer being recharged at any significant rate. Margat and Saad (1984) use the term fossil to describe water that has been present in an aquifer for several thousands of years. Simpson (1985) reports that the principal period for groundwater recharge in the Middle East was either 35,000 to 15,000 years ago or 8,000 to 5,000 years ago. However, merely having a porous rock and a water input does not necessarily result in storage unless a suitable geological structure is present so that the potential aquifer is underlain by an impermeable rock. Nir (1974) lists a number of such 'traps' varying from lava flows underlain by volcanic tuff found in the Yemen, through to clastic deposits alternating with aquicludes such as along the coastal plain of Israel. Wilkinson (1977) adds an example of sand dunes in Oman overlying semi-permeable bedrock producing a water table in the dunes which supports springs issuing from the base of the sands.

Once water has entered an aquifer it is not immobile and can seep out at springs or leak into other aquifers, but the rate of movement may be very slow. Jones and Rushton (1981) report that movement of saline groundwaters in the Indus Plain has still not reached equilibrium after 70 years of irrigation while the saline water wedge in the Nile alluvial deposits is also not in equilibrium after several thousand years of hydrological changes. S.N. Davis (1974) computes that the average groundwater movement in arid lands is less than 0.1 m per day due to the low hydraulic gradients and is a function of low recharge rates. Thus it would take 5,000 years for water to travel a distance of 200 km, although the rate of movement will be greater in areas of recharge where hydraulic gradients are steeper.

The variability of groundwater is not then simply a matter of spatial variation in the location of aquifers but also the rate of water yielded and the recharge rate. Finally, water quality should be added to this life as Nir (1974) maintains that the mineral content of groundwater increases in direct proportion to the depth of the water table and that even in quite shallow aquifers the salinity can be quite high. In Australia values of around 6 g/l total dissolved solids (TDS) are recommended for cattle and 15–18.5 g/l for sheep, compared to around 3.0 g/l for humans. Values of 0.66 were observed in the Northern Territory rising to 6 g/l in Western Australia (Tamala station) (Australian Water Resources Council, 1975). The reasons for the greater salinity in arid lands are (after S.N. Davis, 1974):

1 Evaporation deposits salts on the surface that can be leached into the ground.
2 Surface salts are augmented by dry fallout of soluble salts.
3 Deserts favour deposits of gypsum, halite and other evaporites.
4 Slow groundwater movement results in less flushing of saline waters.

The problems of water extraction and artificial recharge will be discussed in chapter 9 while chapter 5 deals with the assessment of groundwater resources in arid lands.

It has been suggested that the variability of water supplies in arid lands may be no greater than in more humid areas in absolute terms. Certainly rainfalls can be highly localised in temperate regions and groundwater storage is influenced by much the same environmental conditions across the globe. However, groundwater recharge is more uncertain in drylands and surface flows can be more rapidly generated where there is little or no surface covering of vegetation or soil. Discussion of the relative variability of water supplies is impaired by the poorer data base in arid lands. It is clear, however, that the most important consideration is the greater impact that any variation of water supplies has in arid lands. Concern is growing then over the possible changes that will be brought about by climatic change and manifest through increased drought. Chapters 3 and 4 will discuss these in turn before consideration of the problems of data collection in chapter 5.

3

CLIMATIC CHANGE IN ARID LANDS

Hardly a week passes without some reference to climatic change or the 'greenhouse effect', usually with an apocalyptic description of ice caps melting, the ozone layer depleting, CO_2 levels doubling and a corresponding sea level rise threatening vast areas with flooding. This vision is fuelled by headlines such as McKie's (1984) 'The race against the clock and the thermometer' and Lovelock's (1985) 'Are we destabilising the world climate?' while North and Schoon (1989) commented, 'Mankind's disregard for nature threatens to produce climatic changes on a scale not seen since the ice ages came and went'.

While Kondratyev's (1988) recent text on the effect of atom bomb tests in the 1960s clearly shows the impact humans can have on the atmosphere and climate, conversely Bandyopadhyaya (1983) suggests that such concerns are mainly a problem for 'Western Nations' and the northern hemisphere, which has little relevance for Third World arid lands facing famine and poverty. Nevertheless the tide of opinion is that climate change threatens the entire globe and in particular arid areas where life is particularly sensitive to atmospheric conditions.

EVIDENCE OF CLIMATIC CHANGE

The evidence for climatic change tends to be divided into global or regional changes and the discussion below deals with each in turn. When analysing climatic change one needs to recognise that 'climate' is a statistical average of observations over a given period. Hare (1983 and 1985) suggests that we distinguish between climatic noise (short-term weather changes), climatic variability (variation within a given averaging period) and climatic change (when differences exceed noise and variability). Climatic fluctuations are short-term climatic changes (a few decades) when conditions return to the previous state. It is difficult in all regions to establish whether observed changes are short term or long term but this is particularly acute in drylands because of their shortage of observations and inherent variability.

Evidence that the climate of the Earth has gone through several changes in geological time is reviewed in the texts by Grove (1977), Lamb (1977) and Oliver (1973 and 1981). The last major climatic change was an increase in temperatures

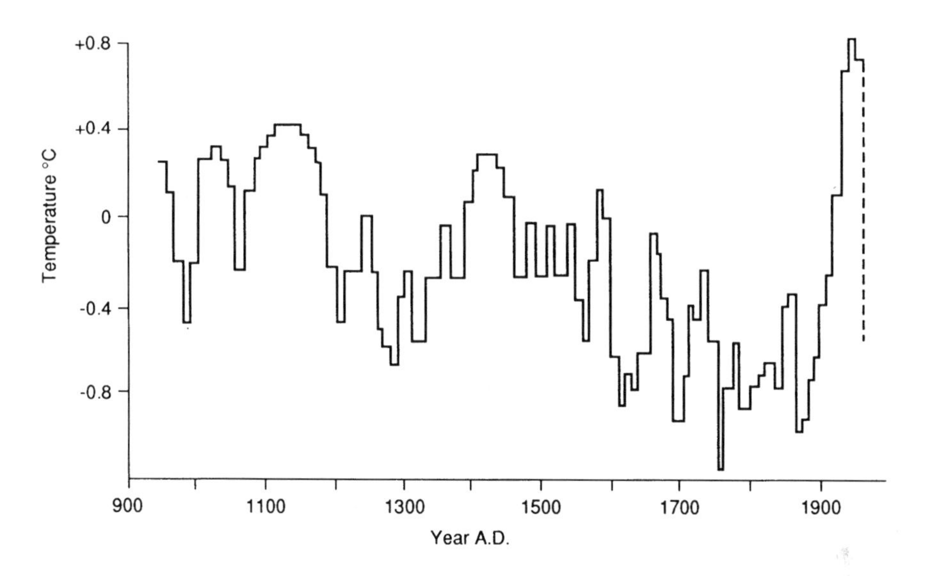

Figure 3.1 Average temperatures in the northern hemisphere after Bryson (1974) based on data from Pall Bergthorsson.

starting 15,000 years ago (McElroy, 1986) and was associated with the disappearance of the ice sheets about 10,000 years b.p. There have been five periods of cooling during the last 10,000 years, superimposed upon a gradual global warming. The last cold period, the 'Little Ice Age', began in the 14th century and continued until the late 1800s (see figure 3.1). Since then, figure 3.2 shows that 19th century temperatures in the northern hemisphere in particular have risen by about 0.8°C (Lamb, 1974 and Kellogg, 1978). Lamb (1973) reports that the extent of Arctic ice decreased by 10 per cent and the growing season in England was extended by 2 to 3 weeks. This period of warming eventually came to an end during the 20th century (Perry, 1984) and was followed by a sharp decrease in temperatures. Hare (1985) suggests the decline in temperature of 0.5°C started in the late 1930s while Perry (1984) reports the decline in temperature was less than this and was only clearly evident by the 1960s. This was followed by a return to a warming trend evident by the 1970s. There are insufficient climatic observations to conclude whether these changes are fluctuations or indications of climatic change. This can only be discerned with knowledge of the causes of such changes. It is notable, however, that 20th century changes have been significantly greater than those in historical times which suggests that mankind may well be influencing and even accelerating these changes. A comparison of the downward trends of rainfall visible at some arid land locations evident in figure 3.3 with the corresponding increase in temperatures shown in figure 3.2 has led to speculation that increasing aridity may result from a continued rise in temperatures.

The climate data base is relatively poor in arid lands and can be inadequate for

Figure 3.2 Changes in the average surface temperature of the earth from 1870 to 1978, showing the warming from about 1890 to 1940 and the cooling which followed, after Lamb (1981).

the investigation of secular changes. Consequently the incidence and persistence of drought has often been used to signify the onset of new climatic conditions but chapter 4 will point to the dangers of this assumption. Nevertheless the widespread and often persistent nature of drought has been noted in many parts of the arid realm (see figure I.1) as a cause for concern, perhaps heralding worse conditions in the future. The examples cited below are merely illustrative of the incidence of drought and possibility of climatic change in arid lands and much of the material is gleaned from reports summarised by Milas (1983) and Hare (1983, 1984a and 1984b).

Asia and Australia

In 1902 drought savaged Australian livestock herds and ravaged crop yields. This was followed by dry spells during 1912–15, 1965–7 and 1972 but the worst drought in two centuries ended in 1983, after crop production had fallen by 31 per cent and farmers' income by 24 per cent. Rainfalls were half the normal amounts and some areas experienced 4 years of drought. Pittock (1988) reports that areas receiving winter rainfall in South West Australia have experienced a 10 per cent decline this century but it is also noted that increases are evident in New South Wales probably due to global warming (see below). In the same text Hobbs (1988) reviews a number of studies concerning climatic change in Australia and concludes that there is much variability, patterns are complex and there is no clear evidence for sustained climatic change. Dhar *et al.* (1979) and Krishnan (1979) noted the frequency with which droughts in India and Australia coincided (e.g. 1965–7 and 1972–5) while Indonesia also suffered at the same time as the recent Australian droughts, prompting speculation over related causal mechanisms. Vines (1986) found a number of cycles in Indian rainfall and believes that they can be explained by sunspot activity. There also appears to be some association with the El Nino Southern Oscillation (ENSO) although Gregory (1988) revealed that the relationship is complex and spatially diverse.

Latin America

North East Brazil has just faced its worst drought since 1583, with an estimated 10 million people destitute out of a population of 24 million and 90 per cent of subsistence crops failing (Hastenrath *et al.*, 1984, Brahmananda *et al.*, 1986, Hall, 1978 and *The Times* 28 March 1984). In Mexico 1982, drought reduced agricultural production by 12 per cent and staple foods were imported as a consequence costing 1.6 billion dollars. Bolivia has also been struck by its worst drought this century with drastic reductions in livestock numbers (*Newsweek* 20 June 1983) and crops (*The Economist* 6 April 1983).

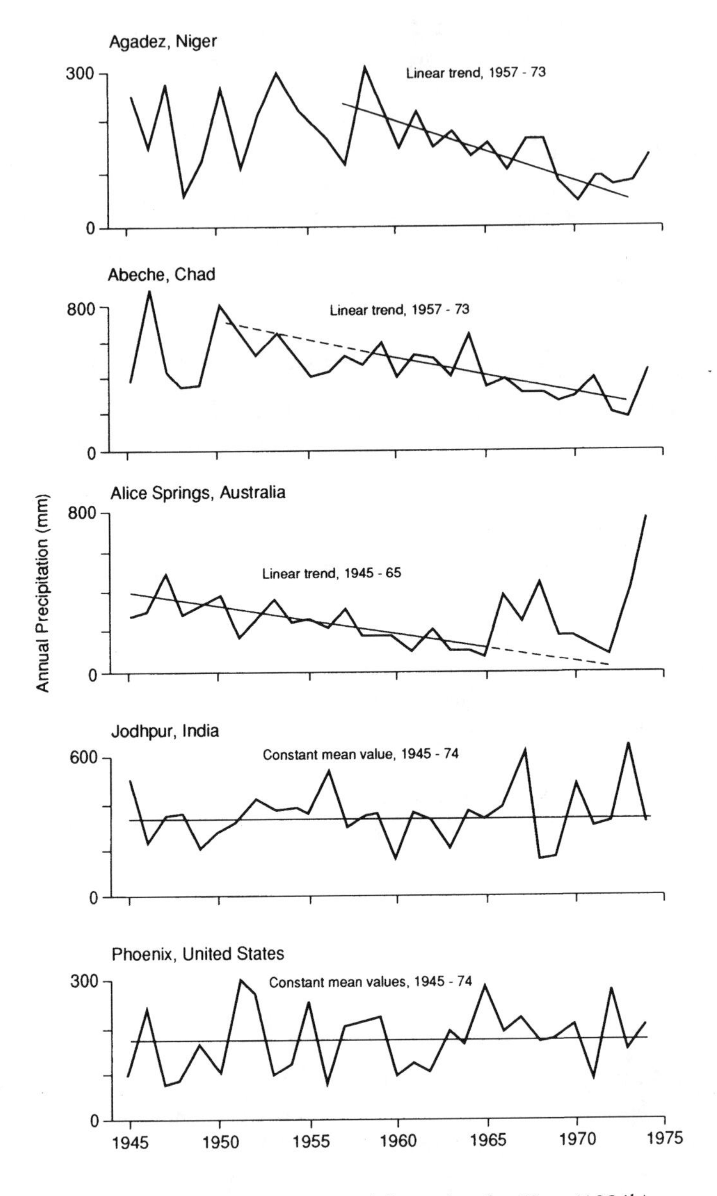

Figure 3.3 Arid lands rainfall trends, after Hare (1984b).

Africa

Glantz and Katz (1987) have reviewed the impact of drought in Africa and listed 13 countries from the Sahel to Southern Africa that were affected by drought in 1986. The states of Southern Africa have been struggling in the 1980s including Zimbabwe (1983 harvest only a third of 1980), Botswana (rainfall has been halved) and Mozambique (rainfall lowest in recent history). Even technologically advanced South Africa found it had to import cereals in 1983 owing to the impact of drought that was predicted by Tyson (1980). To the north, drought has struck Kenya, Somalia and of course Ethiopia (Downing *et al.*, 1987). In Ethiopia 5 million people were estimated to be affected by drought in 1984 rising to 6.8 million by April 1986 (Office for Emergency Operations in Africa, cited by Glantz and Katz, 1987). The Sahel zone of West Africa (which lies between the 200 and 700 mm isohyets, Davy *et al.*, 1976) has perhaps received most attention following the 1968–73 disaster. In a series of articles and papers in the early 1970s, Kelly (1975), Lamb (1973 and 1974) and Winstanley (1973a and 1973b) reported on the possibility of a continuing downward trend in Sahelian rainfall. More recently Lamb (1982), Motha *et al.* (1980) and Nicholson (1979 and 1981a) have commented upon the persistence and continental scale of drought in Africa. In 1984 Tooze reported that the WMO considered that the drought in the Sahel had not yet ended, and in 1986 Sir Crispin Tickell, in a lecture to the ODA, stated:

> The documentary record for the last 70 years shows a slight decline of rainfall from 1955 and acute drought since 1968.

This view is supported by Flohn (1987) who commented:

> The Sahel drought, lasting without interruptions until, (at least) 1985.

Clearly there is a strong body of opinion that the area has experienced and is continuing to experience increasing aridity. Winstanley (1983) suggests that there is now sufficient evidence to identify two rainfall trends in Tropical Africa. Areas such as the Sahel characterised by a single rainfall maximum are experiencing a linear trend towards lower rainfall. On the other hand regions with different climatic regimes are conversely experiencing a trend towards higher rainfalls. Winstanley notes, however, that his view of increasing aridity is controversial. Faure and Gac (1981) for instance suggest through an examination of the River Senegal that the present Sahelian drought should now be at an end, to be followed by a wetter period until severe drought will again be encountered around 2005. After the heavy rainfalls of 1988, reported by Toulmin (1988b), their analysis should be given further consideration. In reviewing the available evidence for UNESCO, Hare (1984b) saw little evidence to support the notion of a worldwide trend of increasing desiccation. There is then general agreement that there has been a rise in global temperatures during the last 100 years. Whether or not this will continue can only be answered by identifying the cause for this increase. The picture for rainfall is less clear. In arid parts of West Africa there is much evidence for a long-term decline in precipitation but rainfalls elsewhere in the arid realm show no clear trends.

CAUSES OF CLIMATIC CHANGE IN ARID LANDS

Given the speculation over global climatic trends and the difficulty of using the incidence of regional drought to reveal climate change, the evidence should be supplemented by an explanation of the mechanisms involved. Explanations for recent climatic changes can be divided into two main groups: external and internal processes. External processes concern changes in the amount of solar energy reaching the Earth while internal processes account for alteration of solar energy once it reaches the Earth through either the composition of the atmosphere or changes to the land surface.

Solar radiation

The amount of solar radiation received by the Earth depends upon factors such as its shape and size and the distribution of land and sea. Changes in these parameters have had a pronounced effect upon the climate through geological time (Budyko, 1978). It is believed that the impact of meteorites lifted dust into the atmosphere 65 million years ago reducing temperatures sufficiently to cause the extinction of dinosaurs. The Milankovitch theory concerns changes in the Earth's solar energy receipts occurring through its orbital eccentricities which alter the input of solar radiation with cycles of 105,000, 41,000 and 21,000 years. Harrington (1987) reports that we are now 6,000 years into the next cooling cycle. Such variables, however, operate over too long a time span to explain contemporary climatic changes (McElroy, 1986) although they may reinforce change through feedbacks. Solar emissions are variable over a shorter time period and sunspot activity has been correlated with droughts in Nebraska by

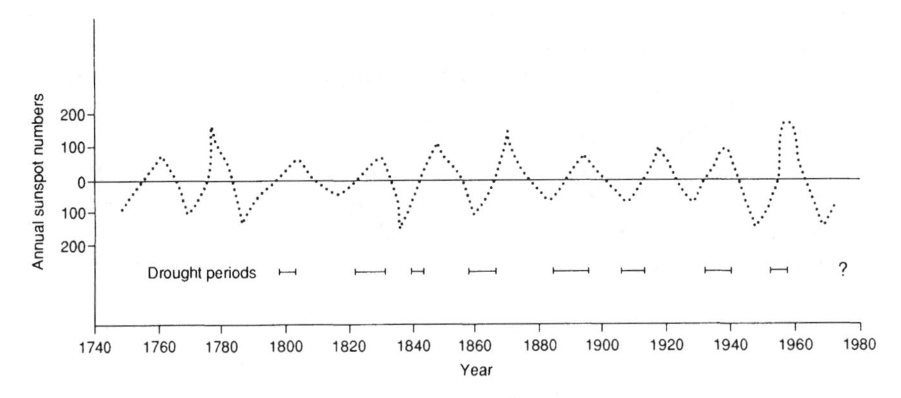

Figure 3.4 Sunspots and drought in Nebraska, after Thompson (1973).

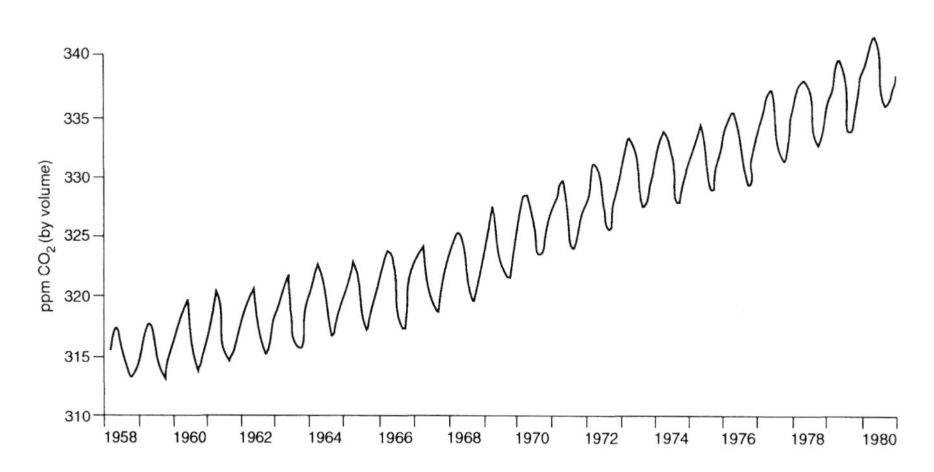

Figure 3.5 Carbon dioxide concentrations, after Royal Commission on Pollution (1984), and Keeling (1990).

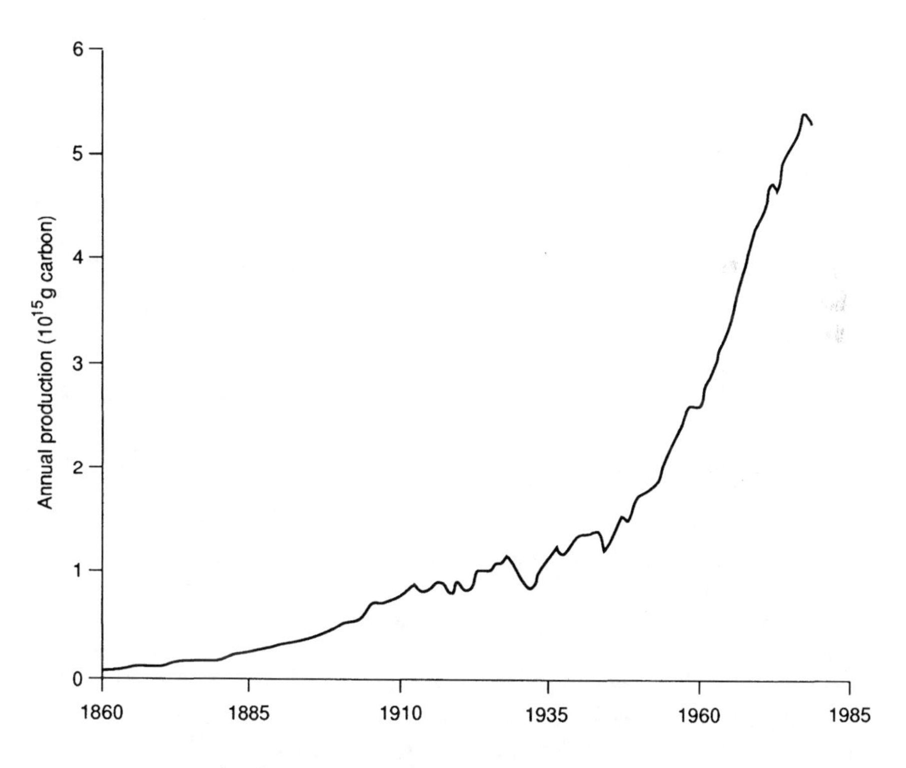

Figure 3.6 Annual carbon emissions, after Royal Commission on Pollution (1984) and Keeling and Whorf (1990).

Schneider (1976) as shown in figure 3.4, and considered by Willett (1976) to explain circulation patterns. Sofia (1984), however, concluded that the record of solar variability is too incomplete and inertia within the climate system means that there is no satisfactory model to explain solar activity and climate change over the last 100 years. Pittock (1983) agreed that many studies were based on inadequate data, but in reviewing other recent research he suggested that progress was being made towards an understanding of the physical mechanisms. Certainly the changes in solar activity during the last 100 years help partly to explain observed variations in mean global temperatures.

The greenhouse effect

This phenomenon has received much media attention with claims of widespread climate warming and associated rises in sea level. The burning of fossil fuels has released large amounts of carbon dioxide (CO_2) into the atmosphere such that concentrations have risen from levels of 180 ppm during the last ice age, (Bunyard, 1987), to around 250 ppm before industrialisation (Bach, 1984a), to 315 ppm in Hawaii in 1958 and reaching 340 ppm in 1982 (see figures 3.5 and 3.6). Predictions based on industrial activities suggest that CO_2 levels will double to 600 ppm during the next century and by 2100 at the latest, without changes in energy usage (Hare, 1984a and 1985 and Mitchell, 1987), though there are dissenters.

Greenhouse gases are also being released by the burning and clearing of forest with the *Guardian* (18 April 1988) reporting that the amount of the Amazon forest cleared has doubled over the last 2 years. Bunyard (1985) points out that man's agricultural activities over the last 150 years have contributed an amount of CO_2 that is comparable to the burning of fossil fuels, while Zimen (1979) conversely reports that some believe this input is negligible. Goudie (1986) notes that 160 Giga tonnes (Gt.) of carbon were emitted between 1850 and 1981 (a rate of 0.7 to 1.7 Gt. a year) through combustion of fossil fuels with 5.3 Gt. released in 1981. In comparison, 2 to 6 Gt a year are released through forest clearing with an additional 2 Gt. from organic decomposition, a substantial contribution but difficult to quantify accurately. Furthermore the Royal Commission on Pollution (1984) warned that not enough account has been taken of the absorption of CO_2 by the world's oceans. Nevertheless CO_2 levels have risen markedly during a period when global mean temperatures have risen by 0.5°C (Wigley and Raper, 1987). Oke (1978) demonstrates how CO_2 and other gases such as methane, ozone and water vapour are effective absorbers of terrestrial radiation with a mean wavelength of 10μm (see figure 3.7). Ramanathan *et al.* (1985) list over 30 greenhouse gases contributing to global warming. It follows that increases in these gases in the atmosphere will then result in higher temperatures. Whether this is sufficient to explain recent changes and predict future conditions can only be ascertained by modelling the effect of $C0_2$ levels on climate.

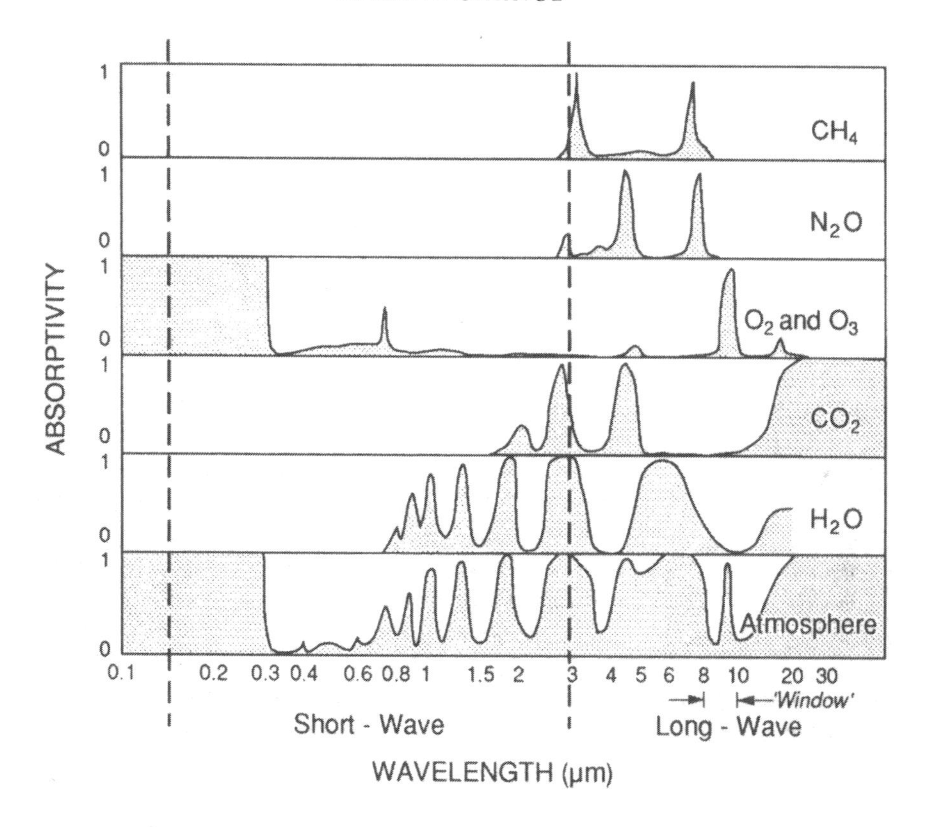

Figure 3.7 Absorption of solar and terrestrial radiation by atmospheric gases, after Oke (1978).

Ocean temperatures

Satellite observations of ocean temperatures have allowed examination of the correlation between temperature and rainfall and a complex pattern has emerged. Glantz (1987) reported that West African droughts might be related to ocean surface temperatures in the Atlantic. Furthermore the warm upwelling off the Peruvian coast known as El Nino (ENSO) appeared to coincide with drought in the Sahel during 1972 and 1982. The presence or absence of El Nino appears to be influenced by the strength of the trade winds in the south-east Pacific with a time lag evident. Caviedes (1988) concurred that prior to an El Nino event there are powerful south-easterly winds which produce a build-up of water in the west Pacific; this is later advected eastwards, affecting the normal upwelling of cold water off the South American coast. Lockwood (1984) reviewed the relationship that was first observed in the 1920s between monsoonal rainfall and sea surface temperatures in the Pacific. It is suggested that equatorial Pacific rainfall is in phase with rainfall over parts of Asia such that changes in Indian monsoonal rainfall are followed by changes in Pacific sea surface temperatures. This may enable forecasts of the El Nino phenomenon and Sahelian rainfall but Glantz and Katz (1987) believe the relationship is still speculative. Nevertheless there is a great deal of interest in this relationship following the work by Folland *et al.* (1986), Owen and Parker (1989) and Parker *et al.* (1988).

Atmospheric dust and volcanicity

Oliver (1981) suggests that the Rajputana Desert (India) remains arid despite moist air aloft because of high concentrations of dust in the atmosphere. During the daytime the dust reduces solar radiation receipts whilst at night the dust cloud cools rapidly to enhance subsidence of the air and suppress rainfall. Atmospheric cooling was noticeable, presumably because of dust following the 1815 Tambora and 1883 Krakatoa eruptions (Hare, 1985). Schonwiese (1988) demonstrates how a combination of volcanic activity indexes, sunspot activity and CO_2 concentrations can explain temperature changes in the northern hemisphere since 1860. Interest has also focused upon dust from agricultural and industrial activities. Middleton (1985) found that dust over the Sahel has increased dramatically in recent years and suggests this may be related to drought and poor landuse practices. Conversely Olsson (1983) working in Sudan found that it has not been proven that dust emanated from agricultural areas or where land degradation is found. Brazel and Brazel (1988) did not find significant relationships between farm abandonment, dust production and precipitation in Arizona. Hare (1984a and 1984b) concludes that there has not been adequate testing of the hypothesis that increasing dust concentrations can lead to more arid conditions.

Surface albedo

Otterman (1974 and 1975) working in the Middle East showed that surface temperatures were lower over areas with little vegetation owing to increased surface reflectivity of solar radiation (albedo). Charney and Stone (1974) suggested the mechanism whereby the overgrazing of vegetation led to higher surface albedos; this in turn reduced the amount of energy absorbed by the atmosphere producing stable air masses and thus decreasing rainfall. Lower rainfalls led to desiccated vegetation which completes the positive feedback producing even lower rainfalls. Unfortunately there was no account of advected energy from other regions and no account of the effect of lower evaporation leading to more energy being available to combat the effects of higher albedos. Dregne (1983) reports that we do not have sufficient knowledge to draw definite conclusions over this process. Studies in the Sahel using remote sensing have in fact suggested that there has not been any significant change in surface reflectivity (Courel *et al.*, 1984). Hare (1983) reports, however, that this process leading to decreased rainfalls has been reaffirmed by recent studies and desiccation is further enhanced by declining soil moisture contents. It appears that in arid lands much of the rainfall may come from re-evaporated rain and therefore declining soil moistures act to intensify drought conditions (Hulme, 1989). Although Kidson (1977) suggests that large-scale perturbations in atmospheric circulation are the more likely cause of changes in rainfall than alteration of land surface conditions, Cunnington and Rowntree (1986) and Walker and Rowntree (1977) have demonstrated the possibility of this mechanism through the operation of climate models, but they do state that there is a lack of 'real' evidence of this feedback process. Changing albedos and soil moisture through land degradation may then be the trigger for climatic change leading to more arid conditions, but there is still a lack of evidence to show that this is actually happening in the arid realm.

There are then a range of possible mechanisms influencing the climates of arid lands although the impacts of a temperature increase through global warming continue to receive most popular attention.

CLIMATIC CHANGE PREDICTIONS

There are a growing number of scientists warning of global catastrophe associated with greenhouse warming:

> The Earth is warming. It has been warming for more than 15,000 years. But the recent warming is accelerating due to continuous accumulation of heat trapping gases in the atmosphere.... Global warming is the greatest crisis ever faced collectively by humankind; unlike other earlier crises, it is global in nature, threatens the very survival of civilisation, and promises ... only losers over the entire international socio-economic fabric.

(Woodwell and Ramakrishna, 1989, also quoting the Conference on Global Warming and Climatic Change at New Delhi, 1989)

> Scientists are now broadly agreed that the greenhouse effect is bringing about the greatest and most rapid climate change in the history of civilisation.
>
> (Lean *et al.*, 1990)

Predictions concerning the conditions in arid lands can be based upon probability calculations of precipitation or the identification of trends in rainfalls. Alternatively models of atmospheric processes can be used to assess the impact of possible changes in atmospheric composition, surface albedos and solar emissions. We have seen in chapter 2 that rainfalls are highly variable and are often non-normally distributed with data often only available over short periods of observation. Probability calculations and extrapolations of trends are therefore of dubious reliability.

Harrington (1987) reviewed work on models of climatic change and reported that temperature changes during the last 100 years could only be explained adequately by a combination of CO_2, sunspot activity and volcanicity. Unless the energy balance and mixing of the oceans is also considered the magnitude of some changes are exaggerated. Investigations have been taking place on the possible effects of changes in these variables.

There are now a number of models that attempt to calculate the effect of CO_2 concentrations upon climate. These have recently been reviewed by Bach (1984b), Dickinson (1986), Gleick (1988), Henderson-Sellers and McGuffie (1987) and Mitchell (1987). It appears that the Earth will experience a rise of 1 and possibly 2°C in mean global temperatures as CO_2 concentrations double. UNESCO predicted that this is expected to occur around the year 2030 while Scroggin and Harris (1981) suggested a later date of 2075. At a meeting of scientists at Villach, Austria, it was concluded that such doubling is likely to occur by the middle of the 21st century (Bolin *et al.*, 1986) producing a sea level rise of between 20 to 140 cm (including thermal expansion of the oceans and melting of ice caps). Of the arid realm, Egypt is perhaps particularly vulnerable to sea level rises which will threaten the coastal populations directly and inland by saline water intrusions and degraded water and sanitation facilities (Tolba, 1990).

In a recent study, Bach (1988) presented energy scenarios ranging from continued usage of fossil fuels until their exhaustion compared to more efficient consumption and declining demands. The envisaged annual carbon emissions by the year 2100 varied from 111.5 Gt to less than 0.1 Gt. It is estimated that this will result in a temperature rise of 0.4 to 3.54°C or a rise of 0.9 to 6.4°C if the climate is more sensitive to concentrations. If all greenhouse gases are taken into account then comparable rises of between 0.9 and 4.51°C and between 2.05 and 8.33°C are predicted. There remains much debate over these predictions of energy consumption and the quantity of gases remaining in the atmosphere (Schware and Friedman, 1981) but this has not stopped investigations of possible impacts.

The effects of doubling CO_2 were calculated by Wetherald and Manabe (1978) while quadrupling CO_2 was investigated by Manabe and Stouffer (1980). They indicate that global warming will be most pronounced in the higher latitudes with only a relatively small increase in the tropics. Bach (1984b) reports that a significant rise in temperature will be found in North East Europe and North Africa. Increasing temperatures and CO_2 may lead to higher yields although they could encourage vegetative growth rather than grain production while weeds may become more of a nuisance. Pittock (1988) reported that for C3 plants (corn sorghum, sugar cane) there could be a 10–50 per cent increase associated with a doubling of CO_2 concentrations. There will be an associated increase in the intensity of the hydrological cycle such that evaporation will increase in most latitudes while precipitation increases will not be as marked at latitudes of 25°N and 30°S. This is associated with a poleward shift in the subtropical high-pressure belt. Mitchell and Warrilow (1987) suggest this will increase winter rainfall in regions lying between 45 and 60°N, summer convective rainfalls will increase due to surface heating, but lower summer soil moisture contents will result due to changes in evaporation and surface runoff. Thus some parts of the arid lands may experience more rainfall while other areas may be facing drier conditions, although consideration of extreme temperature ranges rather than mean values may produce more dramatic results.

Oliver (1981) reviews suggestions advanced by Bryson (1973 and 1974) that drought in areas such as the Sahel can be explained by changes in the position of the subtropical high pressures. These in turn are believed to be controlled by the equator to pole temperature gradients. Emissions of CO_2 coupled with air turbidities may then be causing declining Sahelian rainfalls. This hypothesis does not, however, fully explain changes in temperature over the last 100 years and associated droughts. The models on which many of these predictions of climatic change are based have been assessed by Mitchell (1987) and some are criticised for not taking sufficient account of changes to cloud cover and energy exchanges with oceans. Manabe and Stouffer's (1980) model generally produced reasonable results for temperatures although it failed to reproduce the effects of cold water upwelling in the eastern Pacific and predicted too warm conditions over the Antarctic. Precipitation is overestimated at high latitude and underestimated in the tropics. It was therefore decided to delay hydrological forecasts based on CO_2 until models with a higher resolution become available. Mearns (1988) also concluded that while reasonable results have been obtained in reproducing global temperature changes there has been little success with precipitation and the hydrological cycle.

The UK Channel 4 programme 'The Green House Conspiracy' (Equinox, 1990) argued that major flaws could be found in global climate forecasts including: the lack of historical evidence with data biased towards urban centres in the northern hemisphere; the poor association between CO_2 and temperature over the last 100 years; models have too coarse a spatial resolution; and the role of water vapour and other feedbacks have not been fully ascertained. It is believed that it is not yet possible to provide reliable estimates of regional climatic changes:

> I do not think the approach of those attempting to model the response of the climate to increased carbon dioxide is wrong, but I do think that the efforts have been incomplete and that it is premature to make far reaching prognostications.... A warming may indeed occur, but much more work needs to be done before we should accept current forecasts for the next century, based on carbon doixide doubling, as a basis for action.
>
> (Bryson, 1989)

Bach (1984b) also states

> details of changes at any given location are presently beyond the capability of G.C.M's (General Circulation Models).

Dickinson (1986) agrees:

> it is not yet possible to estimate regional changes with any confidence.

Parry (1990) compares forecasts for the Sahel from the Canadian Climate Centre, the US Geophysical Fluid Dynamics Laboratory and the UK Meteorological Office models, with predicted warming of between 1 and 3°C in summer and rainfalls either unchanged or declining by 5 or 10 per cent. In Australia the same models predict rainfall increases of 5 to 15 per cent but summer temperatures are more consistent with an agreed 2°C rise. Perry (1984) reviews suggestions that studies have yet to separate natural climate change from that induced by CO_2 concentrations and this point is also made in a recent analysis of temperature trends (Boden *et al.*, 1990).

Lovelock (1985) questions the approach used by many climatologists in modelling atmospheric processes which ignores natural feedbacks. Debate over climatic change is further fuelled by suggestions that it may be that this is really a 'Western' problem, or a problem solely for the inhabitants of the northern hemisphere:

> The rapid exhaustion of natural resources, environmental pollution, the increase of carbon dioxide in the atmosphere, the rarefaction of the ozone layer, and other similar environmental problems with which the North is primarily concerned ... are peculiarly related to the industrial and military civilization of the North, and not to the predominantly agricultural civilizations of the South.... There has been no such experience of climate change or variability in the South.
>
> (Bandyopadhyaya, 1983)

Despite general agreement that global warming will continue Hare (1984b) concludes that there is no clear sign that this will result in improved or diminishing moisture supplies in arid lands. Parry (1987) sounds a note of caution that whilst there may not be significant changes in mean global temperature and even precipitation the question of increasing variability has been largely ignored, yet this element is of great consequence for the inhabitants of arid lands.

4

DROUGHT

Drought is a constant threat for the inhabitants of arid lands. Chapter 3 on climatic change listed some of the droughts occurring in the arid realm during the last 20 years and figure I.1 shows the widespread nature of this phenomenon. But does this represent a serious water resources problem? Evidence that drought caused agricultural production to fall by 12 per cent in Mexico, by 31 per cent in Australia and between 20 and 30 per cent in the USA suggests an emphatic answer of yes. Yet chapter 1 demonstrated how many arid lands' inhabitants have evolved strategies that combat drought. An occasional drought is then unlikely to cause widespread disruption unless other factors are operating from persistence of drought due to climatic change through to economic disruption. The main theme employed in this discussion is that the **effects** of drought need to be included in any analysis of its occurrence. Too often drought has been treated as simply a reduction in rainfalls with the assumption that this could explain all other disasters from famine through to desertification. In arid lands it is all too easy to blame all problems on the environment. For example,

> On August the 4th, 1984, the late President Kountche of Niger urged his countrymen to join 'the fight against the advancing Sahara' in order to avoid the humiliation and disgrace of desertification. He used the occasion to crack down on merchants who stole food-aid and slack civil servants, and to sack 30 traffic police. On April the 15th, 1985, in announcing even more draconian measures against desertification, he shelved plans to liberalize the domestic political system 'in the face of the more pressing problem of how to feed the population ...'. 'We cannot talk politics on an empty stomach,' he said. [He] called on Niger citizens to step up their fight against the advancing Sahara desert.
>
> (Agence France Presse)

This abuse of the threat of desertification and drought has been aided by the meaning of such terms becoming vague and confused, as Humpty Dumpty said:

> The question is which is to be master – that is all. When I use a word it means just what I choose it to mean, neither more nor less.
>
> (Lewis Carroll)

The starting point of an analysis of drought is therefore an examination of alternative definitions.

DEFINITIONS OF DROUGHT

The Oxford Concise Dictionary defines drought as 'lack of moisture'. Aridity (chapter 1) is a climatic term concerning **average** conditions that have prevailed for some time. Drought on the other hand is a more ephemeral phenomenon concerning **abnormal** reductions in water supplies. It is then confusing to consider 'permanent drought'. The persistence of drought should be seen to herald the onset of drier conditions due to climatic change, requiring the re-evaluation of what is abnormal. Tannehill (1947) suggested that drought was

> a prolonged and abnormal moisture deficiency,

which was reiterated by Palmer (1965) who also noted that

> drought can be considered as a strictly meteorological phenomenon.

Beran and Rodier (1985) concurred that whilst the impact of drought upon society was important its primary characteristic is a reduction in precipitation. This emphasis upon rainfall has been labelled **meteorological drought** which has been defined as:

> Drought occurs when rainfall is less than 80 per cent of normal levels.
> (Indian Met. Office, Dhar *et al.*, 1979)

> A prolonged period of below normal rainfall.
> (Dracup *et al.*, 1980).

> A period of at least 15 consecutive days without 0.01″ of rain in any one day.
> (UK Met. Office, Goudie, 1985)

> A period of 10 days with a total rainfall not exceeding 5 mm.
> (USSR, Krishnan, 1979)

> We shall consider drought as an extended period of deficient precipitation relative to normal.
> (Namias, 1981)

> A situation when actual seasonal rainfall is deficient by at least two times the mean deviation.
> (Ramdas, 1960)

> A prolonged lack in precipitation, less than average.
> (Russell *et al.*, 1970)

> Precipitation drought due to lack of rainfall.
> (Timberlake, 1985)

Some of these definitions are quite precise and provide a measure against which the extent and persistence of low rainfalls can be examined. There are at least three problems with this type of approach: first of all it is difficult to establish norms or means owing to the variability of rainfall and paucity of data; secondly such deviations may have no significant effect; and thirdly these definitions only consider rainfall. An analysis of reductions in water **supplies** should also take account of rivers, groundwater and lakes while increasing **demand** through population growth or landuse changes can produce water shortages. Drought is multifaceted and Subrahmanyam (1967) suggests there is water supply drought, agricultural drought, climatic drought and hydrological drought; to which Wilhite and Glantz (1985) add socio-economic drought. Faure and Gac (1981) used flows in Sahelian rivers to identify drought conditions, Grove (1986) examined discharges in the Niger and the Nile while Nicholson (1978 and 1981c) also employed the levels of African lakes. These approaches, however, only examine the supply of water while ignoring demand. Some assessment of demand can be obtained by considering **agricultural drought** which has been defined as follows:

> A period of dry weather of sufficient length and severity to cause at least partial crop failure.
>
> (US Weather Bureau, Krishnan, 1979)

> To the farmer drought means a shortage of moisture in the root zone of his crops.
>
> (Palmer, 1965)

> When soil moisture is depleted so that the yield of plants is reduced considerably.
>
> (Tannehill, 1947)

> Drought is a condition of moisture deficit sufficient to have an adverse effect on vegetation, animals and man over a sizeable area.
>
> (Warrick, 1975)

> In agriculture drought means a prolonged shortage of moisture in the root zone of crops and is often related to critical plant growth stages.
>
> (Yevjevich *et al.*, 1978)

But of what use are such definitions for the urban dwellers or pastoralists in arid lands? Namias (1981) states that:

> The definition of drought is partly contingent upon its impact on society and on the economy.

This view is supported by Sandford (1978):

> To the economist it (drought) means a water shortage which adversely affects the established economy.

Hewitt and Burton (1971), however, suggest we should consider

> a period in which moisture availability falls below the current requirements of some or all living communities in an area,

which is similar to Rasmusson (1987):

> Drought implies an extended and significant negative departure in rainfall relative to the regime around which society has stabilised.

There are then a myriad of drought definitions ranging from those concerned with quantifying water supplies, those considering the demand as well as supply of water through to examination of societal impacts. Landsberg (1982) argued that no universal definition of drought exists and is supported by Yevjevich *et al.* (1978) who state:

> In the past there has been no universally acceptable drought definition.

Lee (1980) suggested that,

> there are almost as many definitions of drought as there are people whose livelihood depends upon a regular supply of water.

It is not surprising therefore to find much debate and confusion over the occurrence of drought, as Copans (1983) states:

> A drought becomes a drought when someone wants it to be so.

It is important to resolve the confusion because solutions to water resource problems can only be effective if the causes of drought are known and these in turn can only be understood if the incidence of drought is accurately assessed.

Table 4.1 Countries affected by drought, April 1986 (millions of people)

	Total	*Affected*	*Displaced*
Angola	8.6	0.6 (7%)	0.5
Botswana	1.1	0.6 (55%)	–
Burkina Faso	6.9	0.2 (3%)	–
Cape Verde	0.3	0.1 (33%)	–
Chad	5.0	0.4 (8%)	–
Ethiopia	43.6	6.8 (16%)	0.3
Lesotho	1.5	0.5 (33%)	–
Mali	8.1	0.4 (5%)	0.1
Mauritania	1.9	0.9 (47%)	0.2
Mozambique	14.0	2.1 (15%)	0.4
Niger	6.1	0.4 (7%)	0.2
Somalia	4.7	0.2 (4%)	0.2
Sudan	21.6	6.0 (27%)	0.9
Total	123.4	19.2	3.0

Source: UN Office for Emergency Operations in Africa, 1986

Table 4.2 Food production trends in selected African countries (index based upon 1979–81 average as 100)

	Total food production						
	1982	*1983*	*1984*	*1985*	*1986*	*1987*	*1988*
Ethiopia	107	100	90	100	109	103	105
Mali	115	119	106	110	118	111	122
Niger	101	100	82	102	105	97	124
Senegal	112	85	102	122	127	139	119
Sudan	99	101	94	113	112	97	126
	Per capita production						
	1982	*1983*	*1984*	*1985*	*1986*	*1987*	*1988*
Ethiopia	102	97	86	91	98	91	91
Mali	109	105	94	96	100	91	97
Niger	95	92	73	88	88	78	98
Senegal	111	79	92	108	110	117	97
Sudan	93	92	83	97	94	79	99

Sources: FAO Production Yearbook, 1988 and Quarterly Bulletin of Statistics No 1, 1990

The way in which more rigorous application of drought definitions can aid understanding of the causes of drought can be exemplified by use of the Sahel as a case study.

SAHEL DROUGHT CASE STUDY

Of all the continents, Africa appears to have been the hardest hit by drought. In 1983 the FAO (United Nations Food and Agriculture Organisation) (Milas, 1983) reported that 22 countries in Africa were affected by drought and famine. Table 4.1 shows that by 1987 this figure had been modified to 13 countries but with still 19.2 million people affected. It is evident that over half of the drought-ridden countries of Africa lie south of the Sahara in a band ranging from Ethiopia in the east to Senegal in the west. This region has been loosely called the 'Sahel' although this term is sometimes used solely for the French-speaking West African states.

Table 4.2 reveals the extent of the problem. Food production per capita has been falling, a large proportion of the cereals consumed are imported and the inhabitants of this area are heavily dependent upon food aid.

News reports of disaster in the Sahel first began to appear in the early 1970s with the FAO (11 May 1973) reporting:

> In some areas there now appears serious risk of imminent human famine and virtual extinction of herds vital to nomad populations.

Clarke (1978) reports the contents of a telex sent to UNICEF in 1973 referring to

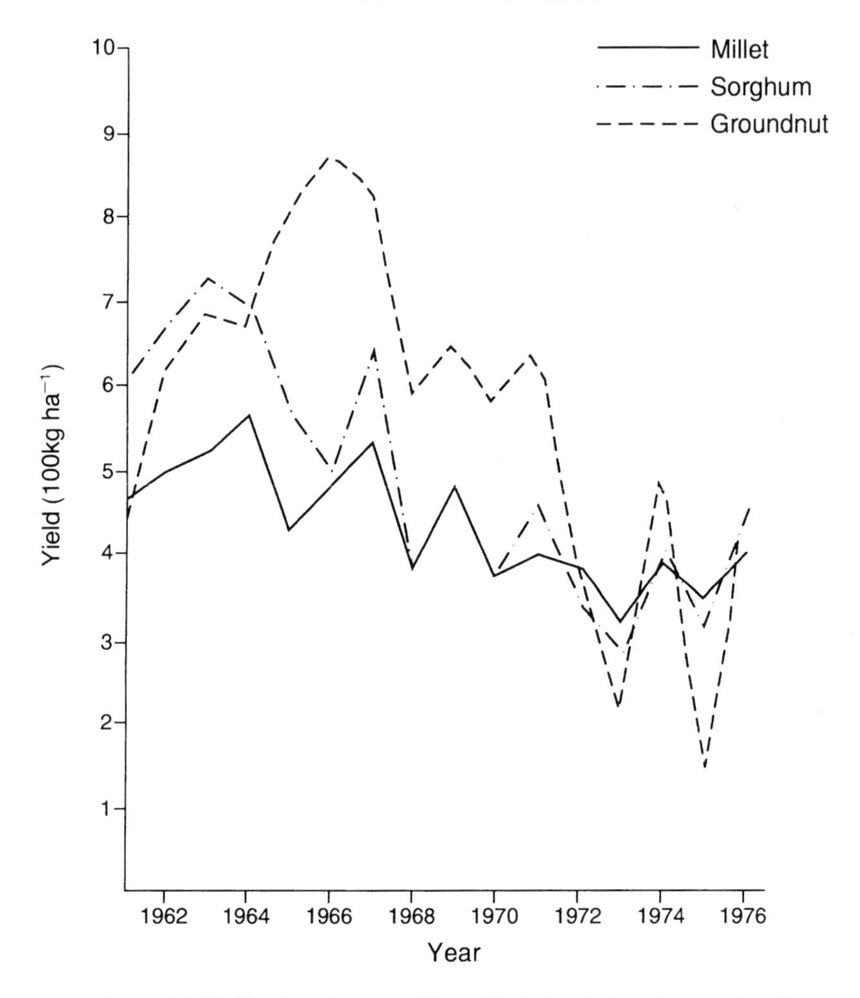

Figure 4.1 Yields of staple crops, Niger (Ministère du Development Rural).

a rush to towns, drought for four to five years, desert is moving southward and six million lives threatened.

Newspapers such as the *New York Times* and *Time Magazine* estimated that some 60 per cent of cattle perished due to drought in the early 1970s (although recent estimates now suggest losses of 30 to 40 per cent) while figure 4.1 indicates the substantial decrease in the yields of staple crops for the country of Niger which lies in the middle of the Sahelian region (figure 4.2). Copans (1983) records drought-related human deaths of 100,000 by 1973. Several years later the *Guardian* (10 October 1977) reported that drought had struck Sahelian crops again and noted that the Mauritanian Government had called upon the international community to 'save human lives from a certain death'.

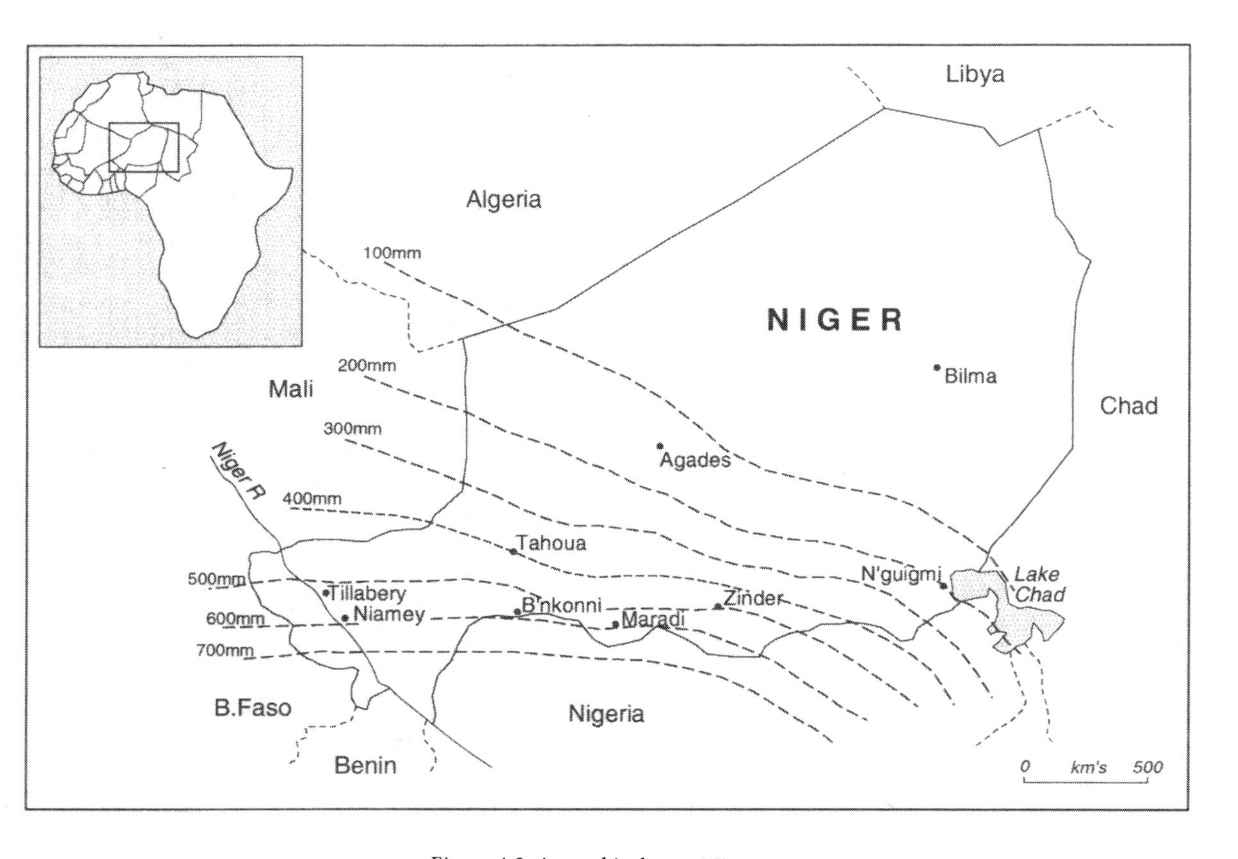

Figure 4.2 Annual isohyets, Niger.

More recently the ODI (1987) announced that drought had returned to the Sahel and Toulmin (1988a) reports FAO estimates of cattle losses as 60–70 per cent in the early 1980s. Some now believe that drought has never ceased but continued to plague this region for the last two decades. Sir Crispen Tickell when he was head of the ODA in 1986 said of the Sahel:

> The documentary evidence for the last 70 years shows a slight decline of rainfall from 1955 and acute drought since 1968.

This view is supported by Flohn (1987), who commented upon

> the Sahel drought, lasting without interruptions, until (at least) 1985,

and Hare (1984a):

> Thus the drought which began in 1968 in the sub Saharan regions never really ended.

Declining rainfalls

These reports and the marked downward trend of rainfall since the 1950s (figure 3.3) suggest that the climate of this region may be changing with increasing desiccation possibly ahead. Nicholson's (1981b and c) analysis of Sahelian rainfalls over a longer time scale using a variety of data sources revealed that there have been several dry periods since 1600 but the 20th century appears to be witnessing one of the most intense and protracted as shown by Farmer (1989) (figure 2.6) and Glantz (1989). Although Hare (1984b) maintains there is little evidence of a worldwide pattern of increasing aridity there is much concern for the future of the Sahel with Vice President Bush (14 March 1986) being urged to give aid to the Sudan to combat the Sahara Desert, which, it is claimed, is advancing at 9 km per annum.

Winstanley (1973) and Lamb (1974) have related rainfalls in the Sahel to general circulations and precipitation in the northern hemisphere, identifying 100 and possibly 200 year cycles. If correct this could mean even more arid conditions in the future. However, the length of record is much too short for such extrapolations without the causal mechanism being explained.

The cause of declining rainfalls may be found in the combined effects of greenhouse gases, air turbidity and solar emissions which (see chapter 3) can explain 20th century global temperature changes. The relationship between Sahelian rainfall and temperature has not yet been established although there has been some success in predicting low rainfalls in 1986 and 1987 using sea surface temperatures (Folland, New Scientist 1 October 1987). Unfortunately the high rainfalls in 1988 were not forecast and there is still work to be done on such models. There is also a possible local mechanism which could explain increasing aridity through land degradation.

Land degradation and drought

Goudie (1986) explains how overgrazing around water points could lead to a large expanse of desertified land. Charney and Stone (1974) argue that removal of vegetation results in higher surface albedos (short-wave reflection) which lead to a reduction in the amount of energy absorbed by the atmosphere, producing stable air masses and so decreasing rainfall. The resulting desiccated vegetation produces even higher albedos and hence lower rainfall. Many aspects of the region's energy balance are ignored, however, and studies in the Sahel using remote sensing by Courel *et al.* (1984) have suggested that there has not been any significant change in surface reflectivity. This model of change has therefore appeared invalid, but other studies based on work by Walker and Rowntree (1977) suggest that rainfall in the region may be largely due to water that is transpired by vegetation rather than transported in by moist winds from the south. If this is the case then removal of vegetation may well have a deleterious effect upon Sahelian rainfalls. Hulme (1989) has reviewed the evidence for and against this explanation and has pronounced a verdict of 'not proven'.

The myth of drought

Explanations for declining rainfalls have been offered which suffer from inadequate models and data, but are we really witnessing climatic change or

Plate 5 River Niger at Niamey. The large seasonal regime of many exogenous arid land rivers offers opportunities for flood retreat farming and other activities along the river banks when dry, but this also presents problems for navigation and water supply.

merely a periodic dry spell in this highly variable environment? Goudie (1986) states

> numerous studies of available meteorological data ... do not allow any conclusion to be drawn to the question of systematic long term changes in rainfall, and the case for climatic deterioration ... is not proven.

The idea of climatic change leading to increased aridity is not advocated by all. Wijkman and Timberlake (1985) have claimed that in many areas there is no significant decline in rainfalls. The evidence for disaster in the Sahel appears overwhelming with reports of livestock deaths, agricultural crops failing and even destruction of the soils in the area. However, when disasters strike arid areas of the world *drought* becomes an easy *scapegoat*. It is all too easy to blame any environmental disaster on the climate (Agnew and O'Connor 1990). Increasingly attention is being given to other non-environmental factors. Food production is being seen as a complicated system involving a highly variable environment that is being opened up to outside economies and thus becoming destabilised. Why then do the inhabitants of arid lands appear to be more susceptible to change? Why can they no longer respond to fluctuations in rainfall? That is, low rainfalls are seen as the 'trigger' for disaster not the root cause. Even more radical explanations ignore the climate altogether and place the blame fully upon economic disruption through bungled development attempts and increased economic competition.

Plate 6 The lack of tree cover on the hills around Gonder in Ethiopia reveals the extent to which deforestation has taken place and this may be having an impact on the low rainfalls recently reported for this region.

Plates 7 and 8 The picture of dead livestock against a village fence taken at the end of the dry season in Mauritania can be taken to indicate the impact of drought. Yet the picture of livestock grazing on the newly emergent vegetation in the valley of the River Senegal was only taken a few kilometres from this village and suggests that it is the lack of fodder, not a lack of water that is responsible for such livestock deaths.

Economic disruption

The pastoral production system in chapter 1 was seen to offer a range of solutions to combat the arid environment in which it operates. The reports of livestock deaths listed above clearly show that this is no longer the case. What then has changed? Analyses by Baker (1974) and Garcia (1981) suggest that attempts to develop these pastoral herds lead to massive increases in animal numbers prior to the reduction in rainfall commencing in 1968. Veterinary care, deep boreholes and the provision of water points coupled with good rainfalls in the early 1960s enabled herds to multiply with the consequent overloading of the range. When drought occurred, livestock herds were decimated, not by lack of rainfall but through a lack of grazing. That is, the disaster in the Sahel was due more to mismanagement of the environment than climatic change. There are also a number of other changes taking place in Tuareg society, for instance:

1. Keenan (1977) reports upon the sedentarisation (settling down) of the traditional camel-herding groups in response to economic and political pressures.
2. There are increasing wage-earning opportunities from tourism (Clarke, 1978) through to uranium mines.
3. Slavery has been abolished.
4. Migration routes now cross international frontiers which are sometimes closed.
5. The constraint imposed by a lack of water is being reduced. Wells and boreholes can tap groundwater resources while trucks can be equipped with water reservoirs and driven into the desert.

Such changes have been taking place while the demand for meat is growing and these previously isolated societies have been opened up to 'Western' economic influences. Such rapid, recent changes have altered the labour force and lead to an intensification of the use of the rangeland which has possibly lead to land degradation and destruction of the herds. This may have been exacerbated by low rainfalls but the cause of the disaster is seen as inappropriate attempts to develop the pastoral economy. Unfortunately there is very little data available to support this explanation of events. Numbers of animals are notoriously difficult to obtain and a recent survey of Niger by Swift (1984) suggests that the claims of decimation of livestock herds may have been exaggerated and there has been some recovery of herd sizes. Norton-Griffths (1989) shows that there has been little growth in livestock numbers in the Western Sahel and the effect of low rainfalls in 1973 and 1984 are clearly visible. In a review on desertification, Warren and Agnew (1988) maintain that there has been an overestimate of its extent and effects, a view that is supported by Ahlcrona's (1986) study in the Sudan. Clearly the pastoral production system is changing under the influence of Western intervention but whether this has resulted in the destruction of the pastoral economy and permanent environmental damage is uncertain.

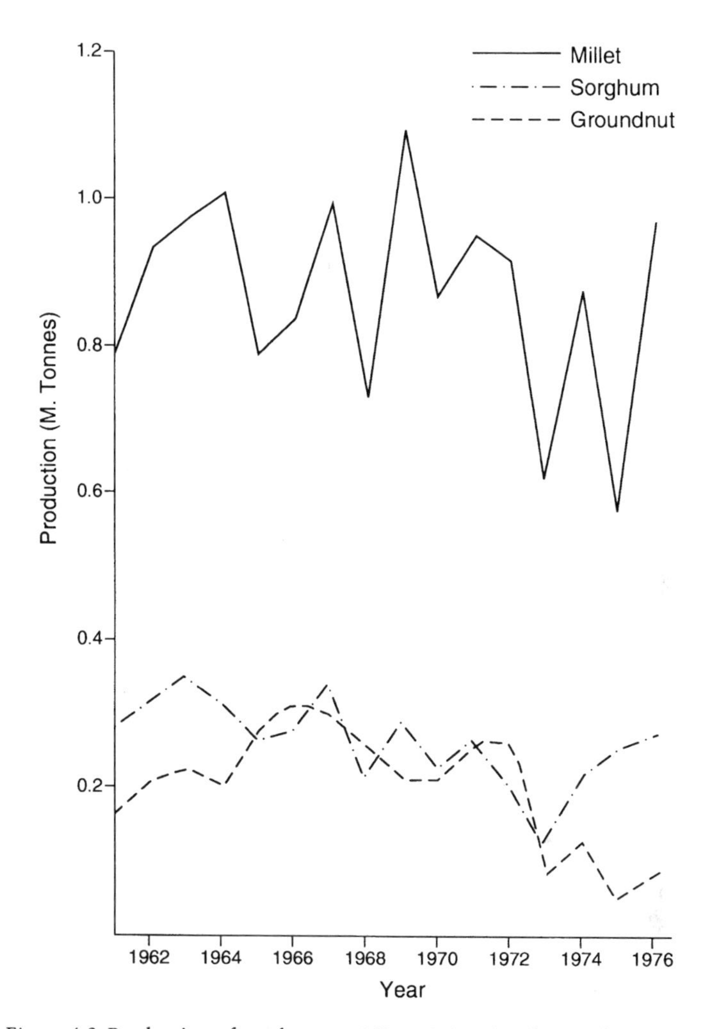

Figure 4.3 Production of staple crops, Niger (Ministère du Development Rural).

Turning from livestock to dryland farming, figure 4.3 reveals that for one Sahelian country, during the early drought years, crop production has generally been maintained but with a marked variability. Comparison of figures 4.3 and 4.4, however, indicate that during this time production has only been kept reasonably static by generally increasing the area under cultivation. Norton-Griffiths (1989) believes this is true for many other Sahelian countries, and shows that the GVP (Gross Value Product) has increased by 37 per cent in the Western Sahel from the 1960s to the 1980s, of which 33 per cent is accounted for by an increase in the areas cultivated. Yields have shown no long-term trend and contributed little to gains in production. The reasons for this increase in cultivated area are threefold:

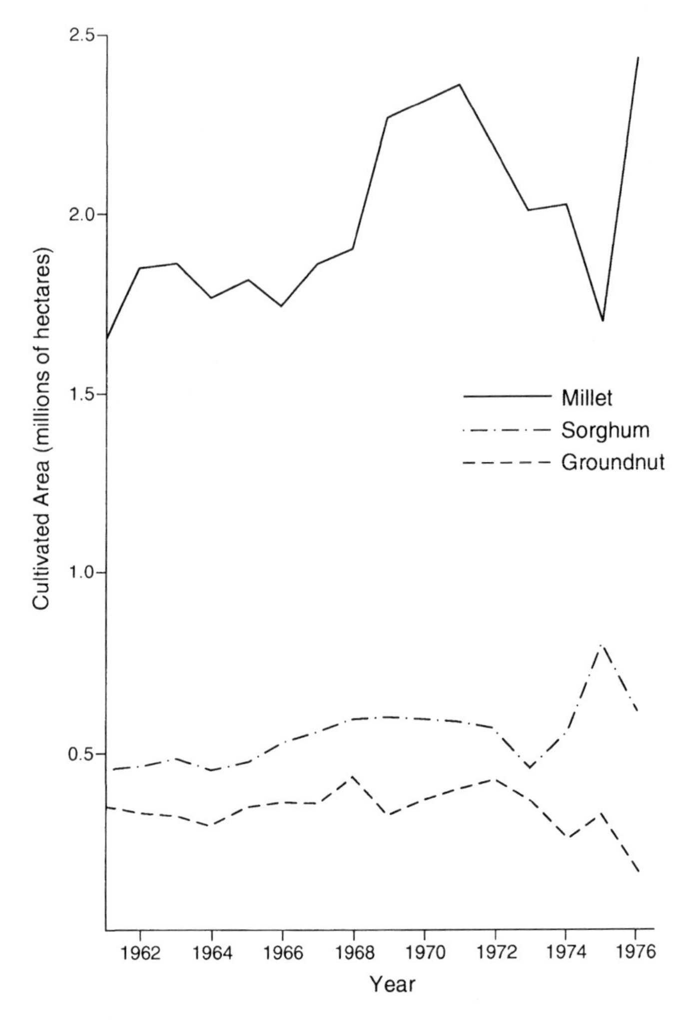

Figure 4.4 Cultivated areas of staple crops, Niger (Ministère du Development Rural).

1 Glantz (1986) notes that higher rainfalls in the early 1960s encouraged a northerly migration of farmers onto the more marginal lands to the north. This increased the competition for land between the pastoralists and farmers, requiring government action to limit the northward limit to which farmers could cultivate. There was also an attempt to prevent any damage to farmers' fields by grazing herds through keeping animal migrations within several metres of major roads, although this then reduced the amounts of animal manure spread onto the fields.

2 The growth of population through health care programmes, cessation of warfare and famine relief has created a growing demand for land. This has

aided the push onto more marginal lands and led to the fallow period being reduced as land can no longer be left 'idle'. Consequently more people are living in drier, northern areas while soil fertilities are declining. The US Office of Technology Assessment (1984) reported that African cities have the highest growth rate in the world with populations doubling every 10 years. The rapid expansion of urban populations increases the demand for food encouraging intensification of the dryland farming system. Curtiss *et al.* (1988) calculate that population is growing at a rate of 2.8 per cent in African countries where famine is a risk which is outstripping food production growth of only 1.3 per cent per annum. Van den Oever (1989) shows that there is an alarming increase in the numbers of Sahelian inhabitants, which accounted for 1.9 per cent of world population in 1950, rising to 3 per cent by 2010.

3 At independence most Sahelian states had poorly developed economies and were consequently heavily dependent upon other nations for assistance. Following independence there was a drive to commercialise agricultural production through the introduction of cash crops in order to purchase foreign goods and also to obtain some economic stability. Peanuts had been introduced to West Africa in the 19th century but now governments were actively encouraging the introduction of new hybrids. The UN yearbook shows that in 1964 peanuts accounted for 79 per cent of Senegal's exports and 63 per cent of Niger's. However, the declining terms of trade during the 1960s through competition with other producers (Franke and Chasin, 1980 and 1981) resulted in more and more land being utilised to maintain economic returns. This meant lower fertilities and competition with staple cereals such as millet or sorghum. Glantz (1989) notes that the best land is usually put into irrigation of cash crops displacing cereal farmers and making self-sufficiency in foodstuffs an unobtainable goal.

It is believed that the data on declining yields and erratic production have more to do with the introduction of cash crops and growing population pressure than drought or climatic change. Overemphasis on cash crop exploitation results in soil impoverishment and expansion onto the marginal lands. Although this interpretation has great popularity there is little hard evidence to support the claims of landuse change and soil impoverishment. Ahlcrona's (1986) study in Sudan could find no evidence of increasing occupation of more marginal lands while figure 4.4 does not reveal a dramatic increase in the area under groundnuts. Perhaps the growth in groundnut production started before the 1960s and the effects were not realised until the 1970s? In fact millet shows the most significant increase which is most dramatic just after the 'drought' started in 1968. But this appears to be in response to lower rainfalls rather than reflecting a massive migration onto the marginal lands.

Interpretations then vary between those who believe there has been a drought whether its causes are natural or human, and those who argue that even though

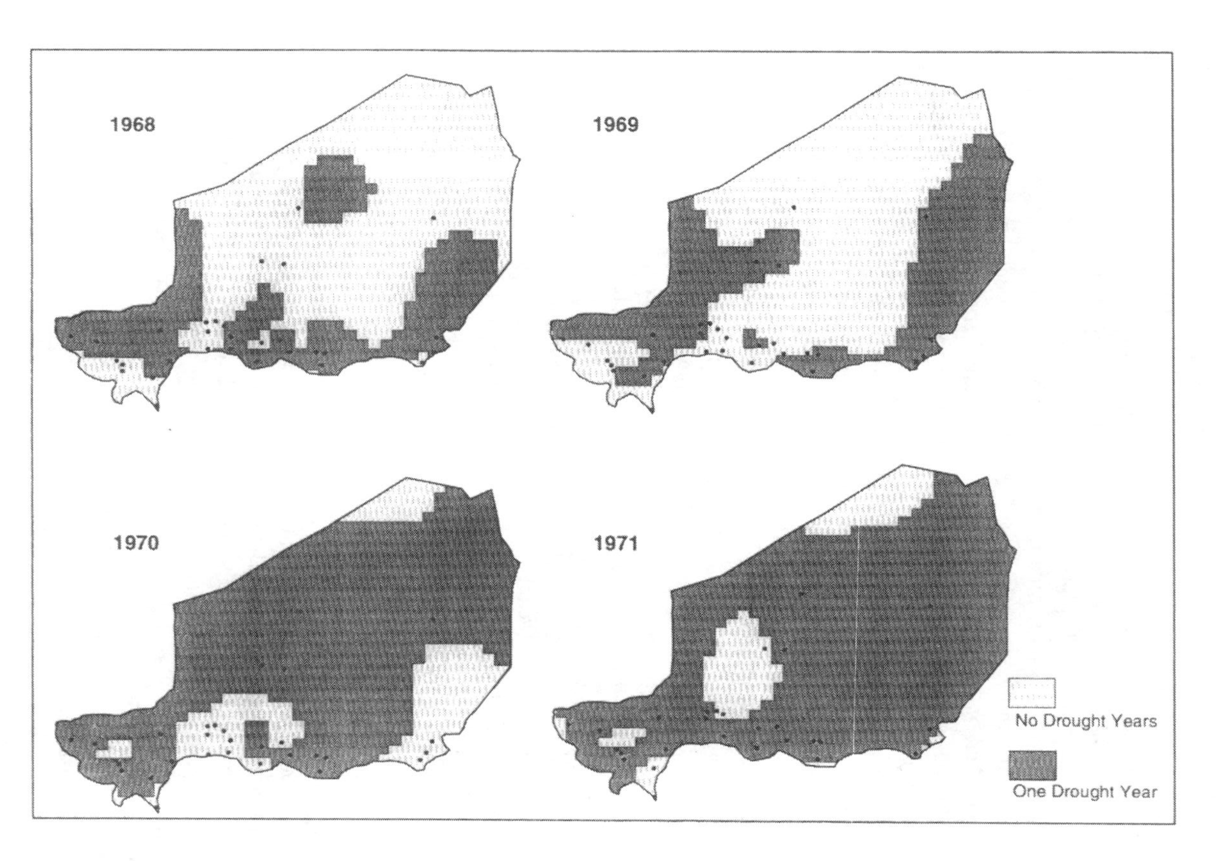

Figure 4.5 Meteorological drought (less than 80% of mean annual rainfall), Niger, after Agnew (1990a).

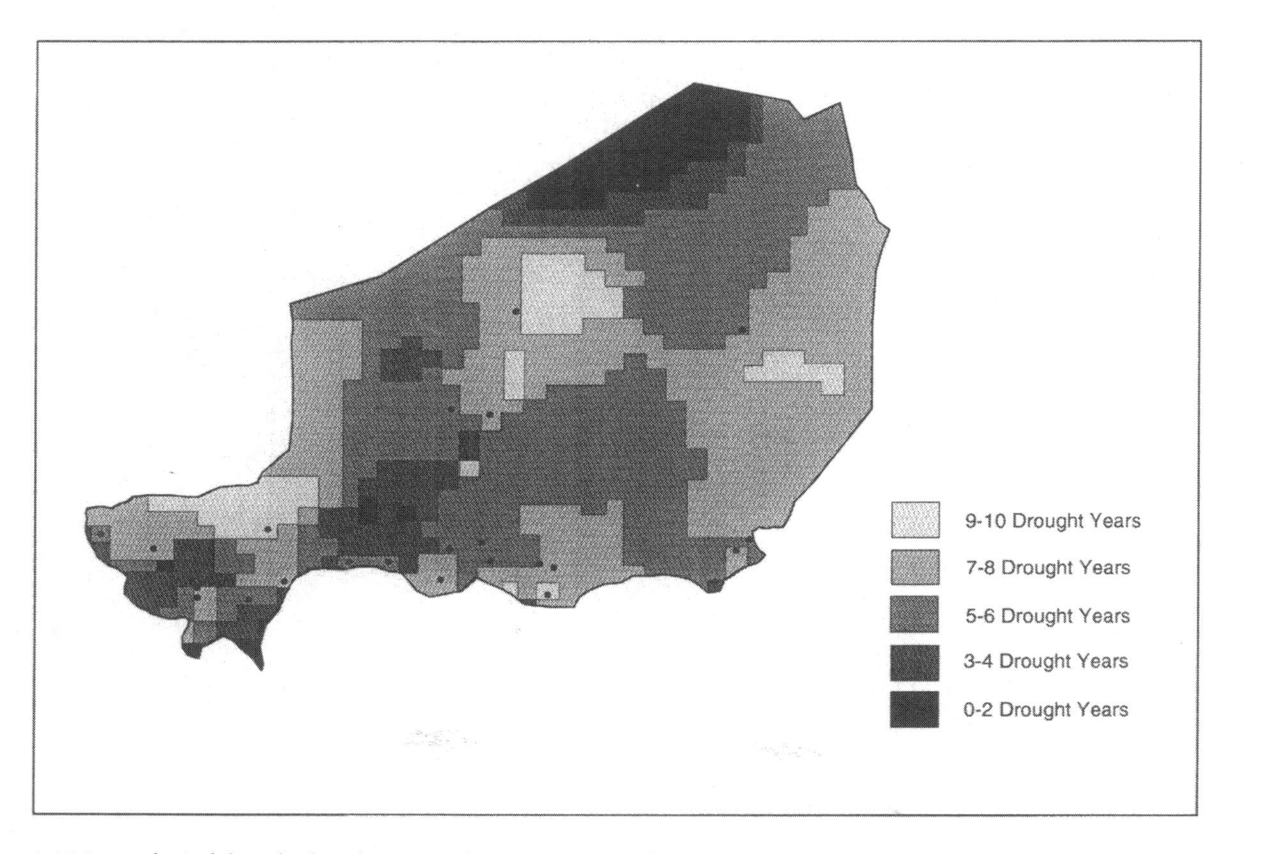

Figure 4.6 Meteorological drought (less than 80% of mean annual rainfall) compilation for years 1967–76, Niger, after Agnew (1990a).

rainfall may have decreased, famine and agricultural disruption is caused more by pressure on the land and economic difficulties. The debate is beset with a lack of data and unprovable hypotheses. In order to gain any further understanding we need to consider the meaning of drought in the Sahel.

Meteorological and agricultural drought

It is evident from the variety of definitions listed above that the difficulty in establishing a meteorological drought is determining when conditions are sufficiently 'abnormal'. The Indian Met. Office (Dhar *et al.*, 1979) suggested a simple calculation of 80 per cent of normal rainfalls. The latter can be applied to data from the country of Niger covering the 10 year period from 1967 to 1976 (33 stations) (see figures 2.3 and 2.4). These results can be mapped by dividing the country into 1,600 grid cells using UCL Mapics software (Battarbee *et al.*, 1983) which generates values for each cell for each year 1967 to 1976. These values were compared with 80 per cent mean annual rainfall (using stations where 30 year means were available, 1947–76). Hence the occurrence of meteorological drought could be mapped. A 4 year period 1968–71 is shown in figure 4.5 revealing that the spatial variability is quite marked throughout the region both in terms of the location and the extent of meteorological drought areas. In 1968, meteorological drought lay roughly between the 300 and 700 mm isohyets whereas in 1970 and 1971 drought conditions are more extensive and only relatively small 'islands' escape abnormally low rainfalls. One must be cautious of the mapped areas above the 100 mm annual isohyet (figure 4.2) because of the lack of climatic data (inevitable given the sparsity of settlement). The effect is perhaps most noticeable in 1969 when a large area of the central region is mapped as 'non-drought' but this result is largely dependent upon the observations from the northernmost station (Iferouane). The vast majority of the population live in the southern area of the country where stations are consequently more plentiful and enable more reliable estimates of the spatial extent of drought to be made.

A map of aggregate meteorological drought conditions for the 10 years 1967–76 can be seen in figure 4.6. Most of the region suffers from meteorological drought for over 50 per cent of the time. A gradient is discernible with more frequent meteorological droughts to the north and east. Figure 2.4 shows annual rainfalls which display a marked downward trend since the 1960s with frequent meteorological drought conditions during the 1970s. This affirms comments made of the enduring drought in the Sahel and possible increasing desiccation. Meteorological drought is not, however, as frequent in the south-west with 2–4 years only of abnormally low rainfalls in many areas. Rainfalls at Niamey (figure 2.3) do not reveal a comparable downward trend and meteorological drought conditions are clearly not persistent. This contradicts reports on widespread drought. It appears, then, that concern for drought and climatic change is most appropriate in the drier, more northern and eastern parts of the country with

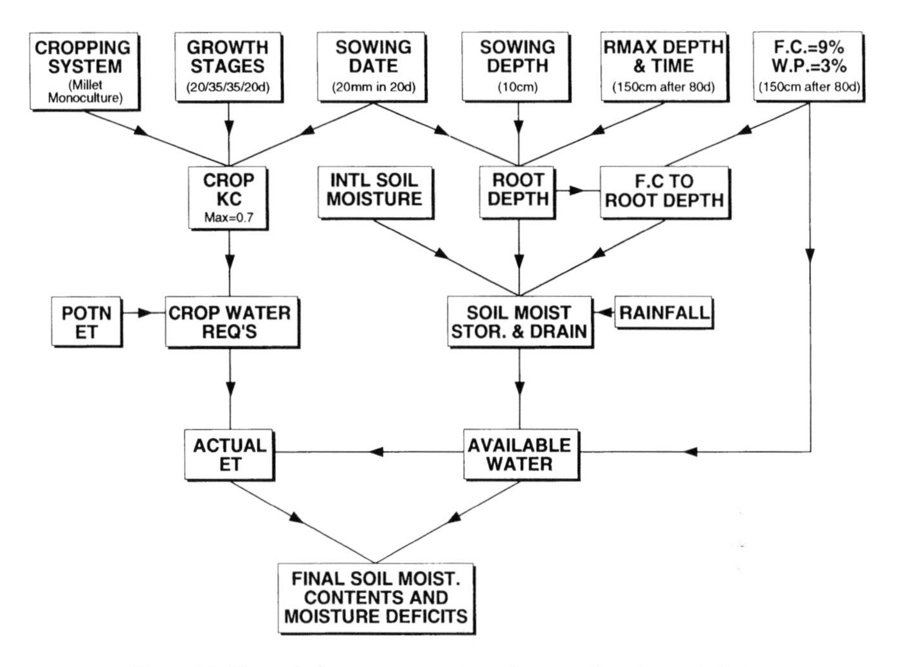

Figure 4.7 Water balance computation of seasonal moisture deficits.

annual rainfalls generally below 300 to 400 mm. But two notes of caution should be mentioned: the above analysis has revealed the spatial diversity of rainfalls so these results should not be extrapolated to neighbouring regions without further analysis; and secondly the above comments have been made for meteorological drought and therefore only concern relative amounts of rainfall and not their effects. One way of assessing the latter is to examine agricultural drought and its effect on dryland farming.

Agnew (1980, 1982) has developed a model that calculates the soil water balance in the Sahel on a 5 day time scale with a mean accuracy of 10 per cent. This model was recently described and used to distinguish between meteorological and agricultural drought conditions (Agnew, 1989, 1990 and 1991). Briefly, figure 4.7 shows the essential elements whereby inputs of potential evapotranspiration and rainfalls are used with crop coefficients to calculate the actual evapotranspiration and soil water balance every 5 days. Data was used from the nine climatic stations (located in Niger) shown in figure 4.2 for the years 1967–78. The crop employed was millet (*Pennisetum typhoides*) because it is the most important rain-fed crop in Niger (Ministère du Developpement Rural, 1976). The model produces a seasonal estimate of moisture deficits summed for the growing period of the crop such that

$$\text{millet moisture deficit} = (\text{potential} - \text{actual})\text{evapotranspiration}$$

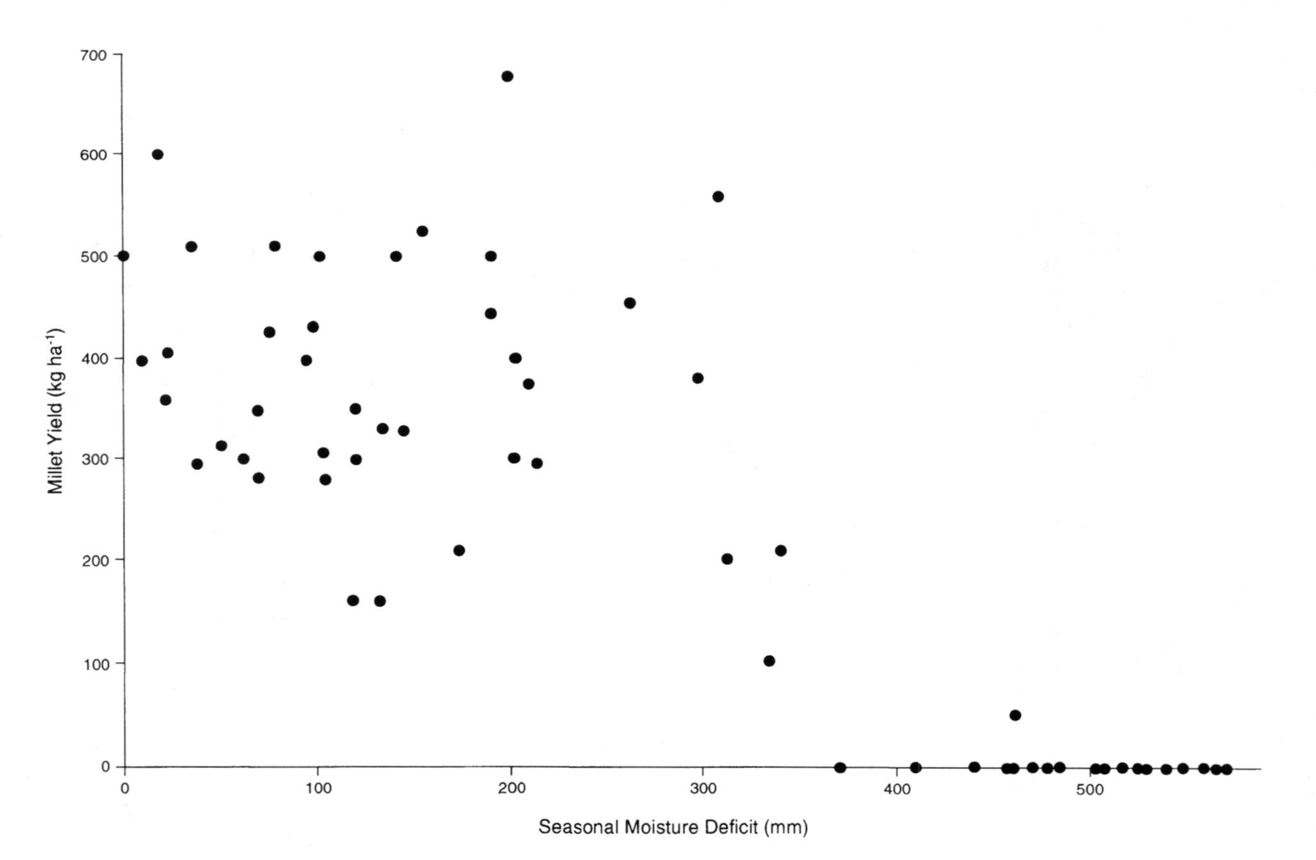

Figure 4.8 Seasonal moisture deficits and millet yields 1967–78, Niger, after Agnew (1990a).

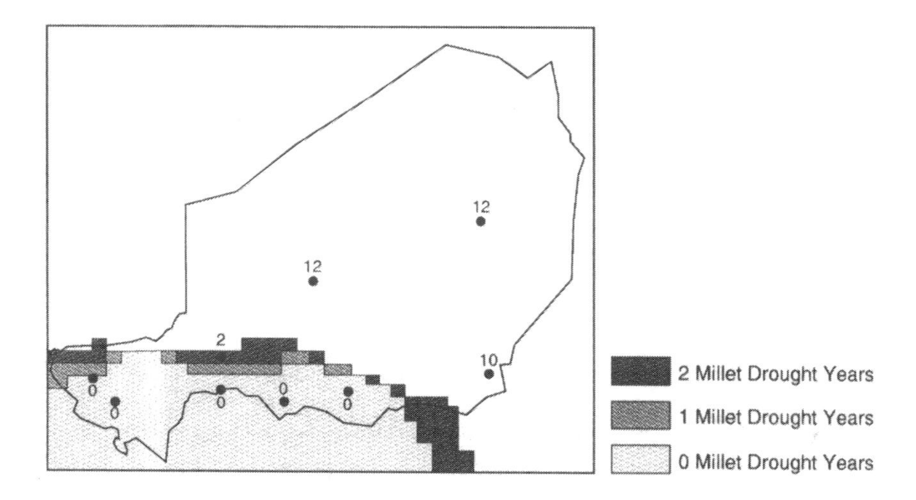

Figure 4.9 Compilation of millet drought years during 1967–78, Niger, after Agnew (1989).

Figure 4.8 shows the relationship between millet moisture deficits and yields. Deficits less than 200 to 300 mm appear to have no effect upon yields with the latter probably affected by soil fertilities, seed variations, disease, etc. Deficits greater than 300 mm seriously reduce yields and this demarcates the limit of successful cultivation and provides a threshold with which to investigate millet drought.

Did an agricultural drought affect Niger in the 1970s? Figure 4.9 has been produced using the Mapics package by classifying all of those areas where 'millet drought' (300 mm seasonal millet moisture deficit) was encountered during the 12 year period 1967–78. It indicates that only parts of the area where millet is traditionally grown to the south of the country suffered severe millet moisture

Table 4.3 Frequency of millet droughts during the years 1967 to 1978

Station	*Seasonal* *300*	*Moisture* *250*	*Deficits (mm)* *200*
Agadez	12	12	12
Bilma	12	12	12
Birni N'Konni	0	1	1
Maradi	0	0	0
N'Guigmi	10	12	12
Niamey	0	0	1
Tahoua	2	4	6
Tillabery	0	1	4
Zinder	0	1	5

deficits, and then for only 1 or 2 years during 1967 to 1978. Table 4.3 reveals that if the definition for millet drought is relaxed to include deficits down to 250 mm or even 200 mm there still remain a number of stations that do not appear to be experiencing frequent millet drought conditions.

There is then little evidence of a 'millet drought' persisting throughout the 1970s in this southern Sahelian region. Explanations of famine and disruption of dryland farming in this area should therefore look to socio-economic changes and the impact of the environment upon pastoralism rather than blaming the cause upon drought. A different conclusion is obtained for the more marginal lands to the north, lying on the Sahel/Sahara boundary. This area appears to be persistently affected by 'millet drought' with possible exceptions in 1974 and 1978. However, these are called marginal lands because they are prone to drought. Hence one should be asking why farmers are trying to settle in this area, competing with the pastoralists, rather than moving south when the rains fail. Drought may be the immediate problem for the inhabitants of this region but the cause of disruption goes beyond examination of climate alone (Agnew, 1984b). In fact Ahlcrona (1986) working in the Sudan reports that there is little evidence for a northward migration of cultivators (1961–83) which reinforces the view that the observed economic disruption and famine may owe little to the desiccation of the staple dryland crop, millet. It is true that most of the stations employed in this study with the exception of Tahoua lie outside of this region of 'marginal lands'. It is not, however, surprising that there are few major settlements in this area and that there is consequently a paucity of climatic observations. The exact positioning of the boundaries of this marginal zone is therefore speculative as they are dependent upon extrapolations from widely spaced stations. Nevertheless it is clear from figure 4.9 that most of the sites were free from persistent millet drought during 1967–78.

Summary

Analysis of annual rainfalls suggests that it is the drier, northern parts of the Sahel that have been experiencing the most noticeable reductions leading to concern over persistent drought and climatic change. The effects will probably be most pronounced upon pastoralism as this is the only reliable landuse in this area. The FAO suggest that the term 'Sahel' should only refer to this drier (less than 300 mm) region. Hence claims of persistent drought and climatic change in the Sahel are probably valid for this environment leading to pastoral disruption aggravated by economic changes. The danger is in assuming that the same changes are taking place in the wetter south which has also been included as part of the Sahel by the WMO (up to 700 mm) and Rasmussen (1987).

In the wetter south there is evidence that meteorological drought has been less frequent than to the north and that agricultural drought may have been quite scarce. Hence declining yields may well be due to non-climatic factors leading to land degradation. In between lie the 'marginal lands' where opportunistic farmers cultivate during 'good rainfall years'. Here agricultural drought is more widespread but this is to be expected given the marginality of the area. If farmers continue to remain in this area when drought strikes it must be asked, what is stopping their migration? The cause of the problem appears to lie in pressure on the land and political difficulties. It is not yet clear whether these marginal lands are also suffering from increasing aridity or merely experiencing climatic fluctuations that are aggravating a rural economy destabilised by other socio-economic changes.

There then appear to be different drought-related problems affecting different parts of the Sahel and it is dangerous to extrapolate from one climatic region to another. Above all drought should be seen as an abnormal reduction in water supplies for a particular landuse, a question of both the supply of and demand for moisture. Drought can then occur both through climatic change and landuse change. But in such a variable environment it is often difficult to distinguish between change and fluctuation. Hence there remains confusion over what is happening and why. This is likely to remain a problem in the analysis of drought in arid lands until more rigorous, landuse-related drought definitions are employed and there is more information on both climate and socio-economic systems.

5

DATA COLLECTION AND ANALYSIS

Many disciplines, from civil engineering to geography, are concerned with the measurement and analysis of hydrological variables. Therefore, as might be expected, there is a wide range of excellent texts (for civil engineers Shaw, 1983 and for geographers Ward, 1975 and Rodda *et al.*, 1976). Additionally, there are specialist manuals, the most complete of which is the *National Handbook of Recommended Methods for Water-Data Acquisition* (USGS, 1984). However, as the vast majority of hydrological investigations relate, for obvious reasons, to humid environments, so the literature displays a similar bias. As a result, while the basic techniques are amply covered, the particular problems encountered when working in the arid environment receive comparatively little attention. In this chapter, the bases of hydrological measurement and analyses are assumed and the focus is exclusively upon those aspects peculiar to the arid zone. While the framework is that common to the standard works on hydrology, the detail is drawn from specialized texts (e.g. Jones, 1981 and Rantz and Eakin, 1971), together with the field experience of the authors.

For each of the hydrological variables determined, their characteristics within the arid zone are discussed. The resulting problems of measurement and monitoring are then considered, together with the particular difficulties for data analysis. The distinctive arid zone hydrometeorological features which give rise to deviations from the humid zone 'norm' are (Jones, 1981):

1. high levels of incident radiation;
2. generally high diurnal and seasonal temperature variations;
3. low humidity a short distance inland from the sea;
4. increased wind effectiveness with frequent dust and sandstorms;
5. sporadic rainfall of high temporal and spatial variability and high intensity;
6. even greater variability of short duration runoff events in ephemeral drainage systems;
7. depending upon the proportion of 'fines' present, a high variation in infiltration rates in channel alluvium;
8. high sediment transport rates;
9. potentially large groundwater and soil moisture storage changes; and
10. high potential evaporation and evapotranspiration rates.

In the arid zone, problems of hydrometry, hydrological analysis and interpretation and water resource assessment and development are severe (Jones, 1981). Indeed, the most difficult problem is the most basic: what to measure and where to measure it. In *Guide to Hydrological Practices* (World Meteorological Organisation, 1974) the following network measurements are defined:

a) optimum network:
by interpolation between values at different stations, it should be possible to determine with sufficient accuracy for practical purposes the characteristics of the basic hydrological and meteorological elements anywhere in the country. By characteristics is meant all quantitative data, averages and extremes that define the statistical distribution of the elements studied; and
b) minimum network:
is one which will avoid serious deficiencies in developing and managing water resources on a scale commensurate with the overall economic development of the country.

For most of the arid zone, the network is either minimum or subminimum. The reasons are partly economic, related to the sparse population density, and partly environmental, related to the extreme nature of the arid environment which results in mechanical breakdowns. Of some concern is the reduction in the number of African recording stations in the 1980s, an alarming trend for rainfall data collection. The imposition of a minimum network can be seen as an initial stage towards the development of an optimum network. The careful construction of the minimum network is obviously vital if long-term records are to be of value.

PRECIPITATION

The key characteristic of rainfall in the arid zone is variability in both time and place. Accurate measurement, monitoring and forecasting all therefore pose problems. There are difficulties on all time scales. Long-term trends cannot be discerned with any accuracy and, as a result, the same statistics can produce totally different interpretations. Furthermore, a trend must be discerned from an oscillation or a fluctuation. Monthly falls are even more variable than annual, but analysis of the alternating pattern of wet and dry periods has yielded some useful results (Jones, 1981). Even within one storm, rainfall intensity varies considerably, so that the length of the rainfall period may mean little in hydrological terms. As a result, the establishment of the most basic statistics, the central tendency, is highly problematical, the more so the shorter the time period. Indeed, even for annual rainfall, the median or perhaps the mode may represent a more meaningful assessment than the mean. Since most rainfall data suffers from a positive skew, it will not always fit a normal distribution probability function and gamma or log-normal functions have to be employed (see chapter 2).

Rainfall can be measured directly by various types of raingauge and can be inferred using a range of other techniques. To relate rainfall to runoff and other important hydrological variables such as agricultural requirement and recharge, a raingauge network is needed. However, given the rainfall characteristics and particularly its generally convective nature, it is only feasible to model for small representative areas, for the basic reasons discussed in Jones (1981):

1 the lack of precision in rainfall measurement, due to the extreme spatial variability in the intensity and amount of rainfall;
2 the short period of runoff production;
3 the complexity of basin hydrological processes (e.g. infiltration, recharge and runoff);
4 the high water retentive capacity of the surface rocks and deposits;
5 the relatively large storage capacities of the shallow aquifers;
6 the low proportion of rainfall which actually runs off; and
7 the range of potential measurement errors.

Within the arid zone, the optimum raingauge network would consist of a relatively sparse distribution of recording and totalising gauges, spread to take into account the major landscape variables and climatic elements. This would be supplemented by a number of dense networks of recording gauges, sited in representative locations. To this pattern would be added, according to logistical possibilities, daily raingauges. An example can be seen in the Batinah of Oman, where there is a thin network of recording gauges in the northern Oman mountains and within each of the major wadi basins. Within reach of each of the main settlements are daily raingauges and a denser network is being established in representative basins.

Within the basic recording network of most arid zone countries, daily raingauges predominate. However, the standard national gauge is determined by the national meteorological organisation and this can cause problems of comparability. It has long been noted (Bull, 1960) that the raingauges used by different nations (with no standard diameters or heights) produce different observations. The arid lands with their varied colonial pasts contain a diverse range of raingauges making comparisons problematical, although the WMO have proposed a worldwide standard gauge with a diameter of 127 mm, the rim located 1 m above the ground and protected by a wind shield. Furthermore, in arid countries, the gauge is commonly sited on a roof top or other secure area, where readings are adversely affected by the local environment. It must also be said that lack of training and experience frequently results in inaccurate readings and inadequate maintenance. Totalising raingauges are read, according to their design, at fixed intervals: weekly, monthly or seasonally. Thus, results can only be used to provide a measure of annual or possibly seasonal rainfall. Recording raingauges, depending upon their design, provide measures of rainfall intensity, besides recording totals, but any mechanical device is prone to failure as a result of dust and sand while data loggers need to be insulated against extremes of both high and low temperatures.

The techniques for measuring rainfall are much the same as for more humid areas and have been reviewed in basic hydrological texts such as Rodda *et al.* (1976), Shaw (1983) and Ward (1975). The particular problem posed by arid land precipitation is obtaining representative measurements given the high intensity and localised nature of rainfall. The development of radar and satellite imagery has helped to provide more reliable areal estimates. Radar beams are attenuated by raindrops as a function of the diameter of the raindrops (WMO, 1974). The diameter of the raindrops is in turn related to rainfall intensity which can therefore be mapped, but this still requires calibration using ground-based gauges to provide estimates of rainfall reaching the surface. Radar cannot therefore be used independently of a ground-based network of raingauges while physical obstacles (towns, hills, etc.) can limit the area covered. Satellite imagery does not suffer from this difficulty and the method is based upon identifying storm clouds by observing cloud-top temperatures (Barrett and Martin, 1981). Not all cold clouds precipitate but the University of Reading Meteorological Department have been successfully developing a technique to estimate rainfall in the Sahel (Dugdale *et al.*, 1991), based upon charting the duration of cold clouds using infrared imagery. They have found that different relationships were established for different parts of the rainy season requiring further investigation before the technique is fully operational. Silsoe College, Cranfield, has been using a similar approach for moisture assessment in arid lands (De Souza, 1990).

In the light of the basic measurement problems, data must be subjected to quality control checks before analysis can begin. In particular, the careful estimation of missing values is important. Estimates, or statistics based upon them, must of course be distinguished in any published record.

The most fundamental product of analysis is the relationship between rainfall and area. In the arid zone, the sparsity of raingauges presents a particular problem, enhanced if the network is not specifically designed to reflect the geographical variables. Although it is time-consuming and involves a large subjective input, the most satisfactory way of illustrating rainfall-area relations is through isohyets. These will illuminate the relations between rainfall depth and altitude, distance from moisture sources and aspect. The other method commonly used for areas of non-uniform raingauge spacing is the construction of shapes, height-adjusted Thiessen polygons or triangles around each gauge. The result is that greater weighting is given to records of gauges in the less dense parts and flatter areas of the network.

Other important measures are concerned with rainfall intensity and probabilities. Rainfall intensity may be expressed as the mean intensity within selected time intervals, or the intensity for any duration. With the latter approach, rainfall depth can be classified by intensity segments, thereby allowing relationships to be established with other hydrological variables such as the infiltration rate. Intensity as a function of rainfall duration and return period is also vital in flood estimation. As a result, many empirical formulae have been derived and it has been shown, for example, that there is a marked constancy of depth–duration ratios and depth–frequency ratios. The probability of daily rainfall occurrences also has hydrological and agricultural importance in the arid zone. These may be shown as the number of days or rainfall periods per year with rainfall greater than the specified level, or the rainfall amount for single rain days, or rain periods.

There are then inherent difficulties in establishing and maintaining a satisfactory network of raingauges in arid lands. Remote sensing and radar may go some way towards solving this problem. Finally, owing to its potential ecological importance, the measurement of dew must be noted. As there are no standard techniques available three basic approaches have been adopted: gravimetric, electrical and visual (Monteith, 1957, Anderson, 1988d and Agnew and Anderson, 1988). During a research programme in Oman, gravimetric sensors proved particularly satisfactory (Anderson, 1988).

EVAPORATION AND EVAPOTRANSPIRATION

Evaporation is affected by the following climatic elements:

1 air temperature;
2 relative humidity;
3 net radiation; and
4 hydrological variables such as water quality and depth and the extent of the water surface.

For open water surfaces, evaporation may be computed by calculations of:

1 the water budget;
2 the energy budget; and
3 the mass transfer.

For the first of these approaches, inflows and outflows are monitored, the residual representing evaporation. The energy budget approach requires the measurement of incoming and reflected radiation, together with the net energy advected and the increase in energy storage of the water body. The mass transfer procedure is based on the aerodynamics of vapour flux and requires measurements of windspeed, air and water temperature and vapour pressure.

The term evaporation has often been interpreted as pertaining to an extensive, open water surface of freshwater, whereas evapotranspiration is used for vegetated surfaces. During the conference on 'Soil water balance in the Sudano-Sahelian zone' (Sivakumer *et al.*, 1991), it was argued that the term evapotranspiration should no longer be used as it led to confusion over the processes taking place, i.e. evaporation and transpiration. Consequently estimates of water losses based upon the assumption of a uniform surface under the same climatic influences would be erroneous. However, the term evapotranspiration has gained widespread usage particularly amongst empirical scientists and it is likely to continue to be used for the time being.

It is necessary to distinguish between actual rates of evapotranspiration (AE_t)

and potential rates (PE_t) when water is freely available and the vegetative surface is therefore at its saturation point (Fuchs, 1973). The primary controls of rates of water loss for PE_t are not hydrological (water supply and quality) but are meteorological, i.e. energy balance, humidity, windspeed and temperature. Crop characteristics also influence PE_t through leaf area, aerodynamic resistances and ground cover. Hence potential evapotranspiration has been defined as

> evaporation (diffusion of water vapour into the atmosphere) from an extended surface of a short green crop, actively shading the ground, of uniform height and not short of water.
>
> (Ward, 1975, p. 98)

PE_t is then a value that is determined from meteorological conditions usually for a grass surface and other plant covers may have different potential rates for the same atmospheric observations. It has proved very difficult to measure actual evapotranspiration. Wallace (1991) reviews the methods that have been employed in arid environments and mentions porometers, the use of tracers and heat pulse techniques for plant transpiration rates. He observes that soil moisture changes are, however, the most widely used approach for evapotranspiration studies, often through the use of lysimeters (a column of soil is isolated and all inputs and outputs are monitored). Lysimeters unfortunately suffer from boundary effects and maintenance problems and are more suitable for research than routine applications (Kohnke *et al.*, 1940, Mather, 1954 and Pelton, 1961). At a broader scale water balance computations (DaMota, 1978, Eagleson, 1978, Hignett, 1976 and Parkes and O'Callaghan, 1980) can estimate AE_t but the reliability of such calculations is too dependent upon the accuracy of measuring and collating all the other variables in the hydrological cycle. The Bowen ratio is a meteorological approach where the gradients of temperature and vapour in the boundary layer of the atmosphere (Oke, 1978) and the soil energy balance are observed in order to calculate AE_t (Black and McNaughton, 1971, Gay and Stewart, 1974, Sargeant and Tanner, 1967 and Van Wijk and Borghorst, 1963). This method can be successful in arid lands when vertical gradients are significant but it requires reasonably sophisticated instrumentation and there may be problems with row crops. Alternatively eddy correlation can be employed which attempts to monitor the changing vapour flux within the boundary layer. Until recently this approach foundered on the lack of suitable instruments (Campbell and Unsworth, 1979 and Anderson *et al.*, 1984). The Institute of Hydrology (UK) has developed a device which will measure AE_t using this method but it is rather sophisticated for normal applications. Hence AE_t rates are often inferred from PE_t estimates.

Various instruments have been devised to measure PE_t of which pans and atmometers are the most common. Atmometers measure the water evaporated from a moist surface, i.e. filter paper (piche type) or porcelain plates (Bellani type) and plates (Livingstome type). They are cheap and easy to use but are reported to be oversensitive to windspeed in arid climates thus overestimating true rates (Baldy, 1976 and Jackson, 1977). However, comparison between piche and PE_t calculated by the Penman formula (see below) using an automatic weather station in Northern Tunisia indicated that similar results could be obtained near aquatic environments where advection was less significant (Agnew, 1984a). Evaporation pans come in all shapes and sizes, installed both above and below the ground (Gregory and Walling, 1973). Thornthwaite (1954) showed that the size of the pan affected water losses, Nordensen and Baker (1962) found that the depth of water was significant while Dagg (1968) indicated that even the brand of paint had an effect. Perhaps the most significant factors include the exposure of the pan, i.e. whether it is surrounded by fallow land or well watered vegetation and prevailing atmospheric conditions. Doorenbos and Pruitt (1975) provide correction factors depending upon these considerations with pan observations reduced by between 15 and 65 per cent.

Dissatisfaction with instrumental measurements of PE_t has led to the use of meteorological formulae. Those based on temperature alone such as the Thornthwaite equation (Thornthwaite, 1944, 1948 and 1954) and Blaney–Criddle (Blaney, 1954, Blaney and Criddle, 1950 and 1962) have been criticised for ignoring aerodynamic factors and their use of empirical coefficients. They also tend to be suitable only for monthly estimates and are therefore best treated as a last resort when air temperatures are the only data available. Nevertheless they remain popular, especially the Blaney–Criddle formula (US Department of the Interior, 1971) due to their simplicity and low data demands. Several formulae based on solar radiation have been developed (Fuchs, 1973 and Doorenbos and Pruitt, 1975) but this meteorological variable is not widely observed and PE_t is influenced by more than the atmospheric short-wavelength energy balance. Dalton suggested in the 19th century an aerodynamic equation incorporating windflow and vapour pressure gradients which has been developed during the 20th century (Oliver, 1961 and Sellers, 1965). Their use of empirical constants, however, limits their applicability to wider locations (Wilson, 1984). The formula that has gained universal acceptance for estimating potential evapotranspiration is that devised in 1948 by Penman (1948, 1949 and 1963), which is considered by Jones (1981) to be most useful for arid environments.

The Penman formula combines both the energy budget and aerodynamic elements affecting evaporation. The rationale of the formula and methods of calculation are provided in most hydrological texts (Shaw, 1983, Ward, 1975 and Wilson, 1984) and irrigation scheduling discussions (Doorenbos and Pruitt, 1975, Hillel, 1982 and Withers and Vipond, 1988). It involves daily observations of temperature; windspeed; vapour pressure or humidity; and solar radiation or day length or net radiation. It has been used to compute daily evapotranspiration rates in a range of arid land conditions. Criticisms concern the complex nature of the formula, the need for a number of meteorological observations and the assumption that energy storage is negligible.

The Penman formula was subsequently developed to include the soil heat flux and wind profiles, but this makes the computation even more complex while the

Plates 9 and 10 Climatic observations for estimates of evaporation should be based on measurements taken over short grass that is actively growing and not short of water. These pictures reveal the difficulty in practice of achieving such conditions in arid lands (*9* in Oman and *10* in Tunisia).

information required is simply not available widely enough (Fuchs, 1973). Penman's original formula has also been criticised for ignoring plant stomatal resistances and aerodynamic conditions (Thom, 1975). The formula was based on the premise that diffusivity of water vapour into the atmosphere is controlled by vertical temperature and humidity gradients and the turbulent transfer coefficients within the boundary layer (Sellers, 1965). Assumptions by Penman concerning these conditions enabled only one set of readings to be required, at a height of 1 metre, providing the evaporating surface was saturated.

Montieth (1980 and 1981) proposed a modified version of the Penman formula to take account of stomatal and aerodynamic resistances, which has subsequently been incorporated into the MORECS system of calculating evapotranspiration rates in the UK, (Thompson *et al.*, 1981) while Brutsaert and Strickler (1979) have also tried to produce an approach for 'non-potential conditions'. Lemeur and Zhang (1990) found the Penman–Montieth approach provided reliable estimates of evapotranspiration in an arid region of China but this conclusion was based upon an analysis of the annual total water balance which is a rather crude means of comparison.

Monteith (1981) noted that there would be problems using his approach in the case of row crops with incomplete ground cover so often encountered in arid lands. Shuttleworth and Wallace (1985) addressed the problem of sparse crop cover by further developing the Penman–Monteith equation, but note that information is required on crop height, leaf area, stomatal and substrate resistance, net radiation and soil heat flux. This data requirement is seldom satisfied and the resulting equation is primarily used only for research. Consequently the original Penman (1948) formula remains the main approach for estimating PE_t especially for irrigation scheduling (Burman *et al.*, 1983), although Shuttleworth (1988) recommends that such a semi-empirical approach can only be used to estimate the average potential evapotranspiration for periods of 10 days or more (which is adequate for most water resource applications). The results are only valid for the well watered vegetated surface over which they have been collected, i.e. grass, although alfalfa is sometimes employed. The PE_t for a particular crop can be estimated by the use of crop coefficients where

$$PE_t(\text{crop}) = PE_t(\text{reference surface}) \times \text{crop coefficient } (K_c)$$

Crop coefficients for a given crop tend to change during the growth cycle and different values are usually employed for phenological stages. A comprehensive listing of crop coefficients can be found in Doorenbos and Pruitt (1975) (see chapter 6). This approach, however, only provides an estimate of PE_t(crop), not AE_t(crop).

The relationship between AE_t and PE_t is influenced by soil moisture availability with plant transpiration reducing as soil moisture declines (Boonyatharokul and Walker, 1979, Kramer, 1952, Lassen *et al.*, 1952 and Slatyer, 1956). The exact form of this relationship has been the subject of debate with Veihmeyer and Hendrickson (1948, 1955) and Gardner (1960) arguing that in sandy soils, $AE_t = PE_t$ from field capacity (upper limit of soil storage when drainage has ceased following saturation) to wilting point (lower limit of soil storage when plants cannot extract any more moisture). This appears to be peculiar to coarse sands as Denmead and Shaw (1962) and Jackson (1977) show that AE_t rates will fall below PE_t at various points between field capacity and wilting point depending upon conditions such as the magnitude of PE_t rates and plant characteristics such as root density. Operational studies generally assume soil moisture is freely available up to a threshold beyond which a linear or exponential function is applied as water becomes limiting and AE_t falls below PE_t. At wilting point AE_t reaches zero. For example, Parkes and O'Callaghan (1980) divided the plant rooting depth into six horizontal layers containing 5, 7.5, 12.5, 25, 25 and 25 per cent of available water. For grass it was assumed $AE_t = PE_t$ until 30 per cent (70 per cent in winter) of the available water remained and then declined exponentially to zero. Penman embodied this approach in his 'root constant' concept which estimated the moisture that could be extracted by different plants before water became limiting, after which there was a steady decline in AE_t rates. The root constant was employed by the Met. Office to compute water balances in the UK (Grindley, 1969) until superseded by MORECS.

Monteith (1991) has, however, criticised this approach as being overly empirical and not based on reality. The crop coefficients do not represent aerodynamic and surface resistances except, perhaps, where water is not limiting. Consequently it is argued that the use of crop coefficients should be phased out and replaced by more reliable estimates based upon the Penman–Monteith formula. While this is undoubtedly correct, this recommendation ignores the lack of available instrumentation for this task and until resources become available it is likely that crop coefficients will continue to be employed.

There is, finally a remaining criticism of this empirical, crop coefficient approach to estimating evapotranspiration summarised in a paper by Morton (1983), who argues that we are assuming that PE_t and AE_t are independent of each other when in fact there is a complementary relationship, i.e. negative slope. If AE_t is high then air humidities will be raised and relatively less energy will be devoted to sensible heat, and hence PE_t will be low; conversely if AE_t is low then PE_t will be high (Kovacs, 1987). As Morton states (p. 12)

> the assumptions used to estimate areal evapotranspiration in the conventional conceptual models are completely divorced from reality. They continue to be used because of their tractability and because of the ease with which unreal coefficients can be optimised when the relevant variables are correlated through their pronounced seasonal patterns.

The use of meteorological data is therefore of dubious reliability except where areas are so small that their evapotranspiration will have negligible effect upon atmospheric humidity and temperature or when moist areas are extensive. The Penman method of estimating crop water requirements should therefore only be used as an initial guide for estimating evapotranspiration.

A recent development that may prove useful in the future is the use of remote sensing (Reginato, R. *et al.*, 1985 and Hatfield, 1983). The temperature of the evaporating surface is an important variable influencing AE_t rates and infra-red sensors can provide such information. Unfortunately their resolution of 0.5 °C is not sufficiently accurate. Nevertheless the spatial pattern of surface temperatures and particularly ground cover can be a valuable aid to the extrapolation of point PE_t and AE_t estimates.

A particular problem concerning arid lands is the evaporation from bare soil or surfaces with a sparse vegetation cover. Interest in water balances for such conditions has been fuelled by the desire to develop agroclimatic models to forecast crop failures from early season growth of crops (FAO, 1986) and the realisation of the importance of soil moisture in climatic modelling (Hunt, 1985). Kondo and Saigusa (1990) have developed a model of bare soil evaporation based on water vapour diffusion but such an approach is too complex for routine use. A view echoed by Markar and Mein (1987) who also note that empirical estimates were too simple and therefore unreliable. Nevertheless there are a number of sources quoting crop coefficients for bare soils. Doorenbos and Pruitt (1975 and 1977) publish a graph relating frequency of rainfall, PE_t and K_c. When applied to Sahelian conditions during 1967 to 1978 K_c values were generally 0.2 to 0.4 with 0.62 during wet phases. Burt *et al.* (1980) reduced these coefficients by 30 per cent for bare sands to take account of the lack of depression storage resulting in less direct evaporation. Wallace *et al.* (1990) report values of K_c of 0.5 to 0.8 for bare soil and 0.65 for fallow bushland, both after heavy rainfall at a station within the Sahel. Wright (1981) used a K_c of 0.3 for corn during early growth but this increased to 0.39 after rainfall, although in a subsequent paper Wright (1982) suggested values of 0.2 for bare soil with a low evaporative demand falling to 0.05 under dry conditions. Agnew (1988a) also found values around 0.05 for drying, bare sands while Jensen (1974) suggested figures of 0.16 to 0.2 for early season growth of cereals. Clearly work still needs to be done on estimating evaporation rates from bare, arid land soils.

SURFACE WATER

Within the arid zone, runoff typically occurs in the form of short, isolated flows, generated by discrete storms. Peak rates of flow occur rapidly and diminish downstream as a result of infiltration and overbank flow. Thus, events are governed by the characteristics of the specific storm and the particular basin and, as a result, general relationships between seasonal or annual runoff and rainfall cannot be derived easily. Indeed, the coefficient of variation of annual seasonal runoff in ephemeral streams typically exceeds 100 per cent. With the variety of other contributing factors, significant among which have been found to be drainage-area size, main-channel slope, basin length, surface area of lakes and ponds and the annual number of thunderstorm days (Benson, 1964), it can be seen that the possibilities of modelling the processes by which flood discharge is produced in arid zones are very limited. Practical requirements are better served by the analysis of the records of peak discharges, irrespective of the causal conditions.

Within the arid zone, there is a vast variety of runoff regimes, but a distinction can be drawn between endogenous and exogenous water courses. Both the nature of the flood peaks and the characteristics of the water courses impose severe restraints upon the siting of gauging stations. Ideally, there should be a minimum of one such station for each active wadi, but if more can be installed, then a range of techniques such as flood routing can be employed. As with rainfall measurement, a sparse network of gauging stations can be supplemented by small, well instrumented representative basins.

Measurement practice for water courses on any scale is well known and covered in all standard textbooks (see Graf, 1988). However, in the arid zone, normal practice is commonly hindered by the following problems:

1. the rapid rise and fall of the flood peak, which itself is frequently of destructive force;
2. the instability of channels, particularly on the lowlands;
3. the high velocities of the water commonly associated with large sediment movements;
4. the occurrence of mud flows, the water content of which cannot be accurately gauged; and
5. the frequent occurrence of pulsating flow.

For small springs and seepages, a current meter, 'v'-notch weir or flume may be sufficient. For larger water courses, float-operated recorders or pressure recorders have been successfully used in the arid zone (Jones, 1981). Also of great importance for flood records and facilitating the development of an inexpensive denser network are primary staff gauges, with crest gauges. Velocity may also be measured by acoustic, float, pressure, tracer, electromagnetic or optical procedures. Under stable conditions, the stage–discharge relations at a gauging station can be plotted to produce a rating curve. In the arid environment in which controls are unstable, such curves will be far less accurate. Water levels have been observed for some time on major arid land rivers; Evans (1990) for example examines records for the Nile dating back to 3,000 BC. The interpretation of such data is not simple, however, with gaps, movement of the gauging station and changes in the channel morphology. It also possible, for arid land rivers, to find the water level recorder has been washed away by a torrent or the recorder pen has stuck during the vital period as a result of inactivity. Thus the one flood event remains unrecorded.

In many arid areas, of course, there are no gauging stations permanently located and discharge must be estimated or calculated by indirect methods. The most basic of these is the drainage area of the basin, with allowances for the various significant geographical and meteorological variables. For individual channels, the slope–area method (Dalrymple and Benson, 1967), for which measurements of cross channel area and water surface slope can be made in the field, is the most flexible. If floods are constrained by culverts, dams or weirs, then other techniques can be applied (Hulsin, 1967 and Bodhaine, 1968). In countries such as Malta and Oman, where the number of gauging stations is severely limited, the slope–area technique has allowed effective flood mapping (Anderson, 1988a).

The estimation of flood peaks and volumes in large arid catchments is best attempted by statistical methods, incorporating whatever evidence is available on geomorphology and flood levels. The simplest statistical method consists of the adjustment of normal or other types of distribution to the annual series by plotting on suitable probability paper or by using the general equation for hydrologic frequency analysis (Sokolov *et al.*, 1976). A more powerful approach than this single-station method is that provided by regional flood frequency analysis, which incorporates all elements of regional hydrology, together with a homogeneity test to ensure that gauging station data is hydrologically similar.

Flood estimation can also be achieved through deterministic methods, particularly the use of the unit hydrograph. However, what is a simple technique in humid climates is distorted by constraints of the arid zone particularly in the determination of 'effective rainfall'. Finally, estimates can be derived from empirical formulae such as the Creagar Envelope Curves (Jones, 1981).

BASIN SETTING

In the landscape, the basin is the basic hydrological unit. It will be contended later (chapter 11) that, in the arid zone, the basin may not always be the most appropriate unit for resource assessment. However, within physical hydrology, measurements are usually related to the basin. Thus, there is no difference between the humid and the arid zone in the variables which are used to describe the basin setting. Nevertheless, it is reasonable to assume that, given the nature of the hydrological system, minor variations may be of greater significance in the arid zone.

Characteristics of the basin setting, for the most part, change only over comparatively long periods and can therefore be considered static. They provide the background against which dynamic measurements such as rainfall and surface runoff can be assessed. They are important, at one extreme, for an understanding of the complete hydrological system, and at the other for providing, in many cases, the only factor available for prediction. Most basins within the arid zone, for example, have few, if any, raingauges. Thus, the basin area may be the only variable available from which a forecast of flow and possibly flood can be made. If humid representative basins are instrumented and monitored, hydrological processes can be modelled, but such detailed work is only of real practical value if the results are transferable. Similar results would only be expected in basins with similar settings. Furthermore, it may be possible to make allowances for dissimilarities. Such procedures depend upon the measurement of a comprehensive range of variables to describe the setting.

The *National Handbook of Recommended Methods for Water-Data Acquisition* (USGS, 1984) provides a detailed reference list of basin characteristics and how they may be measured. The characteristics are grouped into four categories: topographical, soil, geological and landuse. In most of the arid zone, except on valley floors, soil is generally of limited importance, while the variation in landuse is likely to be small. By far the most important and, generally, the easiest to measure are the topographical characteristics. Except for those which vary significantly on a small scale, most can be measured to a reasonable approximation from maps, aerial photographs or satellite imagery.

Depending upon the degree of aridity, the landscape typical of the arid zone varies. In general, it is characterised by sharp breaks of slope where the uplands meet the plain. The highland is generally bare with little or no soil cover, while the lowland is flat and covered with alluvial fans. Drainage is frequently endoreic and, while channels in the uplands are usually clear and incised, those on the plain present an ever-changing anastomosing pattern, extremely difficult to analyse. If they do not lead to the sea or an inland body of surface water, they become increasingly indistinct until they disappear. However, a sharp distinction must be made between these endogenous drainage systems, originating within the arid zone, and exogenous flow, which arises in neighbouring humid areas and merely traverses the arid zone. The latter tends to present less exacting problems for hydrological investigation.

The topographical variables can be classified into those which describe:

1 the total basin;
2 the drainage network; and
3 the channel.

The major factors of importance in the basin are the basin drainage area, the dimensions of the basin, measures of basin slope, basin shape, an index of surface water storage and the hypsometric curve. In a small mountain basin, or the upper part of a larger basin, following rainfall runoff will exceed the infiltration rate and the storage capacity within the channel alluvium. The result is that there is a rapid increase in discharge for every comparatively small increase in catchment area. Downstream, in the lower part of the basin, rainfall is likely to be less, runoff less and channel storage capacity greater. As a result, major increases in the catchment area have little effect upon the discharge. Indeed, in the lower reaches, discharge may decline. Thus the relationship between catchment area and discharge is complex and will differ for each basin. To establish this relationship, discharge needs to be measured at a number of places, from the upper basin to the outlet. However, when, for a particular region, one or two such relationships between catchment area and discharge have been established, reasonable extrapolations can be made for unmonitored basins.

There are several morphometric measures for describing a drainage pattern, including stream order and bifurcation ratio, but the most important variable is drainage density. Channel measurements are more significant for detailed analysis and more precise forecasting. The major factors are the length, slope, storage capacity, sinuosity ratio, pattern and symmetry of the main channel. The key variables of a specific uniform reach used for field investigation and a measure of roughness is indicated by the Manning coefficient (n):

$$\text{Velocity} = \frac{R^{2/3} \times S^{1/2}}{n}$$

where R = the hydraulic radius (in m) and S = the sine of the gradient. The likely value of Manning's n for a typical wadi would vary between 0.025 and 0.040. An indication of the range is provided by table 5.1.

Along channels where there is no standard measuring equipment, and where there are no dams or other obstructions to flow which can be used, the slope–area method has proved very effective. The essential requirement is a measure of the maximum height to which water rose indicated by debris line, marks on walls or rocks, deposits or erosion. The reach selected should be 75 times the channel depth and should be reasonably straight. Using the water marks, three cross sections are measured and the gradient of flow is found by levelling from the water mark at the start of the reach to that at the end.

In Malta the seasonal drainage channels are known as widien and only two of these have been monitored for flow. The 50 year flood in the Wied Qirda, a channel deeply incised into limestone, was calculated by the slope–area method with table 5.2 listing the results.

Since the reach of the wied is relatively straight, similar discharge measurements would be expected. Differences probably result from the combined effects of side-channel inputs and seepage losses into the limestone.

Of the other basin variables, geology is likely to be of greatest use in the arid zone. The number of rock and structural characteristics which could be indicated is large, but the most important in hydrology is porosity. Soils, although largely absent in uplands, are of importance on plains. They may be classified according to their infiltration characteristics into hydrologic soil groups. Other key factors are: soil drainage class, particle size distribution, permeability (hydraulic conductivity), and depth.

Table 5.1 Typical values of Manning's n

Bed Material	*Size (mm)*	n
Concrete	–	0.012–0.018
Firm Earth	–	0.025–0.032
Sand	1.2	0.026–0.035
Gravel	2–64	0.028–0.035
Cobbles	65–256	0.030–0.050
Boulders	over 256	0.040–0.070

Table 5.2 Discharge Estimates for Wied Qirda, Malta (using slope-area method)

	Upstream	*Central*	*Downstream*
Cross-sectional area (m^2)	125.5	68.2	98.6
Hydraulic radius (m)	3.4	2.7	3.3
Velocity (m/s)	4.3	7.2	5.9
Discharge (m^3/s)	530.0	490.0	580.0

The US Geological Survey landuse classification (Anderson *et al.*, 1976) comprises nine basic categories. Of these, rangeland and barren land are undoubtedly the most significant in the arid zone. Additionally, but covering far more restricted areas, would be agricultural land, urban land, forest land and water. The most accurate assessment of landuse is by field mapping, but, on any scale, this is likely to be too expensive and time-consuming. Therefore, measures of sufficient accuracy can probably be obtained by remote sensing. For hydrology, a particularly useful variable is the runoff curve number (US Soil Conservation Service, 1972) which has been calculated according to landuse category and hydrologic soil group.

However, the difficulties involved in using any such techniques in arid lands must be stressed. In particular for Manning's equation there are great problems in measuring the slope of the water surface, in selecting the appropriate roughness variable and in dealing with non-laminar discharges such as debris and mud flows. Furthermore, basins are frequently poorly defined and rainfall may be highly localised within a basin. Walters (1989) remarked that the unit hydrograph method presented difficulties in arid lands owing to the localised nature of rainfall and problems over calculating transmission losses. Hydrographs were, however, developed for South West Saudi Arabia using regression analysis between hydrograph features and catchment characteristics.

SEDIMENTATION

Both aeolian and fluvial sedimentation are of concern in arid zone hydrology. However, in general, arid areas lack easily transportable fine sediments, although, owing to the virtual absence of natural protection, any sediment can be easily eroded. Since, from the hydrological viewpoint, sedimentation results primarily from rainfall and runoff, both highly unreliable in their occurrence, the incidence of sedimentation is in itself extremely variable. It is also greatly influenced by extreme events and, in most arid areas, the lack of base flow.

For the reasons discussed in the context of discharge measurement, the assessment of sediment in arid water courses poses serious problems. Not only is there the fundamental problem of obtaining bed-load or suspended-load samples in what may be a hostile environment, but there is also the question of sampling design. The method recommended for arid zone measurements is to obtain samples from verticals equally spaced across the flow. The basic distinction in sampler types is between those with a valve which controls the intake and those which are open and therefore uncontrolled. The former can therefore collect a sample at a point, whereas with the latter, the sample represents an integration of conditions over the traverse length. More sophisticated techniques include automatic samplers and those with a pumping action which produces a continuous record of concentration. Such samplers are all concerned with the suspended-load. The bed-load, which may include material of very large calibre, is not susceptible to such collection and indeed cannot be measured directly satisfactorily. Methods for the calculation of bed-load include empirical formulae, based on research data. In analysis, sediment rating curves can be produced by developing a regression of sediment concentration or load against other measurable variables. The major difficulty in the arid environment is the lack of data, particularly on two variables simultaneously. Furthermore, and more fundamentally, the availability of sediment is a characteristic of the catchment and water course, not of hydraulic conditions. However, it may be possible to obtain data by the analysis of sediments laid under past floods, on flood plains or related to hydrological structures such as dams and reservoirs.

SOIL MOISTURE

Closely associated with groundwater, but distinct from it, is water contained in the unsaturated zone above the water table – soil moisture. Since this is obviously highly responsive to rainfall and evapotranspiration, it lacks the stability of groundwater. As the arid zone is characterized by sparse, thin or totally absent soils, so moisture measurement is only considered in the more detailed studies. More normally, a value is accorded to soil moisture as a result of precipitation or channel flow assessment.

Soil infiltration is influenced by a host of variables including the rate of water supply, soil texture, porosity and structure, organic matter content and antecedent conditions (Free *et al.*, 1940). Field techniques for estimating infiltration rates are discussed by Dunin (1976), Hills (1970) and Hoogmoed *et al.* (1991). Constant head devices reduce the variability introduced by a falling head but such instruments need to be large for the high infiltration rates of many arid land soils. Thus they are cumbersome, difficult to transport and install. Simple infiltration rings are more practical and Hoogmoed *et al.* suggest that an outer ring is only necessary when there is poor contact between cylinders and the soil. Large amounts of water may be necessary with sandy soils to obtain reliable results.

Formulae have been devised to predict infiltration rates. Gifford (1976) reviewed several well known formulae for semi-arid conditions and concluded that Horton's (1933 and 1940) approach was superior but was still unreliable. Doubts were expressed over the usefulness of such formulae to take account of the many factors influencing infiltration. Morel-Seytoux (1976) tried to tackle the particular problem for the time taken for 'ponding' but the use of too many variables produces an unwieldy approach. Field techniques therefore still have to be relied upon or empirical formulae such as those devised by the U.S. Soil Conservation Service (Graf, 1988).

Gravimetric sampling (Reynolds, 1970) remains the basic method for soil moisture sampling but it is destructive, laborious, especially in hot arid environments, and requires weighing and drying the samples. For small-scale investigations, the Speedy Moisture Tester (Anderson and Cox, 1984) has proved particularly useful. Instruments that sample non-destructively are preferred to monitor soil moisture and most soil textbooks contain a description of the devices available, e.g. Marshall and Holmes (1979). The distinctive features of arid lands that affect soil moisture monitoring are soil salinities, high temperatures and a marked diurnal range, and prolonged dry periods, with most changes taking place in the surface layers. Tensiometers are often unsuitable because of the dry conditions while electrical resistance blocks may be sensitive to salinity changes. The neutron probe offers reliable, repeatable sampling but this cannot easily be automated and the device is unsatisfactory for the top 20 cm of soil. The latter is critical as drying of bare arid soils tends only to extend down as far as 30 to 50 cm (Arora *et al.*, 1987, Shimshi *et al.*, 1975 and Tsoar, 1985). Time domain reflectometry (TDR) can be used for soil surface conditions (Gregory, 1991) while the Institute of Hydrology (UK) has developed a soil capacitance probe.

There are then a number of alternative devices (figure 5.1) with some very recent developments, but as of yet no one device is suitable for all arid environments. Remote sensing is also making a contribution to soil moisture monitoring (Stroosnijder *et al.*, 1986). This approach is most useful in isolated locations with sparse vegetation cover. It is based on monitoring visible and infra-red wavelengths of the electromagnetic spectrum which have lower reflectances from wet soils or monitoring the thermal inertia of wet soils during a diurnal cycle. Alternatively microwave emissions are affected by the soil's dielectric constant which in turn is affected by moisture content and texture (Wang, 1985). However, the reflectance of soils is affected by a number of other variables such as organic matter and colour, while Agnew (1986) found that surface roughness due to animal trampling and the movement of sand along the surface also significantly reduced reflectances. Hence information of surface conditions needs to supplement satellite imagery before reliable estimates can be made.

Darcy's law has been used to model soil moisture movement, relating velocity to the hydraulic potential and hydraulic conductivity. It states the conditions governing the movement of moisture in saturated soils when velocities are low and flow is along the path of potential gradient (Childs, 1969). Modelling soil moisture in unsaturated conditions is more difficult as the hydraulic conductivity changes with soil water potential. To overcome this problem, Richards (1931) combined Darcy's law with the continuity equation (Youngs, 1988). The continuity equation is a water balance of soil moisture changes for small element volumes whose distribution through the soil can be used to determine water movement by employing soil water diffusivity (reviewed by Lascano, 1991). This

Comparison of Field Instruments for Measuring Soil Moisture

	Tensiometer	Gravimetric Sampling	Lysimeter	Neutron Probe	Electrical-Resistance
Sensitive at Low Tensions	Yes	Yes	Yes	Yes	No
Sensitive at High Tensions	No	Yes	Yes	Yes	Yes
Measures Water Content	No	Yes	Yes	Yes	Yes
Measures Water Potential	Yes	No	No	No	Yes
Easy to Install	Yes	Yes	No	Yes	Yes
Laborious to Monitor	No	Yes	No	No	No
Easy to Maintain	Yes	Yes	No	No	No
Mobile	No	Yes	No	No	No
Fragile	Yes	No	No	No	No
Relatively Cheap	Yes	Yes	Yes	No	Yes
Withstand Extreme Temperatures	No	Yes	Yes	Yes	Yes
Requires Calibration	No	No	Yes	Yes	Yes

Figure 5.1 Comparison of instruments measuring soil moisture.

type of approach has been criticised as being too sophisticated for the 'real world' type of applications (Arora *et al.*, 1987). Youngs (1988) is also critical of the applicability of such an equation and states (p. 428):

> To apply classical theory of soil water movement with confidence in hydrological modelling, it needs to be established that the soil conforms as nearly

as possible to the inert uniform porous material that Richards' (1931) equation assumes and that isothermal conditions prevail.

Empirical methods using field capacity as the upper limit of soil moisture therefore dominate attempts to model soil moisture storage and drainage because of their simplicity. Field capacity has been assumed to occur at a potential of −0.33 bars (Richards and Weaver, 1943 and Peele *et al.*, 1948). It can then be determined by using pressure-plate apparatus; by employing equations based upon soil characteristics, (Gupta and Larson, 1979, Salter *et al.*, 1966, Salter and Williams, 1965, 1967, Salter and Goode, 1967 and Vereecken *et al.*, 1989); or by field sampling following the cessation of drainage (Burden and Selim, 1989). Doubts increasingly have been raised whether there is a fixed point at which drainage ceases (Richards, 1949 and Russell, 1973). Mohan Rao *et al.*, (1990) suggest that improvements in such soil moisture accounting models can be achieved by treating field capacities as a more dynamic term related to capillary conductivities. They found this was more likely to be successful in sandy soils where the capillary conductivity is low up to field capacity, at which point the capillary conductivity becomes sufficiently large to maintain rapid drainage. Youngs (1988) describes these conditions for coarse sands and in such arid land environments the use of field capacity to model water will be more reliable.

GROUNDWATER

Of all the hydrological elements considered, groundwater (its characteristics, measurement and data analysis problems) differs least in the arid zone from what would be expected under humid conditions. Since, by definition, water is a scarce resource, water table fluctuations in many arid areas will be greater, but the same measurement principles apply. Above all, perhaps, the importance of accurate assessments is more obvious, as the consequences of inaccuracy can be so serious. The major problem overall is that, since water is so scarce, it is often used at rates exceeding recharge.

Techniques for groundwater exploitation can be found in basic texts (Anon, 1975, Bear, 1979, Lloyd, 1981 and Walton, 1970). Shaw (1983) lists three basic methods:

1 **Resistivity surveys** electrical resistance is proportional to water contents.
2 **Seismic surveys** velocities are affected by the degree of saturation.
3 **Wells and boreholes** expensive to develop and maintain a network.

The essential differences confronting work in arid lands concern studies in remote locations and inhospitable environments. Remote sensing is providing a new source of information for groundwater detection. Essentially the method employs long-standing techniques but uses satellite imagery to provide cheaper or a wider coverage for initial survey and identify areas of interest. Monte and Cooper (1982) present a methodology developed for detecting groundwater in arid regions based upon identification of:

1 terrain features assuming the groundwater is more likely in valleys than mountainous areas and that alluvial fans have a high potential for storage;
2 phreatophytes (water-loving plants) indicating shallow water tables;
3 springs and seeps which reveal sites of groundwater discharge;
4 surface drainage indicating rock type and permeability;
5 faults and fractures which suggest areas of high yields and water accumulation; and
6 soil moisture.

Everett *et al.*, (1984) note that arid lands are 'ideal' for remote sensing due to the availability of cloud-free imagery for parts of the year and Mann (1984) found that costs could be cut by coupling aerial photography with satellite imagery. Sung-Chiao (1984) noted, however, that for reliable results such studies should include traditional ground-based investigations. Although satellite imagery can be used to map areas of groundwater occurrence and discharge, local investigations are still necessary for hydrogeological conditions. Hoag and Ingari (1985) suggest that while unconsolidated sediments are widely exploited for groundwater, consolidated rocks are underutilised, particularly in arid regions, because of the difficulty in identifying areas of potential for exploitation. They suggest a six-phase plan of operation for such areas:

1 Remote sensing analysis of area and collection of all documented information to map groundwater development potential areas.
2 Geophysical reconnaissance to rationalise most promising areas.
3 Detailed geophysical investigation to locate sites of test wells.
4 Test drilling and pumping.
5 Production well drilling and pumping.
6 Production well system design and implementation.

The importance of geophysical surveys for groundwater exploration was stressed by White (1987) who reported the success rates for borehole siting in Victoria Province, Zimbabwe, during 1983–4 as: geophysical 90 per cent successful; hydrogeological reconnaissance 66 per cent; aerial photointerpretation 61 per cent; and logistical location 50 per cent.

Consideration also needs to be given to recharge rates (see chapter 2). Houston (1987) notes that most aquifers are recharged by surface (rainfall) percolation that can be assessed by:

- examination of base flow hydrograph,
- monitoring well levels,
- water balance computations,
- examination of the chloride contents of rainfall and groundwater.

Each approach contains, however, sources of error and more than one method should be employed. Chidley (1981) is more emphatic over the problem of estimating recharge rates and states (p. 149):

The subject of estimating groundwater recharge is at present not very precise. There are too many factors involved which do not lend themselves to easy and economical measurement.

WATER QUALITY

Surface water quality is subject to rapid change, exhibiting a wide range of values. In contrast, groundwater remains relatively uniform, with values over only a narrow range, since its characteristics are controlled very largely by the local geological environment. Particular problems commonly encountered in the arid zone include downvalley increases of dissolved materials, as a result of irrigation water reuse. Contamination also occurs in coastal zones where abstraction exceeds recharge rates, allowing seawater incursion.

Procedures in water sampling are no different in the arid zone from elsewhere and tests to indicate biological, chemical and physical quality are detailed in the *National Handbook of Recommended Methods for Water-Data Acquisition* (USGS, 1984).

WATER RESOURCE ASSESSMENT AND MANAGEMENT

The assessment and management of water resources has already been discussed in the socio-economic context. With regard to physical measurement, optimum development of resources, bearing in mind requirements and legal aspects, requires a catchment (where appropriate) investigation. This overall reconnaissance is then followed by more detailed investigations.

A reconnaissance survey will include the following aspects:

1 a description of the hydrogeology;
2 an assessment of the chemical and thermal quality of the water;
3 an analysis of water development already undertaken by man;
4 an isohyetal map of average annual precipitation and runoff;
5 a calculation of the time distribution of precipitation and runoff, computed for the nearest recording station;
6 a flood frequency analysis;
7 an assessment of the average annual water budget;
8 a determination of the source, occurrence and movement of groundwater; and
9 a determination of perennial yield and potential development (Rantz and Eakin, 1971).

This level of investigation provides an approximate estimate of the water supply potential and the lack of detail can be overcome by the implementation of a comprehensive observation programme. This will involve, in particular, the establishment of precipitation and stream gauging networks, together with the provision of appropriately sited instruments for measuring evaporation and evapotranspiration.

These key elements can be used to construct the hydrologic equation:

inflow = outflow ± storage change

However, in the arid zone, storage itself poses problems of measurement, since surface and groundwater resources are not easily separable. Surface storage presents particular difficulties as a result of the high ratio of storage to mean annual runoff volume required to produce the degree of control to allow effective management. It is also obviously hampered by high evaporation and sedimentation rates. In the case of groundwater, the effects of runoff variability are reduced as are the influences of evaporation and sedimentation. Thus, as indicated by socio-economic considerations, the key to arid zone water management is groundwater and the aquifer.

The groundwater inventory is difficult to construct because many aspects are not directly measurable. *The American Society of Civil Engineers Manual of Engineering Practice No. 40* lists the following items within the groundwater inventory:

a) supply elements: surface inflow, subsurface inflow, precipitation, imported water and sewage, decrease in surface storage, decrease in soil moisture storage, and decrease in groundwater storage; and
b) disposal elements: surface outflow, subsurface outflow, total evapotranspiration, exported water and sewage, increase in surface storage, increase in surface moisture storage, and increase in groundwater storage.

From the groundwater inventory it is possible to estimate the safe yield for a particular aquifer. This is a measure of the amount of water which may be extracted annually without producing undesired results such as:

1 impaired water quality;
2 an excessive withdrawal over available supply;
3 violation of water rights or law; or
4 the reasonable costs of pumping.

However, since conditions change, the safe yield is itself variable. Safe yield may be determined through a knowledge of the water storage capacity for various elements of the water table of the aquifer. It may also be inferred from a knowledge of the coefficient of permeability, hydraulic gradient and cross-sectional area of the aquifer. These factors are used in the equation of hydraulic continuity. Neither these approaches nor direct methods are precise and the result would be, at best, a best possible estimate. Therefore it is vital that the conditions under which the estimate was made are listed.

Part III

ENHANCING WATER SUPPLIES

It is evident from chapter 1 that arid lands are not totally devoid of water but water resources are highly variable and consumption is increasing. Many arid nations now consume more water than is replenished naturally within their borders. The imbalance between supply and demand is accentuated by uncertainty over the availability of water resources due to the vagaries the environment and the difficulties of assessing those that are known. The need to regularise water supplies, to store that available during periods of excess and to exploit new resources is then only too apparent. Part III examines the techniques that have been used to enhance water supplies in arid lands. Although much of the discussion focuses upon modern, 'high-tech' and often large-scale techniques there is an attempt to balance this with reference to more traditional, often small-scale measures.

6

IRRIGATION

Much has been written on 'making the deserts bloom' through the use of irrigation. Boyko (1968, p. ix) quoted the USDA yearbook of 1948 which stated:

> Sand deserts alone occupy an area seven times as large as the total agricultural area of the US.

He suggested that a major objective facing scientists was to turn these 'vast wastelands' into agricultural productive land.

Biswas (1978) reported UN recommendations that irrigation should be used to increase food production. The same sentiments were voiced again for the continent of Africa in the 1980s (Moses, 1986). The technology is appealing given the potential increases in production suggested from experimental farm results in table 6.1.

Shalhevet *et al.* (1976) report that wheat yields in Israel were maximised by two irrigations; compared to 86 per cent achieved by early irrigation, 75 per cent was achieved by late irrigation and only 51 per cent with no irrigation. Despite many reservations over using a package of irrigation, hybrid seeds and chemicals this technology has led to significant increases in the yields of wheat and rice since its introduction as shown by table 6.2.

Montague Keen (*Financial Times* 18 April 1990) reporting from Tel Aviv's

Table 6.1 Crop yields achieved through irrigation (kg/ha)

	Max. yields (irrigation)	*Average yields*
From India (White, 1978 and Worthington, 1977)		
Rice	10,000	1,600
Maize	11,000	1,100
Wheat	7,200	1,200
Potato	41,100	8,000
From Near and Middle East (El-Gabaly, 1976)		
Corn	1,470	400
Wheat	1,240	500

Table 6.2 Average wheat and rice yields (kg/ha)

	Wheat		*Rice*	
	1961–5	*1981*	*1961–5*	*1981*
India	850	1,650	1,500	2,000
China	900	1,950	2,800	4,250
Mexico	2,100	3,500	2,300	3,150

(From Byres *et al.*, 1984, compiled from 1976 FAO Production Year book and March 1982 FAO Monthly Bulletin of Statistics.)

international irrigation conference noted that modern irrigation practices have enabled farmers to raise output by 10 times and increased farmers' income by 13 to 15 times. The optimism fuelled by such reports has led to a massive expansion of irrigated lands during the 20th century and it is clear that through the 'green revolution' irrigation has alleviated hunger in many parts of the Third World (World Resources Institute, 1989, p. 51). Table 6.3 shows the dramatic recent increases in the area under irrigation. At the end of the 18th century the world's irrigated area is estimated to have been 8 million ha, increasing fivefold by the end of the 19th century to around 40 million ha (FAO, 1973) with 17 million ha in India, 4 million ha in the USA and 2.4 million ha in Egypt. During the 20th century the world's irrigated area appears to have increased further to over 270 million ha by the 1980s with an average growth rate of 21.6 per cent which the figures in table 6.3 show to be accelerating. However, World Resources Institute (1990) report that the growth rate of irrigation is tailing off with a figure of 271 million ha forecast for the year 2000 rather than being achieved during the 1980s. The figures in table 6.3 should therefore be treated with some caution and used only as a very general guide.

Nevertheless the growth in irrigation this century has been staggering and accounts for 28 per cent of World Bank expenditure during the 1980s (Johnson, 1990). In the USA irrigation has increased over 300 per cent since 1939 (Stewart and Musick, 1982) and by 1974 18 per cent of US cropland was irrigated supplying 27 per cent of agricultural production. The irrigated area in the Middle East increased by 32 per cent during the 1960s and 1970s from 15 million ha (1961) to 20 million ha (1985) with increases of 191 per cent in Tunisia; 163 per cent in Morocco; 170 per cent in Yemen PDR; 91 per cent in Israel; and 90 per cent in Libya (Beaumont, 1988). World Resources Institute (1986 and 1987) estimate that irrigation in India increased from 28 million ha to 37 million ha in the 1960s and by the 1980s had reached 50 million ha; and that irrigated agriculture now (1987) covers 18 per cent of the world's cropland supplying 30 per cent of world food production. Kay (1986) suggests figures of 20 per cent of all cultivated lands are irrigated producing more than 40 per cent of agricultural output. It is difficult to assess how much of this growth has taken place solely in arid lands because irrigation can be found in nearly all climatic conditions. Heathcote (1983) suggests that the irrigated lands in the arid realm expanded by a third in the 1960s and 1970s and that they now contain 80 per cent of the world's irrigated lands (FAO, 1973).

Irrigation is a thirsty technology. Only half the water supplied to agriculture is available for reuse compared to the 90 per cent that is available from water supplied to industry and homes (Postel, 1986). Over 70 per cent of global water use is devoted to agriculture and most of this is required for irrigation. From 1900 to 1950, world water use doubled and it is estimated that it will double again before the end of the century mainly because of population growth and the expansion of irrigated agriculture (World Resources Institute, 1987). Table 6.4 shows some examples of the significant national water demands due to irrigation.

The rapid growth of irrigation and optimism over raised food production is being replaced by a more pragmatic evaluation of irrigation prospects. A quick comparison of tables 6.1 and 6.2 reveals that, for example, in India yields remain below expectations. Many irrigation projects around the world are beset with problems of salinization, water logging, poor harvests, excessive water use and indebtedness. Disappointment with the performance of irrigation has led UNESCO (White, 1978) to suggest that more emphasis should be placed on improving existing irrigation schemes rather than expansion while Heathcote (1983) poses the question, 'Irrigation; panacea or pandora's box'? The costs of irrigation are high with US$1,900–4,500 per ha in South East Asia and US$10,000 in Mexico and sub-Saharan Africa (Johnson, 1990). In the same article it is reported that Pakistan collects Rs1 billion for public services associated with irrigation but costs are in excess of Rs 2 billion (US$80 million).

The reasons for the apparent failures of irrigation lie in a range of explanations from poor planning, overoptimism and cultural factors, yet some irrigation systems have operated successfully for thousands of years and can be considered to be sustainable. There has been much debate over the meaning of this term and Turner (1988, p. 5) notes:

> The precise meaning of terms such as 'sustainable resource usage', 'sustainable growth' and 'sustainable development' has so far proved elusive.

Table 6.3 Areal extent of world irrigation (millions ha)

1900		*1930*		*1955*		*1968*		*1978*		*1985*[1]
44		80		120		163		201		271
Annual increase (Mha/year):										
	1.2		1.6		3.3		3.8		10.0	
Growth rate (%):										
	20.1		16.4		23.8		23.3		43.6	

(From Heathcote, 1983 and [1] from World Resources Institute, 1986)

Table 6.4 National water demands

	Total withdrawn (million m^3)	*Irrigation (%)*	*Public supplies (%)*	*Industrial supplies (%)*
USA	525,053	39.5	8.9	50.1
Turkey	29,928	77.8	12.5	9.7
Australia	17,800	57.1	12.1	4.6
Greece	6,945	82.7	10.8	2.1

(From L'Vovich, 1979)

Pearce (1988) describes non-sustainable practices as leading to land degradation through overexploitation of non-renewable resources and in this discussion on irrigation it is employed in the same sense. That is, sustainable irrigation is perceived as sustaining economic use of resources without environmental degradation. We will first examine some of the traditional practices for supplying water to crops that might be considered to have been sustainable due to their longevity before turning to modern developments.

TRADITIONAL IRRIGATION PRACTICES

The precise nature of irrigation can be difficult to establish because there are so many different techniques and irrigation is used for a variety of purposes from supplying water to crops through to sedentarising nomads (Rydzewski, 1987, p. 5). Barrow (1987, p. 199) states:

> Irrigation is much discussed but seldom clearly defined.

Most definitions pay attention to the supply of water to plants, e.g. Ruthenberg (1976, p. 163) states:

> Irrigation describes those practices that are adopted to supply water to an area where crops are grown, so as to reduce the length and frequency of the periods in which a lack of moisture is the limiting factor to plant growth.

Similarly the FAO state (1973, p. 1):

> Irrigation and drainage are processes which aim at the maintenance of soil moisture within the range required for optimum plant growth.

But this could include cloudseeding for instance or rain harvesting, so the FAO further qualify their definition by stating

> when the moisture available is not sufficient ... artificial application of water to the land ... is called irrigation.

The key words of 'artificial application of water' and implicitly 'avoiding plant water deficits' reveal the scope and purpose of irrigation taken in this discussion. As Coward (1980, p. 15) states:

Plate 11 In Oman, water originating from natural springs, is carried above flood damage by ancient aqueducts constructed on valley sides and fed into a date garden at the mouth of the valley.

> A system of irrigated agriculture can be defined as a landscape to which is added physical structures that impound, divert, channel or otherwise move water from a source to some desired location.

Water can be diverted and applied in a variety of ways and by using 'traditional' we make the distinction between 20th century irrigation systems relying upon engines and machinery powered through mainly fossil fuels, as opposed to irrigation systems whose techniques can be traced back to ancient times, relying upon human labour and/or animal draft power. Given that traditional systems have evolved over long periods of time and can be viewed as sustainable due to their longevity, it is suggested that they may offer lessons for modern irrigation designers and managers.

A world map would be dotted with examples of ancient irrigation systems and cultures. Heathcote (1983) mentions sites as old as 5500 BC in Iran, 2300 BC in Mesopotamia and 1200 BC in Peru. Basin irrigation was introduced to the Nile about 3300 BC (FAO, 1973) and Egypt claims to have the world's oldest dam for irrigation built 5,000 years ago. Obeng (1977) notes that irrigation in the Khuzestan province in Iran can be traced as far back as 3000 BC while Cantor (1967, p. 12) mentions ancient examples from China and the Americas and includes the following biblical quotes:

> And the river went out of Eden to water the garden; and from thence it was parted, and became into four heads.
>
> (Genesis 2:10)

> And he said, Thus saith the Lord, Make this valley full of ditches: For thus saith the Lord, Ye shall not see wind, neither shall ye see rain; yet that valley shall be filled with water, that ye may drink, both ye, and your cattle, and your beasts.
>
> (II Kings 3: 16–17)

Many of these irrigation cultures have long since perished due to environmental problems such as salinization (Hills, 1966) but some continue to flourish in arid parts of the world, employing a variety of water extraction techniques. Irrigation farmers may respond to opportunism by planting after flood waters recede on river banks or in seasonal ponds and saturated soils. Chesworth (1990) notes that archaeological evidence suggests that seeds were sown in soils flooded by Nile waters as far back as 5200 BC. Today, women from Mauritanian villages bordering the Senegal River plant rice in heavy clay soils relying solely upon rainfall and overland flow, and Hills (1966, p. 257) mentions such opportunistic planting on an alluvial fan in Eritrea fed by the River Gash. In the strict sense of 'artificial application of water' (Ruthenberg, 1976) this is not irrigation which can be considered to start with the impounding or diversion of surface runoff. Heathcote (1983) suggests that this may be the oldest form of irrigation, using a diversion weir (or seil – Arabic for flood) to direct flood waters onto fields and he mentions an example to be found at Marib, Yemen. This technique encourages the deposition of the fertile silts carried by the flood waters while excess waters can be drained away back into the river. Clearly such an operation is subject to the vagaries of environmental conditions and is only successful where flows are reasonably predictable or frequent. In addition the system only allows one crop to be grown where there is just one annual flood although in parts of Saudi Arabia and the Yemen there may be more than one flood each year that can be diverted and impounded. Consequently a number of techniques have been devised to extract water (rather than diverting) from rivers or to tap groundwater resources for more reliable supplies.

Until the advent of steam pumps, human and animal power had to employed and Cantor (1967), Nir (1974) and Heathcote (1983) provide a catalogue of designs from the simple rope and bucket through to the Archimedes screw which

Plate 12 Irrigation in these date gardens of Oman is controlled by carefully observed water diversions underpinned by the social organisation of the village. Although many such settlements are now threatened through the loss of the market for dates and the out migration of young men.

Plate 13 Nestled at the base of the hills, quite large settlements can be maintained through irrigation using water from wadi gravels and qanat supplies.

only allows a lift of 1 m. The FAO (1973) report that 5 million wells in India were being used for irrigation with lifts from 1 m to 30 m (increasing to 50–150 m for modern tube wells). A shaduf uses leverage but consequently the height of lift is limited and Cantor (1967) reports the area irrigated in India was limited to under a hectare. Larger irrigation rates can be obtained by a saqia (or sakiya) when animals pull a horizontal wheel which drives a vertical conveyor on which buckets are attached. Perhaps the most efficient technique is a vertical wheel (noria) with buckets attached which is driven by the flow in a river and can lift water some 10 m. Working examples of noria can still be found in Syria and South East Asia. Although ingenious these devices can only irrigate relatively small areas and require fields to be only slightly higher than river or groundwater levels. These designs can find modern equivalents driven by electrical or diesel engines but perhaps the most perfected and efficient traditional irrigation system is the qanat.

Qanats can be found operating today in Oman and have been described by Wilkinson (1977). Man-made channels conveying water to an irrigation settlement are called aflaj and the term ghayl falaj refers to a system relying upon perennial surface flows in wadi gravels. Where surface flows are unreliable or absent villagers have sometimes constructed underground channels that tap into an aquifer. This subterranean channel, often several kilometres long, is called a qanat, foggara or karez. Bemont (1961) reported that there were 40,000 qanats in Iran while Goldsmith and Hildyard (1984) report 22,000 remain, which cover 270,000 km of canals. Drake (1988) suggests 4,000 qanats are still in operation in Oman. Ron (1985) cites examples of foggaras in North Africa from Morocco to Libya and qanats in Iran, Azerbaijan, Afghanistan, Pakistan and the galerias of Mexico. Nir (1974) believes this technique originates from Central Asia and Iran and has spread to other semi-arid areas including the Sahara, China and even South America. The qanats in Oman are believed to have been introduced by Persian settlers (Allen, 1987) around 600 BC. Ron (1985) makes a distinction between irrigation systems in the Judean Mountains based on tapping groundwater through using springs or short tunnels (generally less than 50 m) when water tables fall, as opposed to qanats where a 'mother well' is used to tap into the groundwater, horizontal channels are longer and vertical shafts are used for excavation. This distinction has been criticised and the ensuing debate has raised questions over the age and development of such systems (Rowley, 1986 and Ron, 1986).

The construction of qanats is impressive when one considers the technology and social organisation required. During construction a reliable source of undergroundwater has to be found, possibly by some form of water divining. Figure 6.1 shows that a series of vertical shafts (Drake, 1988, mentions some as deep as 60 m in Oman) are excavated in order to build the underground channel and to maintain it from roof falls and siltation. The qanat or channel will have a slight incline to reduce siltation and to allow gravity to bring the irrigation water to the village some distance away. Numerous physical obstacles can be overcome, e.g.

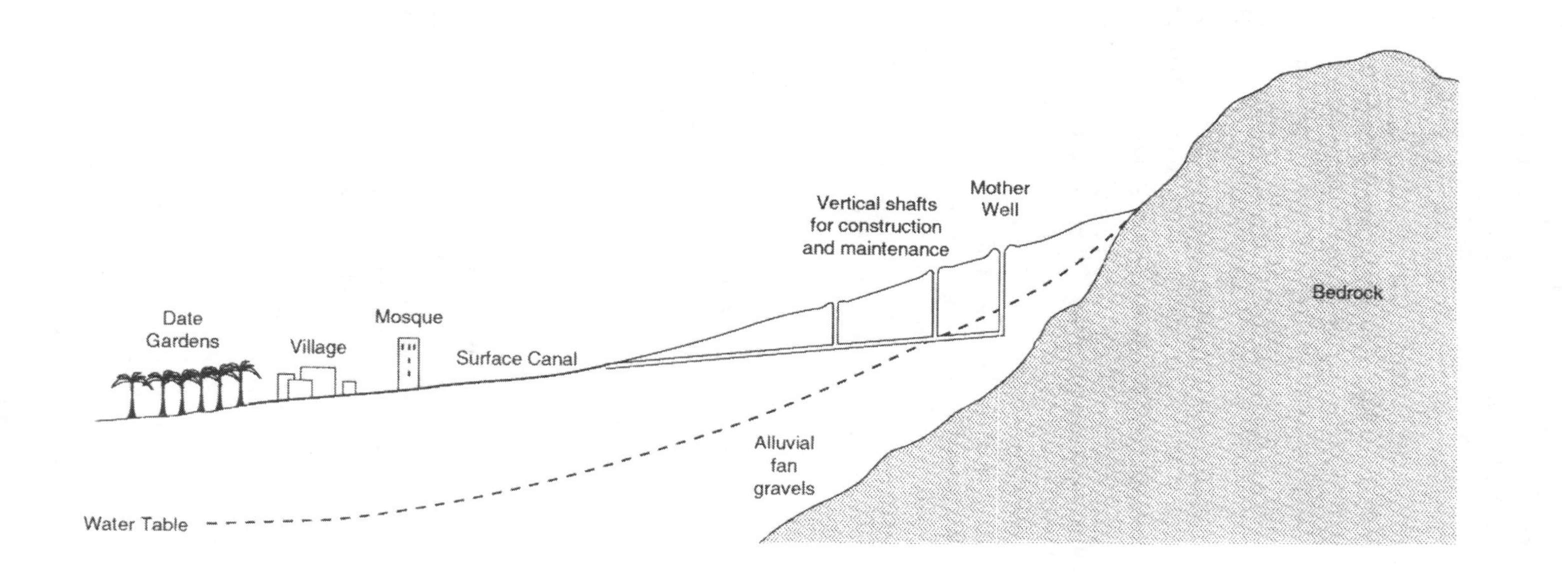

Figure 6.1 Qanat irrigation, Oman.

by taking the qanat underneath a wadi floor that crosses its path or using an aqueduct while raised channels run along valley sides above the level of flood damage. Wilkinson (1977) reports flow rates of 40 l/s from qanats which could support a village of 1,000 people and larger settlements are possible when several qanats have been constructed. A high degree of social organisation is required for the construction, operation and maintenance of these traditional systems. The water is carefully controlled and distributed with a recognised cycle of diversions. Qanats are a communal enterprise using local technology and expertise which may explain their successful operation in many parts of the world. In addition they only supply water that is generated by natural recharge which results in water of constant quality and no degradation of the aquifer. Unfortunately there are now indications of decline and abandonment of qanat irrigation settlements.

Nir (1974) noted in the 1970s that it was unlikely that any new qanats were being constructed and that those in the Sahara in the 1960s were clearly deteriorating. In Oman it is believed that most qanats are ancient and that the ability to divine new water sources and to construct qanats may have been lost. Wilkinson (1977) reports that the average yield of date palms (fed by irrigation) had fallen in Oman from 45 kg to only 20 kg per palm and that 80 per cent were showing signs of ageing. The reasons for such a decline often at first appear to be environmental with sand dune encroachment (Harrison, 1987), the lowering of water tables and salinization but the real causes often lie in social and economic changes. Allen (1987) reports that overextraction of groundwater by new diesel-driven pumps has resulted in falling groundwater tables and consequent seawater intrusion along the Batinah coast, Oman. The situation has been exacerbated by increasing world competition for dates, Oman's loss of its trade routes between East Africa and Asia and the migration of young men to seek employment in the new cities and oil fields. The lack of labour to operate and maintain qanat irrigation is perhaps its greatest threat.

In the Sahara, following 19th and 20th century European colonial expansion, traditional irrigation systems were disrupted through the abolition of slavery (Keenan, 1977). Slaves were important in maintaining wells and channels while the cultivators were kept on a sharecropping basis, retaining only one-fifth of the harvest, the remainder going to the Tuareg 'landlords'. Nir (1974) suggests that during this century these Saharan qanats became 'commercialised' and were no longer operated by the community resulting in overexploitation and poor maintenance. Abandonment followed when the financial returns were insufficient. Hills (1966) claimed that salinization was the major cause for the abandonment of many ancient irrigation systems. While this is true the actual cause of such environmental degradation is probably a result of the breakdown of the irrigation society rather than environmental change. These traditional systems rely upon human resources and when the community no longer acts coherently or with a common purpose it proves impossible to maintain water flows and to combat environmental hazards. Contrary to these reports of declining qanat systems, however, at a recent conference in London (SOAS, 1987), it was apparent that throughout the Middle East qanats continue to function and, far from being fossilised, they are capable of social change. The message for 20th century irrigation designers is clear: that for a sustainable system the community must be incorporated into construction, maintenance and operation.

MODERN IRRIGATION PRACTICES

Modern irrigation systems can be classified in a variety of ways, e.g. gravity or pressurised, diversion or reticulation, food crops or cash crops, etc. The aim of this section is to compare the advantages of different techniques for applying the water and this can be done by considering surface, aerial, subsurface and finally trickle irrigations.

Surface irrigation

Surface irrigation allows water to flood over the surface of the soil and enter the root zone through infiltration. This includes wild flooding, level or basin flooding, graded borders and furrow methods. Surface irrigation is the most common irrigation system and accounts for more than 95 per cent of world irrigation (Kay, 1986) despite recent increases in trickle and aerial methods. Rawitz (1973) describes wild flooding whereby water is fed along temporary ditches,

Plate 14 Surface irrigation can be possible even in sandy soils with high infiltration rates if basins are made small enough but there is a danger of soil erosion.

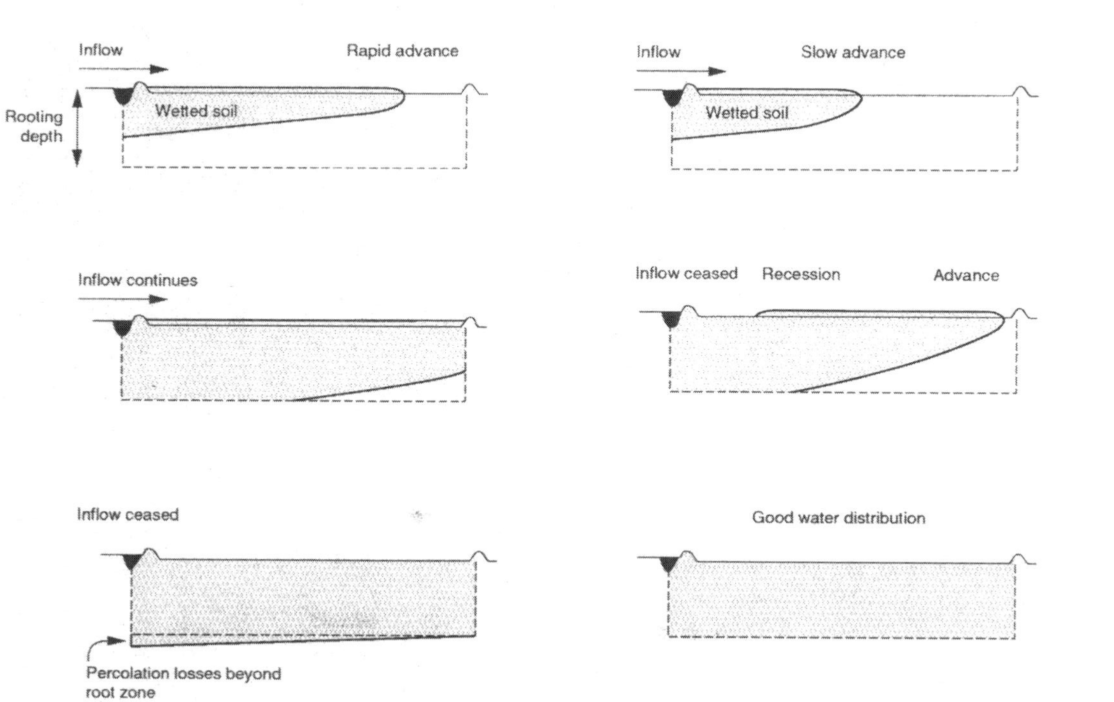

Figure 6.2 Basin and border irrigation, after Kay (1986).

perpendicular to the slope, out of which it flows down the slope and over the fields. This is the least efficient method but involves the minimum of preparation.

Basin flooding is perhaps the most ancient of irrigation applications where water is allowed to inundate fields and to seep into soils. It requires level sites and is typically associated with fine-textured soils although it can be adapted to a variety of soil conditions. Basins are level areas surrounded by small walls of earth or bunds typically 15–30 cm high. The size of a basin is dependent upon soil infiltration rate and the magnitude of water inflow. Coarse-textured soils with high infiltration rates will need small fields to ensure flooding is extensive or a very high inflow rate which can present an erosional hazard. If the time taken to inundate the field is too great then percolation losses may be excessive near the point of water inflow. Figure 6.2 demonstrates the balance between field size, water advance and infiltration, and Kay (1986, p. 28) suggests basins range from 0.01 ha for a sandy soil with a low inflow rate of 15 l/s to 1.6 ha for a clay soil with a high inflow rate of 240 l/s. Although evaporation losses can be high, Dedrick *et al.* (1982) argue reasonably that high water efficiencies can be obtained as there are no runoff losses and deep percolation can be minimised if the system is properly managed.

Surface flooding is widely used for rice paddies on flood plains but water-logging and soil erosion are associated environmental hazards. An alternative surface irrigation technique is to use graded borders where fields with a slight incline are enclosed by levees made of earth. Minimum slopes of around 0.1 to 0.2 per cent are necessary to ensure water moves over the land surface. Rawitz (1973) considers slopes should not exceed 0.5 per cent to minimise the dangers of erosion although Kay (1986) suggests slopes up to 2 per cent in humid areas and 5 per cent in arid areas are possible if a good cover crop is grown, e.g. alfalfa. The size of the field is a function of the soil infiltration rate and water input. Graded borders tend to be narrow (3–30 m wide) and long (100–800 m) which means that quite large farms may be necessary. If infiltration rates are so high that fields of less than 60 m are required then an alternative method of application may be considered. Border irrigation can appear quite similar to basin irrigation with similar crops being grown but there is an important difference in the method of water application evident in figure 6.2. For basin irrigation, water continues to flow across the field giving fairly constant infiltration providing the initial water advance is rapid enough. There are bound to be some percolation losses but with a balanced field size and inflow rate these can be minimised. For border irrigation the inflow of water is stopped after a period of time and before the entire field is inundated; hence infiltration is not continuous. This method reduces percolation losses at the top of the field but requires careful control of the point at which inflow ceases: if too soon the water will not reach the bottom of the field; if too late the water will overflow the border. Kay (1986) suggests inflow should cease when water has reached 60 per cent of the field for clay soils and almost the bottom of the field for sandy soils.

Of considerable importance for both basin and border irrigation is the grading of the land to ensure the flood waters move slowly across the field and do not stagnate in pools. As this is difficult to achieve in practice waterlogging and over-application of water is common. This results in poor aeration in the root zone and often machinery cannot be used on the saturated fields. An alternative is to corrugate the field into a series of rows down which the water travels. This allows better aeration and access by machinery and furrows can cope with much steeper slopes. FAO (1973) suggest furrows can be used on slopes up to 2 per cent for sandy soils, 3 per cent for clays and up to 15 per cent with careful design and management. The length of run varies from 50 m up to 300 m depending upon high or low infiltrations. Furrow irrigation is used typically for row crops, including corn and cotton, while this technique is also suitable for orchards and vegetables. Furrows require more land preparation but can produce good water efficiencies and even handle saline irrigation water. The spacing and shape of furrows is important. Sands, for example, have a high infiltration rate but there is relatively little lateral soil moisture movement. Hence furrows in sands need to have a deep 'v' shape and be closely spaced to ensure the water reaches the end of the field with good lateral wetting.

Although pressurised systems can be used in surface irrigation, water is usually conveyed to the fields by gravity through a network of open ditches which may or may not be lined. This requires careful design and construction to ensure flows are adequate and that water levels in the channels are appropriate for adjacent fields. Water enters the fields by either siphons, sluice gates, weirs or simply breaking earth walls. Consequently the degree of control of irrigation application is highly variable but it is possible to monitor inputs quite closely. The advantages of surface irrigation are that after the land has been prepared, i.e. levelled or graded and borders built, there is comparatively little maintenance and that required can be done by human labour. In flat flood plains where fine-textured alluvial soils are found there may be very little work and these areas are often favoured for this type of irrigation. There are then low maintenance costs, fields can be reasonably large and machinery is not essential. This then can represent a low-technology system although it is possible to automate irrigation scheduling and field applications and introduce quite sophisticated controls. The major disadvantages include the requirement for low slopes and low infiltration rates, i.e. surface irrigation is quite site specific. Water efficiencies can be quite low and World Resources Institute (1987) suggest that water losses could be easily reduced by up to 40 per cent. Doorenbos and Pruitt (1975) suggest water efficiencies of 55 to 70 per cent for furrow irrigation, 60 to 80 per cent for level flood and 60 to 75 per cent for graded borders (see table 6.6). Earth moving may remove the fertile top soil while uneven surfaces can give rise to puddling and percolation losses. Surface irrigation is the most commonly used system in developing countries because it has low capital and maintenance costs and is perceived to require the least amount of training for farmers. It is often then selected for large-scale irrigation systems but this results in a fairly rigid system of fields and channels which cannot be easily modified in the light of technological changes or experience.

Aerial irrigation

Aerial irrigation attempts to reproduce rainfall by spraying into the air water which then breaks down into droplets that cascade onto the land surface. Rawitz (1973) lists sprinkler systems as either static or mobile, which can use perforated pipes (spray lines) from which water issues under pressure or more commonly sprinkler heads can be elevated above the ground to water specific plants or areas. Static systems are characteristic of ornamental gardens, parks or orchards where the vegetation is fixed. They require less labour than mobile systems but are initially more expensive because of the need to invest in fixed pipework and application devices. Alternatively aerial systems can be carried by tractors and can rotate around a centre pivot or travel laterally across fields as a long moving boom. Mobile rainguns have been developed with a large rotary sprinkler operating with high pressures that can irrigate up to 4 ha at a time. Rainguns have been criticised for poor uniformity of water application and for producing large water droplets which can result in significant crop and soil damage. Kay (1983) suggests that these disadvantages may be offset by lower costs and labour requirements compared to other aerial systems. Centre pivot and side-moving systems can overcome the erosional drawbacks of rainguns by using lower water pressures with lower intensity sprays. These mobile systems can, however, get bogged down in heavy clays and are therefore best suited to well drained soils. World Resources Institute (1987) report that rotating sprinkler arms can irrigate up to 50 ha and that Saudi Arabia has installed 12,000 in order to increase grain production. The application rate and area irrigated depends upon the water pressure, the number of sprinkler heads, drop size and movement of the system. It is therefore possible to design systems to take account of crop water requirements, water supply and soil type to ensure that good water efficiencies are obtained and soil erosion and salinization hazards are minimised.

The significant advantage of sprinklers is that they can accommodate steep slopes and consequently do not require initial earth moving. By controlling application rates and drop size, runoff and drainage losses can be reduced and there is no need for canals and ditches which can take up to 5 per cent of the land area. Soils with a high infiltration rate can be more easily irrigated than by surface flooding and mobile systems can be switched from field to field increasing the flexibility of operations. Kay (1983, p. 8) states:

> Sprinklers generally need less water and labour than surface irrigation and can be adapted to more sandy, erodible soils on undulating ground.

Sprinklers have therefore gained great popularity where farmers can afford to buy such systems (table 6.5). The major drawbacks for many arid lands must be the higher costs of these systems compared to flood irrigation, for not only must the equipment be purchased (pumps, pipework and sprinklers or spray lines) but water must also be pressurised. The operation is more complex and therefore requires training and maintenance. The nozzles are prone to clogging if the water

Plate 15 Aerial irrigation using static spray lines can be useful for areas where flood irrigation is unsuitable but there are higher capital and maintenance costs to consider.

Plate 16 Mobile aerial sprays can be automatically controlled to move across fields with uneven topography, producing a reasonable areal coverage at predetermined application rates.

Table 6.5 Application rates for aerial systems (adapted from Withers and Vipond, 1988)

Pressure	*Wetted diameter (m)*	*Rainfall amount (mm/h)*	*Areal uniformity of wetting*
Low	9–25	2.5–25	Fair to good but possibility of large drops
Medium	20–45	3.3–45	Good to v. good; hence suitable for most applications
High	35–120	7.5–48	Poor to good with large drops possibly leading to erosion

is not filtered, electricity needs to be available and high winds can disperse the aerial spray giving uneven watering. Citrus crops can be adversely affected by wetted foliage while the use of water sprays can enhance fungal and bacterial problems. The attraction of sprinkler irrigation systems is the good water use efficiency of around 70 to 80 per cent and their flexibility but they are unlikely to be used in arid lands where surface irrigation is already established and may be more attractive in new areas where soils and slopes prohibit surface flooding. The maintenance costs should not, however, be overlooked.

Subsurface irrigation

The principle of subsurface irrigation is that by controlling the height of the water table, plants can be irrigated from below through capillary rise. The attraction of this system is that it minimises evaporation losses while providing a high degree of control over drainage and soil moisture. Water in irrigation channels can then be used to augment the water table or drain excess water away. Alternatively water can be pumped through perforated pipes installed in the root zone. Rawitz (1973) reports a 30 per cent reduction in evaporative losses but two subsurface systems installed at the Sacremento River delta in California and Huleh Valley in Israel have turned to using sprinklers due to dissatisfaction. The problem with subsurface irrigation is that it requires specific site conditions, i.e. a water table near to the surface and underlain by an impermeable layer and soils with a high hydraulic conductivity so that water can freely move within the root zone. Soils with low permeabilities would require a large number of closely spaced subsurface pipes making the cost prohibitive. Pipes are also prone to damage by rodents and machinery. The soil surface needs to be fairly level so that the groundwater table is uniform (FAO, 1973), while the introduction of salts in the upper soil horizons can require leaching through some surface applications of irrigation.

Trickle irrigation

Trickle irrigation could be classified as an aerial application but it operates at much lower pressures and by water seeping slowly out of surface and subsurface pipes and nozzles rather than by aerial sprays. Trickle irrigation is characterised by a continuous but low rate of water applied at or near to the plant roots. It has tended to be a fairly intensive form of water application developed during the late 19th and early 20th centuries for horticulture. Perforated plastic pipes can be laid alongside rows of plants or individual nozzles can be directed to specific stems and roots. The system is designed by varying the application rates through the diameter of the pipes, the number and type of nozzles and the operating pressure (Heller and Bresler, 1973, suggest applications of 4 to 10 l/h for orchards and 1 to 21 l/h for vegetables). There has been a recent rapid expansion in this technology with an increase in the world areal extent from 56,370 ha in 1974 to 416,660 ha by 1982, i.e. an eightfold increase (World Resources Institute, 1987). This growth is paralleled by expansion in the USA from 4,000 ha in 1972 to 175,000 ha by 1980 (Bucks *et al.*, 1982), reaching 185,300 ha by 1982; with 81,700 ha in Israel; 44,000 ha in South Africa; and 22,000 ha in France (Nakayama and Bucks, 1986).

A number of advantages of trickle irrigation are listed by Caswell (1989) and Heller and Bresler (1973) of which perhaps the most important is that it has a high water use efficiency because water is placed where it is needed by the plant and overwatering is limited. There is a continuous supply of water adapted to plant needs so soil moisture fluctuations are minimised and salinity problems due to evaporation are reduced and there is unlikely to be any surface crusting. Because water is fed at low pressure the soils need not become saturated and good aeration is maintained while plant nutrients can be applied in the irrigation water. Nakayama and Bucks (1986) report that fruit trees are the main crops selected for trickle irrigation. Although most soil types are used, sandy soils are preferred. The main reasons for using trickle irrigation are that water is expensive, limited or saline, that labour is expensive or that local conditions are not suitable for other systems. Unfortunately the orifices can be prone to clogging and therefore the irrigated water must be filtered while reasonably high soil permeability is necessary to ensure water disperses into the root zone from the point of application. Salts can accumulate at the edge of the soil-wetting zone so leaching may have to be included in irrigation applications. Nevertheless good water efficiencies can be obtained although the higher cost than conventional methods is a severe drawback. Caswell (1989) notes that although this form of irrigation serves 185,000 ha in the United States, this only represents 2 per cent of the total area irrigated; and subsequent analysis shows that the quality of land and value of the crop are significant factors influencing adoption of this system. Trickle irrigation is then best suited to reasonably good soils that can ensure an even dispersal of water, where water is expensive to warrant the capital investment and when high value crops or vegetables are to be grown.

Plates 17 and 18 Trickle irrigation can be used to supply carefully controlled applications of water to individual plants although cost is a major obstacle.

Table 6.6 Irrigation application efficiencies (%)

	Barrow (1987)	*Doorenbos and Pruitt (1975)*	*Withers and Vipond (1988)*
Border/Furrow	40–60	55–75	50–70 (medium)
Basin	30	60–80	50–90 (good)
Sprinkler	60–80	60–80	65–85 (v. good)
Subsurface	–	up to 80	70–95 (excellent)

Figure 6.3 compares the relative merits of the above irrigation methods while table 6.6 indicates a general increase in irrigation efficiency from surface through aerial to subsurface applications. But the range is quite marked for each method and there is some disagreement, for instance over basin flooding. This reflects opinions over what can be achieved and what occurs in practice, i.e. the significant role of management and operation of irrigation which will be raised below.

ENVIRONMENTAL HAZARDS

The environmental costs involved in irrigation include excessive use of water, waterlogging, salinization and health risks. The environmental costs of irrigation have been forcibly stated by Hillel (1982, p. xiii):

> Since its inception seven or more millennia ago in the river valleys of Eastern, Southern and Southwestern Asia and North East Africa, irrigation development has in some places become ultimately self destructive. All too often the short term gain in production, leading to intensive settlement, was followed inexorably by long term loss in the form of water resource depletion and pollution, as well as soil erosion and degradation.

Excessive use of water resulting in low water efficiencies has been one of the major causes of environmental degradation.

Low water efficiencies

Irrigation is a thirsty technology with low water use efficiencies arising from evaporation and seepage losses and overapplications. A problem in discussing water use efficiencies is the lack of a universal definition for irrigation systems. Shmueli (1973) suggests the following:

Irrigation efficiency volume of water consumed by crops compared to the volume of water applied;
Water application efficiency as for irrigation efficiency but also including leaching requirements and effective rainfall;
Consumptive use efficiency evapotranspiration compared to the amount of water depleted from the root zone;
Field efficiency water stored in the root zone compared to water applied.

Summary of the Major Methods of Irrigation

	Bases of Comparison	Basin-Floor Irrigation	Furrow Irrigation	Sub-Irrigation	Aerial Irrigation
1	Total capital costs	Low	Low	High	Medium
2	Total annual costs	Low	Medium	Very low	High
3	Crops	Pastures, grain	Row crops e.g. corn, cotton, potatoes, vegetables, orchards	Annual root crops, vegetables	Pastures, vegetables, orchards, nurseries
4	Soil Type	Adaptable to most soils but fields may be small if infiltration rate high.	Adaptable to most soils having good lateral moisture movement characteristics	Must be capable of capillary rise into the root zone. Must allow good lateral movements of moisture	Adaptable to most soils.
5	Topography	Slopes capable of grading to 1% max.	Slopes varying from 0.5% to 12.0%	Fairly uniform slopes preferred.	Adaptable to most slopes which can be farmed.
6	Water required	Large streams.	Fairly large streams are sometimes necessary depending on the number and size of furrows.	Small flows may be utilised	Small flows may be utilised.
7	Soil or crop damage	Over-watering may cause salting, puddling or surface crusting.	Furrow erosion is a big problem	Extremely unlikely.	Both are possible if water drops are large.

Figure 6.3 Comparison of major methods of irrigation, adopted from Withers and Vipond (1988), Table 7.8; based upon Weisner, C.T. (1970) *Climate, Irrigation and Agriculture*, Angus and Robertson, London.

These definitions pose particular problems. The calculation of crop evapotranspiration in particular has vexed hydrologists throughout this century (see chapter 5). The crop rooting depth is required to calculate the volume of water available at different growth stages, a notoriously difficult variable to monitor. Russell (1973) and Durrant *et al.* (1973) explain that rooting depth is very dependent upon water supply and hence irrigation rates will affect the development of roots (Gregory and Squire, 1979) requiring field monitoring rather than relying upon published figures. Mechanical devices have been invented to sample roots (Welbank and Williams, 1968) but Nye and Tinker (1977) maintain that there is no real alternative to excavation and washing. Unfortunately this is both laborious and destructive and has lead to attempts to use radioactive sources (Boggie and Knight, 1962 and Ellis and Barnes, 1973) or the monitoring of water extraction by tensiometers (Stone *et al.*, 1973) and neutron probe (Agnew, 1980, McGowan, 1974 and Tennant, 1976). Such analyses are rarely performed routinely and rooting depths are normally estimated from soil conditions and plant characteristics; for example, Hignett (1976) suggests for wheat:

$$\text{rooting depth (mm)} = 32W^{1.3} - 100.0$$

where W is the number of weeks since sowing.

The ratio of crop yield to evapotranspiration has been used as a measure of water use efficiency by Arkley (1963), Begg and Turner (1976) and Russell (1973) but this ignores drainage and other water losses. The economic value of the crop can be compared with the costs of water to indicate irrigation efficiency but 'crop value' can be difficult to establish given government pricing policies while 'water costs' can be calculated in a variety of ways. More commonly irrigation efficiencies are calculated by incorporating field, conveyance and system losses as suggested by Doorenbos and Pruitt (1975) and Jensen (1967). Table 6.7 is compiled from observations by Bos (1977) where:

e_p = project efficiency = V_n/V_t = $(e_a \times e_d)$
e_f = farm efficiency = V_n/V_f = $(e_a \times e_b)$
e_d = distribution efficiency = V_a/V_t = $(e_b \times e_c)$
e_a = field application efficiency = V_n/V_a
e_b = farm ditch efficiency = V_a/V_f
e_c = conveyance efficiency = V_f/V_t

V_n = crop irrigation need (crop water requirement − rainfall)
V_t = total water supplied to project
V_f = water delivered to farm
V_a = water applied to cropped area.

Table 6.7 shows that reasonably high conveyance and farm ditch efficiencies can be found of up to 80 per cent but as water progresses through the system there is a cumulative loss of water. Distribution efficiencies are 35 to 70 per cent, field application efficiencies range from 40 to 75% and overall project and farm

Table 6.7 Selected irrigation efficiencies (after Bos, 1977, p. 354)

Country	e_p	e_f	e_d	e_a	e_b	e_c
Egypt	0.3	0.46	0.46	0.66	0.70	0.66
Iran		0.29				
Iran				0.76		
Iran				0.50		
Israel	0.51					
Rhodesia	0.32	0.45	0.57	0.47		0.71
India	0.40	0.57	0.58	0.70	0.82	0.70
India	0.14	0.20	0.34	0.40	0.50	0.67
India	0.25	0.32	0.47	0.53	0.60	0.78
India	0.16	0.24	0.34	0.47	0.51	0.67
USA	0.26		0.66	0.40	0.80	0.83
USA	0.33		0.70	0.58	0.80	0.88
USA	0.28	0.53	0.52	0.55	0.97	0.54
USA	0.33		0.50	0.59	0.80	0.63

efficiencies are generally below 50 per cent and may be less than 20 per cent.

Kovda (1980) suggests most modern systems only have water use efficiencies around 30 to 35 per cent. Stewart and Musick (1982) believe U.S. on-farm efficiencies were around 50 per cent which means twice as much water is supplied as required by the crop. Conveyance losses were so high in India that for every 100 m^3 supplied at the headwaters only 40 to 50 m^3 reached the field (Ruthenberg, 1976). Similar seepage losses of 40 per cent were found in Algeria (White, 1978) while UNESCO (1977a) report that on average 50 per cent of irrigation water is lost by drainage and evaporation.

The causes of low water efficiencies have been blamed on poor control of water resources and bad water husbandry due to either poor training or artificially low prices encouraging overapplication. What is drainage for one farmer becomes waterlogging and salinization for another downstream. The failure to used lined channels either because water is so abundant that it is not perceived as being important, lack of materials and know-how or through cost cutting has led to significant seepage losses. Hillel (1982) suggests the problem may originate in the design stage when insufficient attention is paid to the utilisation of water in the field and the role of individual farmers. Ruthenberg (1976) believes the sharing of water leads to its uneconomic use and mentions that water metering may help the situation, although the problem extends beyond the control exerted by individual farmers and it is difficult to envisage how metering could be cost effective in many rural areas of arid lands.

Salinization

Overuse of water is not only wasteful but also contributes to the problem of salinization. Kovda (1980) states that salinization has been associated with irrigation for thousands of years in the basins of the Tigris, Euphrates, Indus and Nile but the problem is growing. Parts of Iraq that in ancient times supported a flourishing civilisation today have no fertile non-saline soils. The problem was recognised as being serious in the USSR during the 1930s with 2 million hectares affected which had increased to 3.5 million hectares by 1940 (Kovda, 1946). Salinization has affected about 20–30 million ha of irrigated land or 7 per cent of the total since the beginning of modern irrigation (Kovda, 1983) including vast areas of India, Pakistan, Egypt, USSR, Iraq and Syria and Patagonia. Postel (1989) suggests the problem is even greater with an estimate that 24 per cent of all irrigated land suffers from salinization, amounting to 60 million ha.

Around 1 to 1.5 million ha are added each year to the total area affected by salinization (World Resources Institute, 1986). In the USA 20 to 25 per cent of all irrigated land suffers from salinization (El-Ashry, 1985). The water table in the Indus Valley, Pakistan, was 30 m below the surface before irrigation but due to overwatering and seepage the water table rose to within a few centimetres of the surface in low-lying areas. By 1960, 3.2 million ha (58 per cent of the total) were affected by waterlogging; 4.5 million ha suffered from salinization; and 1.4 million ha (26 per cent of the total) went out of production (Jackson, 1977). In India, 6 million ha are affected by waterlogging; 4.5 million ha by salinization; and 2.5 million ha by alkalinization (Dogra, 1986).

Salinization is a problem because of the deleterious effects of salts upon soil structure and fertility as well as its toxicity for plants. Brady (1974) describes how the high concentrations of sodium salts in sodic soils lead to dispersal of soil colloids making soils impermeable and difficult for plant growth. Kovda (1946) reported that yields of cotton were reduced by 50 per cent for slightly saline soils.

Table 6.8 Effects of saline irrigation waters

Dissolved solids (ppm)	*Conductivity (mS/cm^{-1})*	*Detrimental effects*
<500	<0.75	no significant effects
500–1,000	0.75–1.5	salt-sensitive crops affected (peas, potatoes, cabbage, apples)
1,000–2,000	1.5–3.0	many crops affected adversely
2,000–5,000	3.0–7.5	salt-tolerant crops only (cotton, rape, kale, sugar beet)

(From Marshall and Holmes, 1979 and Brady, 1974)

Table 6.8 indicates the effects of saline irrigation waters upon crops.

Cantor (1967) provides the following graphic description of the effects of salinization upon fields in the Punjab:

> If you look down at the Punjab from an aircraft, you will see . . . patches of grey decay. This is salinisation, a rotting of the land caused by a rise in the level of the undergroundwater table, which forces the salts of the earth to

> the surface and gradually turns field back to desert. It does not steal inexorably forward, like an encroaching sand dune, but erupts unpredictably in disconnected patches . . . nibbling away at the welfare of the Punjab.

Salinization affects not only soils but also the surrounding ecosystem and Ojiako (1988) reports on the adverse effects upon fish stocks downstream of irrigation schemes on the Anambra River in Nigeria. Salinity levels in the Kent River, Western Australia, increased from 0.5 g/l in 1940 to 1.2 g/l in 1980 due to irrigation; concentrations in the San Joaquin River, California, rose from 0.28 g/l to 0.4 g/l over the same period (World Resources Institute, 1990). Oster (1984) describes the soil–salt–water system as the following balance:

$$\hat{S} = r + g + i + S_m + S_f - d - S_p - S_c$$

where,

- $\hat{S}$ is the salt balance
- r salt content of rainfall
- g salt content of any upward movement of the water table
- i salt content of irrigation water
- d salt content of drainage water

- S_m salt from soil minerals
- S_f fertilizer additions of soluble salts
- S_p salt that precipitates in the root zone
- S_c salt absorbed by the harvested crop.

The causes of high salinities can then be blamed upon natural occurrence of salts, using low-quality saline irrigation water, lack of drainage, overapplication of water and farmers' lack of understanding of the leaching requirements of irrigation. El-Ashry (1985) cites the example of an underground salt dome that is responsible for introducing 200,000 tons of salt each year into the Dolores River, Colorado. Arid land soils and rocks have a natural salinity which may be concentrated in the upper horizons by infrequent wetting followed by evaporation and by plants drawing water out of the soil but leaving salts behind. The presence of soluble salts may be augmented by aeolian transport and surface runoff. Often these salts pose no particular problem as they are leached out and are not allowed to accumulate within the crop rooting zone. When leaching is ineffective through insufficient water being applied then salt deposition can occur. Kovda (1946) describes an Egyptian farm satisfactorily using groundwater with a salt content of 3–5 g/l but when the crop area was expanded the water supply was insufficient resulting in salinization of the upper soil horizons. Waterlogging followed by evaporation can exacerbate the accumulation of surface salts while overapplications of fertilisers can increase the salinity of soil waters. Kovda (1980) reported that in the USA the area of irrigated agriculture doubled between 1949 and 1973 while fertiliser applications had increased fourfold.

Water percolating through soils can remove soluble salts which may ultimately contribute to the salinity of groundwater. If the irrigation system is utilising such groundwater there is a positive feedback with increasingly saline irrigation water. Another possibility is that saline groundwaters will affect the crop rooting zone by capillary rise in fine-textured soils or rising water tables through overwatering. It is estimated (El-Ashry, 1985) that during the 1990s the water table in the San Joaquin Valley will rise to within 1.5 m of the surface and 650,000 ha will require drainage. Saline water discharged by irrigation may also drain into streams and channels producing salinity problems elsewhere in the catchment. White (1978) reports that following the establishment of irrigation the mean annual discharge in the Rio Grande fell from 1.33 to 0.251×10^9 m^3 while salt concentrations increased from 221 to 1,691 ppm. A desalination plant had to be constructed on the Colorado River to reduce salt concentrations from 3,100 to 115 ppm due to the effects of saline irrigation drainage waters (Channabasappa, 1982).

There are a number of solutions for salinization of which good water husbandry, i.e. preventing overwatering and seepage losses, is one of the most effective. This involves lining irrigation and drainage channels with impermeable coatings. Several materials are available but they all cost substantially more than just excavating earth ditches. Better training for farmers and more careful water applications are obvious measures to call for but difficult to achieve in practice. Johnson (1988) reviews attempts in Pakistan to tackle salinization and reports that 8,000 tubewells were installed by 1977, increasing to 12,000 by 1979, but problems were identified over the lack of education and training of farmers coupled with poor overall organisation and management. Even if farmers have good knowledge of irrigation need and leaching requirements this is only useful if they have a reliable and manageable supply of water. Salts can be leached out if sufficient extra water is applied (Oster, 1984 and Shalhevet, 1973). White, (1978) suggests the following regime:

Salt concentrations (g/l)	Flushings
0.5–1.0	every 1 to 2 years
1.0–2.0	once or twice annually
2.0–3.0	several annually
3.0–5.0	each irrigation

Water used for flushing must be free from silts to prevent clogging and poor drainage, and must be low in sodium or hydrolysis may result with associated high pH and impoverished soil structure (Brady, 1974). The best results are obtained for saline soils containing soluble calcium and magnesium salts but sodic soils (high in sodium) present a particular problem. Leaching of sodic soils releases sodium ions which cause dispersal of soil colloids, thus destroying the soil structure and creating impermeable conditions in which plants cannot grow. This can be avoided by first treating a sodic soil with gypsum which will prevent hydrolysis (Loveday, 1984) and then applying the irrigating water or using sulphur where sodium carbonate abounds. The excess water percolating through the soil, however, must not be allowed to produce a dramatic increase in the groundwater table. Salt flushing therefore requires good drainage and the installation of horizontal drains that can feed away percolating waters while also preventing a rising water table from reaching the rooting zone. Wells can be used to lower groundwater levels but there are problems in controlling drainage when soils are fine textured and subsurface movement is limited. The saline water that is collected from these drains must be carried in lined channels to ensure it does not seep back into the soils but this still presents a problem for disposal somewhere in the catchment.

Kovda (1980) reports that the Government of Pakistan has installed tens of thousands of wells to reduce groundwater levels but the scheme lacks horizontal drains and a unified network. The Iraqi Government has invested in both wells and subsurface drains approximately 2–3 m deep and 100–200 m apart with 12–20,000 m^3/ha of water applied for flushing. In Tadzikistan (USSR) the extent of salinization has been reduced from 35 per cent of the irrigated land to 9 per cent by 1972 using drains and leaching. In China, water was diverted from the Yellow River via the 'Peoples Victory Canal' to irrigate 40,000 ha in 1952; and total diversions now supply 3 million ha (Ronghan, 1988). Lingen (1988) reports annual diversions are of the order of 500 MCM (million cubic metres of water) with a further 150 MCM from 5,000 wells. As a consequence the water table rose substantially and vast areas became salinized. A massive programme of new drainage ditches, canals and tubewells resulted in a lowering of the water table and substantial reclamation of the land. Nickum (1988) notes that, despite reservations over the lack of technical expertise and poor management, since the 1970s the saline area has declined and crop yields have increased.

The use of sprinklers or trickle irrigation may help tackle salinity, especially in sandy soils, through their improved water use efficiencies but only if the applied water is free from salts, otherwise evaporation will lead to salt deposition. The use of furrow irrigation as opposed to surface flooding can also reduce salinity problems through reduced depression storage. A more passive approach is to plant crops that are not sensitive to salts such as date palms, cotton, rice and barley (Brady, 1974 and Ruthenberg, 1976). Boyko (1968) refers to successful experiments growing crops in sandy soils with natural seawater for irrigation. Other authors in the same publication cite examples from Tunisia to Kuwait.

There are then a number of techniques for combating salinization but flushing of salts with improved drainage is the main approach. Letey (1984) demonstrates that this approach was established by the end of the 19th century. The continuation of the salinity problem is not then so much a lack of knowledge or technique but has more to do with the exploitive nature of irrigation practices where sustainability is not the principle goal and environmental damage is condoned.

Excessive water extraction

The high water demands of large-scale irrigation operations and low water use efficiencies require the construction of large dams and even interbasin transfers with their consequent environmental effects (see chapter 9). The result of extraction beyond naturally renewable resources is falling water tables, declining water quality and increasingly difficult water extraction with the possibility of subsidence. Irrigated cotton has expanded in the Kazakhstan and Uzbekistan regions of Soviet Central Asia from around 2 million ha in 1950 to some 7 million ha today (M. Walker, 1988, Slow death of the Aral Sea, *Guardian*, 18 April 1988). Overwatering and a lack of lined channels have led to significant percolation losses. The drainage waters leach out some of the naturally occurring salts, augmented by fertilisers leading to salinization problems and the production of a new man-made salt lake called Sarakamyush. The impact has been to reduce dramatically the amount of water reaching the Aral Sea which has reduced by at least one-third since 1960 according to R. Cornwell (1989, Catastrophe on Aral Sea ranks with Chernobyl, *Independent*, 20 April 1989) and at this rate will disappear by 2010. It is estimated by the Institute of Geography, USSR Academy of Sciences, that lake levels have fallen by 14 m, of which only 20 per cent is due to climate variation (World Resources Institute, 1990). Micklin (1988) reports that the lake volume has decreased by two-thirds with salinity rising threefold, resulting in salt deposition on surrounding farmland (Postel, 1989).

High water demands coupled with increasingly powerful pumps have led to groundwater resources becoming severely depleted by irrigation schemes. Postel (1984 and 1989) provides worldwide evidence of this phenomenon including examples from India (Tamil Nadu) where groundwater tables have fallen 25–30 m and China (Beijing) where the rate of fall is reported at 1–4 m each year. Allen (1987) reported overpumping in coastal sites in Oman leading to seawater intrusions inland. World Resources Institute (1986) report that in the USA this problem is widespread and severe. Lean *et al.* (1990) note that the water table of the Ogallala aquifer beneath the United States' Great Plains is falling at a rate of 1 m a year; while water tables at Bangkok have fallen 25 m since the late 1950s. The irrigated area in California increased by 38 per cent between 1950 and 1980 with a consequent rise in water demand of 121 per cent leading to overextractions of groundwater. In the Santa Clara Valley water tables have fallen 37 m since 1910 with associated land subsidence of 4 m near San Jose and seawater intrusions; in the San Joaquin Valley water tables have fallen 60 m (Coe, 1988).

Health problems

Irrigation is often to be found in warm areas which may provide fertile breeding grounds for diseases and their vectors, especially if there is abundant stagnant or slow-moving surface waters, e.g. rice paddies or poorly maintained drainage ditches (FAO, 1973). Obeng (1978) reported the following increases in the infection of schistosomiasis following the introduction of irrigation in Egypt:

	Before	After
Sibaia	10	44 (% infection)
Kilh	7	50
Mansouria	11	64
Binban	2	75

Hillman (1987) lists similar increases from zero or low levels of infection to a prevalence of 30 to 60 per cent in the Gezira scheme by 1940, infections of 53 to 86 per cent between 1937 and 1967 in Tanzania (Atusha Chini project), 90 per cent in Ghana (Lake Volta) and 45 per cent in Nigeria (Lake Kainji). There is, however, a problem over the reliability of such medical statistics including the lack of data available before projects are established. Amin (1977) reports that the incidence of schistosomiasis associated with the Gezira scheme in the Sudan was only around 2 per cent (1940s) through analysis of urine samples from visitors to dispensaries. Village surveys produced adult infections (*S. haematobium*) of 21 per cent and 45 per cent in children, which by 1952 had fallen to 8.86 per cent infection, whereas *S. mansoni* infection increased from 5 per cent to 8.77 per cent during the same period.

The range of diseases associated with irrigation schemes is immense. Proliferation of the *Anopheles* mosquito in flooded fields and canals has contributed to a threefold increase in malaria between 1961 and 1970 (Chapin and Wasserstrom, 1983) and it is estimated that 160 million people are now infected worldwide (Goldsmith and Hildyard, 1984). The same publication notes a corresponding increase in schistosomiasis of 114 million people infected in 1947 rising to 200 million by 1984. The cause of high infection rates is not only the presence of an aquatic environment and temperatures encouraging the spread of disease through the presence of vectors such as snails and mosquitos, but also poor sanitation. Farid (1978) suggests poor hygiene is a major factor influencing the incidence of diseases associated with irrigation projects. Large-scale irrigation often involves the migration of huge numbers of people encouraging the transmission of disease. If potable water is not widely available and if human waste is not adequately treated and kept separate from agricultural waters then the occurrence of disease will be higher. Solutions are therefore to be found in health and hygiene programmes through education and better sanitation, through the provision of medical facilities and by improved water management. Barrow (1987) mentions that if channels are kept mud and weed free snails can be discouraged while some plants also have this effect.

LARGE OR SMALL SCALE?

The lack of control of water resources and poor reliability of flows prompts overwatering when supplies are available. These causes are bound up in the belief that irrigation schemes need to be large scale in order to organise water supplies efficiently and in order to attract sufficient investment and to produce significant increases in agricultural production. Largescale irrigation projects have the attraction that savings can be made through economies of scale and that effective management of water resources requires control over a sufficiently wide area. Large projects are also highly visible and therefore can satisfy political desires, they can more easily attract sufficient investment and they can produce a significant change to agricultural production (Oosterbaun, 1985 and Ruttan, 1986). Worthington (1977, p. 16) suggested that projects of less than 1,000 hectares should be avoided for efficient operations. The move to large-scale production has required the involvement of governments and public funds. This is highlighted by Heathcote's (1983) observation that one of the last major private irrigation projects in the USA originated in 1902!

The Gezira scheme was a large-scale attempt to irrigate parts of Sudan to grow cotton as a cash crop. It covers over 800,000 ha and is run by the Gezira Board (GB) who control the landuse, field rotations, crop planting date, irrigation schedules and fertiliser applications. The farmer acts as a tenant and provides the labour. All other inputs are provided by the project which therefore receives a proportion of the cotton harvest in payment. Ruthenberg (1976) states that the lessons of the Gezira scheme are that one organisation must be responsible for all supplies and water resources and that close supervision is essential. Barnett (1977) is more critical in his discussion of actual practice. Concerning water supplies he mentions that the cumbersome administration of the Gezira scheme is unable to respond flexibly enough to changing environmental conditions and often false initial assumptions over water needs and soils are included in the design. Through using 'average' supply rates of 30 m^3 per day per plot with watering every 15 days there is inadequate knowledge of varying local conditions which can lead to under or overwatering. The 'code' was to water fields for 4 days yet Barnett found the average time required was over 7 days. The result was that crops were underwatered for parts of the year and overwatered at other times. It would appear then that, far from promoting greater efficiency, a centralised administration could produce low water use efficiencies through poor communications, inadequate design and a rigid adherence to initial organisation without continual monitoring and modification. The infrastructure and organisation of large-scale projects tends to be established early and then becomes fossilised and resistant to modification.

Dissatisfaction with the operation of such schemes has led to suggestions that no new land should be opened up to irrigation until existing operations have been improved (White, 1978). Large-scale projects are plagued by high costs due to their complex nature and associated environmental problems and led Balek (1977) to propose:

> Priority should be given in the African tropics to small scale village type irrigation schemes that can be easily managed and financed and in which farmers can fully participate.

Adams and Carter (1987) have discussed the attractions of small-scale community-based operations where irrigation is organised and controlled by the landholders. This 'grass roots' development, it is argued, is more effective at integrating human and physical resources, utilising local skills. The first section in this chapter showed how small-scale traditional irrigation could support communities with little environmental degradation. Problems do exist, however, with small-scale irrigation including inadequate control of water resources, difficulty in attracting investment and capital and the need for the development of infrastructure and marketing possibly without government assistance. Irrigation then faces difficulties of poor water efficiencies, finance and organisation at all scales although mistakes involving large-scale projects may result in more serious consequences.

CROP WATER REQUIREMENTS AND IRRIGATION SCHEDULING

The water required by crops needs to be computed as part of irrigation design to ensure peak and total water demands can be satisfied. There is also the need to monitor water requirements during irrigation operations and to make adjustments in water schedules given possible changes in atmospheric conditions and technological changes not foreseen during the design phase.

Crop water requirements

Crop water requirements are usually calculated as plant evapotranspiration where evaporation from soil and plant surfaces, intercepted water and plant transpiration are combined due to the difficulty of separating these components for routine calculations. Chapter 5 discussed the problems of estimating evapotranspiration, with either the Penman formula or evaporating pans the most widely adopted techniques providing potential evapotranspiration (PE_t) for a reference surface of grass, water or alfalfa. The PE_t for a particular crop can be estimated by the use of crop coefficients where:

$$PE_t(\text{crop}) = PE_t(\text{reference surface}) \times \text{crop coefficient } (K_c)$$

(providing soils are kept moist). Crop coefficients for a given crop tend to change during the growth cycle and different values are usually employed for phenological stages. A comprehensive listing of crop coefficients can be found in Doorenbos and Pruitt (1975), Jensen (1974) and Wright (1981 and 1982) with values ranging from 0.1 for bare soil to over 1.0 when full ground cover is reached.

The use of meteorological data or evaporating pans to calculate irrigation need can be misleading as they should be obtained from a site where evaporating surfaces are moist, that is a climatic station surrounded by well watered, short, green vegetation. Otherwise there can be a significant change in the energy and moisture balances of the atmosphere once the irrigation system is installed. The Penman method of estimating crop water requirements or the use of evaporating pans should therefore only be used as an initial guide and monitoring of crop water requirements during irrigation becomes essential.

Irrigation water requirements

The following sequence (after Basu and Ljung, 1988, Burt *et al.*, 1980, Doorenbos and Pruitt, 1977 and Burman *et al.*, 1983) for estimating irrigation water requirements is fairly typical:

1 Select crops, establish planting dates and rotations
2 Calculate PE_t (crop) on a monthly/weekly basis
3 Compute water balance taking account of other hydrological inputs and outputs such as rainfall, surface inflow and drainage, hence deducing field irrigation need (I_n)
4 Determine salinity leaching requirements
5 Determine field application and conveyance efficiencies
6 Calculate total irrigation need, peak water needs and weekly/monthly demands.

The water balance computation will depend upon the data available, e.g. groundwater depth, capillary rise, infiltration rates, rainfall intensities, etc. which will involve some reconnaissance fieldwork. The water required for salinity leaching can be obtained from tables in Doorenbos and Pruitt (1975) or empirical formulae such as

$$\text{Leaching water (mm)} = n_1 \times n_2 \times n_3 \times 400X \pm 100$$

X = mean % readily soluble salts in 2 m depth of soil
n_1 = Coefficient of mechanical composition: 0.5 sand to 2.0 clay
n_2 = Coefficient of depth of groundwater: 3(1.5–2.0 m), 1(7.0–10.0 m)
n^3 = Coefficient for mineralisation of groundwater: 1(weak), 3(brine).

(From Kovda, 1980, p. 212.)

Irrigation efficiencies have been discussed in an earlier section. Water needs can be calculated by

$$PE_t(\text{crop}) = PE_t(\text{grass}) \times K_c$$
$$CWR = PE_t(\text{crop}) - P + D + R_o - CR - R_i \pm SW$$
$$I_{ef} = E_a \times E_b \times E_c \text{ or } = E_a \times E_d$$
$$I_n = (A \times CWR)/(LR \times I_{ef})$$

where

I_n = irrigation need for area A (m^3)
PE_t = potential evapotranspiration (m)
K_c = crop coefficient
CWR = crop water requirements (m)
P = effective rainfall (m) (Dastane, 1974)

D = drainage beyond rooting zone (m)
CR = capillary rise into the rooting zone (m)
R_o = surface runoff losses (m)
R_i = surface inflows (m)
SW = soil moisture storage in rooting zone (m)
I_{ef} = irrigation system efficiency (%)
LR = salinity leaching requirement (%)
A = irrigated area (m^2)
E_a = field application efficiency
E_b = field canal efficiency
E_c = conveyance efficiency
E_d = distribution efficiency.

Irrigation scheduling

During the design phase, field and project water requirements are estimated so as to ensure that sufficient water resources can be made available at the right times. In practice data is often not available or is of questionable reliability and such calculations should therefore only be used as an initial guide with subsequent monitoring of actual water demands necessary. Jarvis (1975) reported that most crops only store 10 per cent of their daily transpiration needs and therefore when soil moisture becomes limiting plant growth is soon affected. Begg and Turner (1976) reviewed the possible morphological effects of moisture stress upon plants (cell division and enlargement, root growth and leaf area) and the physiological effects (photosynthesis, respiration and translocation) while Kozlowski (1972) has edited a comprehensive series dealing with moisture and plant growth. The impact upon yield is complex and dependent upon type of crop, the degree of moisture stress, timing of deficit and previous history. The economic yield may not be adversely affected as Burr *et al.* (1957) found moisture stress increased the sugar content of sugar cane and Van Bavel (1953) found it improved the quality of tobacco leaves. It is widely reported that cereals are particularly sensitive to moisture stress during flowering and early grain formation (Salter and Goode, 1967). This was reported for wheat (El-Nadi, 1969), for sorghum (Kaliappa *et al.*, 1974), for maize (Fuehring *et al.*, 1966 and Wrigley, 1969). Yet Day and Interlap (1970) conclude that moisture stresses at any time are important while Seetharama *et al.* (1984) examined millet and sorghum and found variable responses: with grain yield reduced by stress occurring during seedling stage and particularly just before flowering yet stress occurring after crop establishment had little effect. Moisture stresses may improve yields in particular cases, but it is best avoided for most crops at all times.

The timing of water applications can be computed from environmental conditions where

$$I_f = (AWC \times RD)/(AE_t \times 100)$$

I_f = Irrigation frequency (days)
AWC = Available water capacity (%)
RD = Rooting depth (mm)
AE_t = Actual evapotranspiration (mm).

AWC is the difference between field capacity and the soil moisture content at which water becomes limiting for plant growth and moisture stresses begin to affect plant yields (see below).

The difficulties of estimating AE_t, rooting depth, field capacity and the point at which water constraints plant growth limit the usefulness of this approach. An alternative is to monitor the crops directly and Shimshi (1973) lists a number of techniques including relative water contents, pressure chambers and devices monitoring stomatal apertures, but mostly these are scientific methods which have proved unsatisfactory for routine irrigation applications. Bielorai (1973) lists a number of visual plant indicators that can be used including colour, leaf angle, stomatal movement and stem sizes, but this approach may be too subjective although it can be argued that with good training this can be effective.

Soil moisture contents can also be used to indicate plant water status (Visser, 1966, Hearn and Ward, 1964 and Jensen, 1968). Slatyer (1956) and Gardner (1960) both found that crop transpiration fell at soil moisture potentials of −0.8 to −1.0 bars. This is confirmed in reviews of such studies for a wide range of crops by Bielorai (1973) and Hagan and Stewart (1972). Irrigation water should therefore be added before soil moisture reaches 1 bar suction.

Reviews of the devices available to monitor soil moisture for irrigation have occurred from time to time (Richards, 1957, Gairon and Hadas, 1973 and Campbell, 1982). Comparisons of techniques are tabulated in figure 5.1. Tensiometers (Hillel, 1971, Richards, 1949) appear the most suitable as they measure soil moisture potential directly which can be related to plant water potential. Tensiometers are only moderately expensive, they provide continuous monitoring and they can be automated; but they only measure at one depth and hence several are needed and in arid environments they require frequent refilling with de-aerated water. Gravimetric sampling (Reynolds, 1970) is cheap and simple but is too destructive for continuous monitoring. Fibre glass electrical resistance blocks (Colman and Hendrix, 1949) have proved inaccurate in wet sandy soils (Agnew, 1980) while gypsum blocks (Bouycous and Mick, 1941) may respond too slowly. Neutron probes (Eeles, 1969, Holdsworth, 1971 and Bell, 1976) do provide repeatable, reliable results but they are expensive and cannot reliably be used for the top 20 cm of soil. Most of these techniques are calibrated against soil moisture content and therefore the soil moisture characteristic curve (Childs, 1969 and Marshall and Holmes, 1979) needs to be established in order to estimate soil moisture potentials. A combination of tensiometers and neutron probe will provide successful monitoring of a wet, irrigated soil in both surface layers and deep horizons.

Apart from the frequency of irrigation, the speed at which water enters the soil

Table 6.9 Soil infiltration rates (from Gairon and Hadas 1973)

Infiltration Class	*Infiltration Rate (mm/h)*	*Comments*
Very low	<5	Flood irrigation recommended
Low	5–15	Sprinklers possible, low intensity
Medium	15–25	Flood only possible for small fields
High	25–50	Flood impossible, sprinklers suitable
Very high	>50	All irrigation difficult

also needs consideration to avoid surface ponding and evaporation losses. Techniques for measuring infiltration rates were discussed in chapter 5. The significance for irrigation is that the infiltration rate determines the amount of water that can be applied before large evaporation losses occur due to depression storage. Gairon and Hadas (1973) provides table 6.9 as a guide to infiltration rates and suitable application techniques.

IRRIGATION DESIGN AND EVALUATION

There has been considerable criticism of the low efficiencies of irrigation systems and their failure to satisfy crop production objectives. Blame is often directed at inappropriate design and the inflexibility of large-scale projects. Yet it is clear from the above discussion that irrigation projects involve numerous complex calculations for which data is rarely available. The situation is exacerbated in arid lands by the variability of the natural environment making water resource estimates particularly difficult.

Physical design

Withers and Vipond (1974) supply a list of studies that are necessary for irrigation system design:

Physical landscape
soil survey; topographic survey; water resource evaluation; climatological analysis; crop ecological evaluation.

System design
site conditions; materials and methods; application techniques; efficiency calculations.

Economic consideration
transport and marketing; funds for construction; cost–benefit evaluation.

System operation
current farm practices; land tenure; culture; organisation of operation; training and extension services.

The FAO (1973) produce a shorter list of:

- water supply
- topographic survey
- soil survey
- power sources and costs
- cropping patterns
- labour supply.

In both lists the physical environment and engineering considerations predominate, with social and cultural factors being considered at a later stage during the operation of the irrigation scheme. System design then becomes hydraulic engineering rather than an evaluation of human and physical resources.

Social considerations

Adams and Grove (1984, p. 143) state:

> There are rarely adequate socio-economic surveys at the early stage of irrigation planning.

De Wilde (1967) describes the failure of the Niger River irrigation scheme in Mali to appreciate the cultural dislocation created by transporting in farmers from Upper Volta (Bukina Faso) resulting in disruption of the labour force through subsequent outmigration. Barnett (1977) reports on problems in the Sudan Gezira scheme through treating farmers as automatons, assuming they will work in a totally predictable manner. Ehlers (1978) reports that the Dez irrigation project in Iran caused severe disruption to local communities due to an emphasis upon economic and not social goals. There is a need to involve local inhabitants and farmers in projects at an early stage and for projects to take more account of cultural and social constraints. Carruthers (1978, p. 302) demonstrates this by listing farmers and 'local elites' along with engineers, economists and agronomists as planning interest groups involved in irrigation.

Clarke and Anderson (1987, p. 240) show appreciation of social factors by including sociological studies and an investigation of land tenure as part of the pre-feasibility stage of design along with the physical resource surveys, as they state:

> The social and political aspects of irrigation developments are of paramount importance. Irrigation developments require new social structures and attitudes, and changes involved may not be easy to bring about.

In the same publication Gordon (1987, p. 161) states:

> The development of water supplies for whatever use is related to social conditions; because of the complex nature of irrigated agriculture the social implications need to be understood and anticipated from the outset.

It is difficult and probably dangerous to provide a 'check list' of social factors that require consideration although attention will need to be given to labour resources, the role of men and women in the community, land tenure, present agricultural practices and local decision making. Baillon (1987) reviews the experience of irrigation in Sudan and argues for a social anthropological perspective when planning such schemes.

Evaluation

The success of irrigation projects is difficult to evaluate because there are often a variety of aims. Rydzewski (1987, p. 5) suggests the following have been used to initiate irrigation:

Maximising returns: on capital invested/water used/cultivated area
Maximising the value of agricultural output
Increasing food production
Increasing export crops
Increasing household income
Creating employment
Redistributing income
Satisfying political ideals
Establishing social stability
etc.

Worthington (1977, p. 53) lists:

- national economic efficiency
- gaining foreign exchange
- sedentarisation of nomads
- drought prevention
- stabilisation of agricultural systems
- modernisation of rural economy.

As projects are best evaluated in terms of achieving their aims it is often difficult to be clear whether projects are successful or not and comparisons between them can be misleading when their objectives are different. An example of such difficulties is provided by Adams (1978) writing on the Senegal's experiences establishing irrigation along the River Senegal commencing in 1946. The aim was to establish a rice irrigation pilot scheme using mechanised agriculture in a sparsely populated area with little local participation. Adams suggests in this respect that the scheme was a 'mitigated success' as the area under irrigation expanded to 5,400 hectares by 1960 supplying a tenth of Senegal's consumption. But it was a financial failure for the farmers with a state subsidy required every year. In 1965 it was decided to expand the scheme to 30,000 hectares. By 1976 20,000 people had joined the scheme but 19,000 hectares of land was found to be too salty to be used and many farmers increasingly found themselves in debt to the project. Mounier (1986), writing about irrigation at the mouth of the Senegal River where it is argued that peasant farmers have become salaried workers, states (p. 112)

> Irrigation might be a failure but it is a highly profitable failure for the French companies involved.

An additional problem with project evaluations is that they tend to occur at just one point in time and therefore may obtain an unrealistic impression of performance. Successful irrigation design is based upon a system of monitoring changing criteria and employing feedbacks in order to alter the initial structures and operations in the light of actual performance and more detailed local knowledge through experience.

Pilot studies

Given the complex nature of irrigation it is not possible to design projects involving more than a few households without using pilot schemes. Pilot schemes are vital for establishing both environmental and human responses to proposed changes. Rydzewski (1987, p. 4) proposed a sequence of:

- Identification and pre-feasibility studies
- Feasibility studies
- Implementation
- Monitoring and evaluation.

The essence is flexibility and the success of a project depends upon being prepared to modify plans in the light of new information obtained through monitoring. However, Palmer-Jones (1981) forcibly demonstrates that pilot studies do not necessarily provide 'correct' responses. A case study from Nigeria is cited where problems were reported due to the uncooperative and lazy farmers. This was tackled through land control, taxation and other coercive measures. An alternative interpretation is that farmers were reluctant to partake in an irrigation scheme that would lead to financial difficulties in the case of wheat production and crop failure due to poor water control for rice fields. The results of monitoring and pilot studies can then be misleading.

SUMMARY

The growth in the world's irrigated area has been dramatic; 20 per cent of the harvested land in developing countries is now irrigated using 60 per cent of fertiliser supplies and supplying 40 per cent of agricultural production (Barrow, 1987). The demand for food is growing in many arid lands, such as the Western Sahel where the IUCN (1989) report population was growing at 2.39 per cent (1960–5 to 1965–70) but is now increasing at 2.79 per cent (1980–5 to 1985–7) with an increase in total numbers from 60 to 107 million. During this time food self-sufficiency fell by −1.9 per cent per annum. It is estimated (World Resources

Institute, 1987) that arid areas such as parts of Nigeria, the Sahel, Egypt and India need to double their present 2 per cent growth rate in irrigated area to meet this demand for food. Of these countries only India is showing any signs of being self-sufficient in food by the end of this century. Kovda (1980) reports that 30 to 40 per cent of the world's food is produced in irrigated areas but that this will have to double by the end of the century to meet the forecasts. In 1973, Shmueli calculated that the potential area for irrigation based on water resources was 500 million hectares (twice current usage) and could be even greater with technologies such as desalination. It is clear that irrigation is expected to help alleviate world food shortages through increased agricultural production and that arid lands have scope for development.

Aerial spraying and deep groundwater extraction can enable hitherto unsuitable arid terrains to be brought into production. This is already evident in the Middle East. Care needs to be taken over excessive groundwater extraction with powerful pumps given the experience of land subsidence in Arizona, where Larson and Pewe (1986) report water tables have fallen by 90 m since the mid 1950s with over 1 m of associated subsidence. The recent exponential growth in the areal extent of irrigation may, however, tail off as more environmental problems become evident and the most suitable sites are used up. Some have therefore argued against further expansion of the world's irrigated area.

Biswas (1978), Houston (1978), White (1977 and 1978) and Worthington (1978) all suggested that it is better to focus upon improving irrigation practices and training rather than initiating new projects given the low efficiency of most existing schemes. The FAO (1973) reported that there was considerable room for economy in water use and that more attention should be paid to 'permanent' irrigated agriculture. More recently, Barrow (1987) and the FAO (1979) have repeated this objective, arguing that most of the 'suitable' sites have already been utilised and there is a greater need to focus upon water use efficiency and recycling. Concern over water efficiency has grown with the realisation of the environmental degradation associated with irrigation, e.g. soil salinisation and erosion. Soil salinisation can be tackled through leaching and improved drainage, seepage losses can be reduced by lining irrigation channels while evaporation losses can be minimised by better field preparation and water application. The cause and hence the solution of poor water husbandry is not purely a technical concern. Greater emphasis is being placed on the social organisation of irrigation systems, the need for better training and education as part of an irrigation development package.

There is some debate over the most suitable scale for irrigation. The need to control water resources over a wide area suggests a river basin size of approach (Saha and Barrow, 1981), a large scale of operation and governmental involvement. The FAO (1973) reported that the participation of governments had become necessary for optimum exploitation of water resources as the scale of operation had increased. In the USA and Australia for instance, governments only had a passive role in early irrigation developments but have become increasingly involved. Uruguay, it is reported, was one of the few countries left where irrigation was exclusively due to private enterprise. The involvement of governments was thought necessary in many countries owing to the complex demands of irrigation involving ownership and occupation of the land and the large capital expenditure on infrastructure and provision of water resources. Cantor (1967, p. 13) stated:

> Irrigation developments were the products of complex civilisations which had progressed beyond the subsistence stage of agriculture. They required the construction of enormous public works, control of the water supply for irrigation and an inevitable development of bureaucracy.

Realisation that irrigation requires more than just supplying water has been recognised by several authors. Withers and Vipond (1988) state:

> Irrigation is far more than an exercise in agriculture and engineering. Major projects have a profound effect on the farming community.... The economic and social patterns of life may be radically altered.

The notion of hydraulic communities has been noted by Kovda (1980) who argued that ancient irrigation systems which survived, managed to integrate water, botanical and human resources. The cooperation and participation of local communities is being recognised as essential for full and efficient use of both human and natural resources. This is easier to say than to do. Griffiths (1984) notes that in the 1970s Nigerian irrigation schemes continued to ignore local farmers but believes that gaining their participation is no simple matter, especially where projects are ill-conceived. Chambers (1980, p. 29) argues that the reasons why 'people' have been ignored by irrigation engineers lie partly in the preoccupation with capital investment, construction and settlement plus the reluctance of many to cross disciplinary boundaries. In addition the difficulty of organising irrigation at lower levels of administration should not be ignored. One of the chief failures of irrigation systems remains social disruption as exemplified by the experience of the green revolution.

The green revolution (GR) is a biochemical package involving new hybrid crop varieties giving high yields in response to favourable environmental conditions supplied through additions of irrigation, fertilisers, pesticides and herbicides. Byres *et al.* (1984) examined the history of the GR in India and observed that following failed attempts to tackle food shortages through reforming land tenure and the feudal system (i.e. social change) it was decided to adopt a technical approach. Although there were substantial initial increases in the production of rice and wheat, problems were encountered through the lack of fertilisers, lack of new variety seeds and shortages of capital. It has also been suggested that the majority of small-scale peasant farmers have not benefited from the technology which has made rich farmers even more wealthy. That is, it is a socially divisive strategy. While rice and wheat production has increased it has been at the expense of land devoted to traditional cereal crops and pulses requiring a

Table 6.10 Crop production in Kenya (1,000 tonnes) (from Woldesmiate and Cox, 1987)

Crop	*1973*	*1983*	*1984*
Pineapples	48	160	145
Coffee	71	92	126
Tea	57	120	116
Sugar	171	350	375
Wheat	150	205	80
Corn	1,600	2,070	1,700
Millet	320	300	195
Rice (Paddy)	36	36	38

change in diet and commercialisation of the food production system. This explanation has been reviewed in several books including Conway and Barbier (1990), Dahlberg (1979), Glaeser (1987) and Pearse (1980). The ensuing social disruption in India erupted in local unrest and riots and resulted in the following governmental announcement reported by Byers *et al.* (1984):

> The new technology and strategy having been geared to goals of production with secondary regard to social imperatives have brought about a situation in which elements of disparity, instability and unrest are becoming conspicuous with the possibility of increase in tensions.
>
> (Indian Government 1969)

Harrison (1987) suggests that the GR has largely ignored Africa because of its more variable water supplies, political instability and lack of investment. This is no longer the case with various attempts to develop suitable hybrids. Kenya for example has experienced quite a significant increase in the yield of some crops as shown by table 6.10.

It is quite clear that Kenya's agricultural growth has been in cash crops which, it is argued, are mainly cultivated by the larger farmers benefiting from irrigation. Food crops are grown without irrigation in smallholdings and were particularly badly affected by the 1984 drought. The general trend, like India, would appear to be that the GR technology is causing greater polarisation of wealth with the richer farmers benefiting disproportionately. Hayami (1984) suggests that this is misinterpreting the situation. Criticisms of the GR have focused upon dualistic rural economies and argued that wealthier farmers benefit disproportionately because of access to information and credit, while the technology is labour displacing resulting in a landless peasantry. Hayami argues that there is very little evidence of this being the case with hybrid seeds being used both by large-scale and small-scale operations, scale of enterprise does not affect yields and there has been an increase in the demand for labour. Hence it is argued that the social disruption due to the GR is caused by population growth and land ownership difficulties. Hayami concludes that more emphasis should be given to population control and land reform, factors that will need to be considered by those designing irrigation systems.

Arid lands' inhabitants are familiar with uncertainty and have evolved a number of strategies to combat the vagaries of their environment. Irrigation is designed to reduce much of this uncertainty for agriculture but the inherent flexibility of arid land cultivators should not be lost through rigid planning and operational practices. Flexibility is the key to success for arid land irrigation with respect to scale, water application systems and labour requirements. This can be achieved by proper training and a design that allows systems to evolve with improved practices and environmental information.

7

DESALINATION

Desalination is the separation of pure water from a saline solution (brackish or seawater). This should not be confused with desalinisation, i.e. the reclamation of saline soils (Worthington, 1978) (see chapter 6). For most water supply purposes a concentration of solids in suspension of less than 500 ppm is required (World Health Organisation standard) although in arid zones water supplies may contain in excess of 1,000 ppm (Porteous, 1975). As 97 per cent of the world's water is saline the potential freshwater resources that could be produced by this process are immense but seawater typically has concentrations of 35,000 ppm which requires a separation of around 98 per cent of total dissolved solids to produce potable supplies.

The technique has long been practised and Popkin (1969) refers to the writings of Aristotle describing the way Greek sailors used evaporation to desalinate seawater some 2,500 years ago; Romans passed seawater through clay soil to reduce the salt content while wool was also used for the same purpose. Khan (1986) also mentions descriptions of desalination by Pliny and Arab, Byzantine and Renaissance writers. Large land-based enterprises had to await the adoption of steam power and 19th–20th century technological developments, leading to the construction of commercial plants supplying over 5,000 m^3 of water per day (approx. 1 mgd (million gallons per day)) by the 1960s. By 1985 the world's desalting capacity had reached 6.8 MCM per day (see Table 7.1) (World Resources Institute, 1987).

Table 7.1 The development of world desalination capacity (capacities in 1,000's m^3 per day)

	1968[1]		*1977*[2]		*1985*[3]
	Plants	*Capacity*	*Plants*	*Capacity*	*Capacity*
Middle East	78	314 (28%)	329	1,822 (49%)	3,785 (69%)
U.S.A.	313	234 (21%)	681	645 (17%)	1,135 (12%)
Rest of World	321	565 (51%)	562	1,241 (33%)	1,892 (19%)
Total	712	1,113	1,572	3,708	6,812

Notes: 1 From Leeden (1975, p. 479) only listing plants producing more than 100 m^3 per day.
2 From Heathcote (1983, p. 71).
3 From World Resources Institute (1987) where 'Middle East' also includes North African countries.

NB: Hornburg (1987) estimates worldwide capacity was much greater by 1985 at 10 MCM per day.

Table 7.2 National water balance for Saudi Arabia (based on Ukayli and Husain, 1988)

Demand	7,430 MCM a year	Agriculture
	1,400 MCM a year	Domestic and other
Supply	900 MCM a year	Surface water
	950 MCM a year	Groundwater – renewable
	6,480 MCM a year	Groundwater – non-renewable
	100 MCM a year	Reclaimed waste water
	400 MCM a year	Desalination

For comparative purposes, 6.8 MCM per day equals 79 m^3 per second which corresponds to the average flow of the River Thames through London (UNESCO, 1969).

Desalination has increased sixfold since the 1960s and two-thirds is located in the oil-rich Middle East countries where desalination makes an important contribution to national water supplies as shown by tables 7.2 and 7.3.

Desalinated water is supplied mainly by the processes of distillation and the use of membranes although other methods such as freezing and solar humidification have been developed and continue to receive attention (Delyannis and Delyannis, 1980, provide a comprehensive listing of the technical developments and patents for desalination and Spiegler and Laird, 1980, present modern engineering considerations and designs).

DISTILLATION

The principle of distillation is that by supplying energy to a seawater/brine solution, pure water is evaporated and the water vapour is condensed and captured. The first type of distillation plants (**boiling**) used boiling of the saline solution through the direct application of heat to produce vapour which was cooled by the incoming saline water (see El-Sayed and Silver, 1980). This technique has been surpassed by **flashing**, i.e. the seawater is heated but is prevented from boiling through maintaining conditions above the solution's saturation pressure. This solution is then fed into a flash chamber at a lower pressure resulting in rapid evaporation releasing water vapour that can be condensed. In order to increase production efficiencies, multiple boiling and multiple flashing were

Table 7.3 Construction of selected desalination plants (from Kishi and Inohara, 1985 and Ukayli and Husain, 1988)

1969	230 m^3 per day	(Dhubai, Saudi Arabia)
1969	885	(Kharg Island, Iran)
1970	18,925	(Jeddah, Saudi Arabia)
1973	28,400	(Al-Khobar, Saudi Arabia)
1978	430,000	(Jeddah, Saudi Arabia)
1985	393,120	(Doha, Kuwait)

developed from ideas in the 1930s, whereby the thermal energy of the brine and distillate is used to produce more distillate through a series of linked chambers. MSF (Multiple Stage Flash) was developed from MEB (Multiple Effect Boiling), the former relying upon flashing while for the latter only 10–20 per cent is obtained in this manner. With steam power it proved possible first to construct distillation plants on board ocean liners in the 19th century and then capacity increased to make land-based plants viable. By the end of the 19th century steam distillation plants could be found where demand was sufficiently high enough to overcome costs. Khan (1986) reports that the first steam-powered, land-based distillation plant was constructed nearly a hundred years ago at Aden while the first large land-based plant was built at Aruba (Netherlands Antilles) in 1930 (Hornburg, 1987); but the most significant growth in this industry has been since the 1950s.

The cornerstone technique for distillation is Multiple Stage Flash. The patent was first filed in 1957 and the first commercial plant of 5,000 m^3 per day capacity (1 mgd) came into service in 1960 (Silver, 1978). By 1970 there were 59 desalting plants of at least this size, 57 employing distillation techniques (Leeden, 1975). There are a number of alternative designs for MSF involving cross tube and long tube arrangements that influence the diameter of the heat transfer tubes and flow velocities. The prime importance is the energy requirement of the process. Silver (1978) calculated that distillation technology required 100 kJ/kg of pure water produced, assuming energy production efficiencies of 38 per cent. He calculates that in the future the best that is likely to be achieved by distillation is 60 kJ/kg compared to 90 kJ/kg for freezing and 40 kJ/kg for reverse osmosis (see below). MSF does have high energy costs but it has had (until recently) a clear advantage over other processes, that is its ability to desalinate water with a very high dissolved salts' content and its capacity for large-scale production. Before MSF no unit exceeded 2,000 m^3 per day (0.4 mgd) but from the first 5,000 m^3 per day (1 mgd) plant in 1960 capacity has grown exponentially, with forecasts of 2 MCM.

MSF distillation operates by a temperature difference created by lowering the saturation pressure, as opposed to vapour compression (VC) whereby evaporation takes place by the creation of a higher saturation vapour pressure through compression (El-Sayed and Silver, 1980). Vapour compression, like MSF, can tolerate high salinities and Heathcote (1983) suggests similar energy requirements but slightly lower operating costs. Silver (1978) is quite emphatic, however, that vapour compression is not an attractive alternative compared to MSF because of the greater complexity of construction for the former plant and hence higher maintenance costs and estimated energy requirements of 168 kJ/kg for VC as against 100 to 60 kJ/kg for MSF.

MSF distillation does suffer from some drawbacks. The costs of construction and operation are significant. Evans (1979) estimates capital costs of US$1,125 m^3 per day and operating costs of US$1.2 m^3 per day, compared to US$225 and US$0.33 for reverse osmosis respectively. The high temperatures and salt concentrations lead to corrosion problems requiring the use of expensive corrosion-resistant materials in pumps and pipes. Salts also accumulate producing scaling, especially if temperatures are raised above 120 °C. Porteous (1975) distinguishes between alkaline salts (from bicarbonates) and calcium sulphate scale, the former being tackled mainly through scale-inhibiting compounds or acid treatment although scale can be encouraged to form on easily removed surfaces or, once present, removed by mechanical or chemical means. Ion exchange is a further alternative where the incoming saline water is passed over a resin bed which replaces the calcium and magnesium ions with sodium. Unfortunately many of these methods are expensive and chemical treatment is the most common solution. Calcium sulphate is not responsive to acid treatment but can be reduced by avoiding high temperatures.

MSF continues to be the work-horse for desalination and Fair (1987) clearly sees it continuing in this position although alternatives are increasingly competitive. Multiple Effect Boiling through vertical tube evaporation or horizontal falling film is arousing renewed interest while technological developments have made significant improvements in the capacity of reverse osmosis.

MEMBRANES

Saline solutions exert an osmotic pressure of around 2.5 million Pa (seawater). In order to separate pure water from the solutes (salts) the saline solution is passed, under pressure, through a semi-permeable membrane which rejects low molecular weight salts but allows water to pass through (**reverse osmosis**). Alternatively an electrical current is applied to selective ion permeable membranes such that dilute and concentrated solutions are produced on opposite sides (**electrodialysis**). The use of membranes was reported by ancient Chinese writers and mention is found in Hebrew scrolls (Dickey, 1961) but commercial development of this technique was limited by the lack of suitable membranes.

Reverse osmosis requires membranes that allow a high filtering of salts and a fast flow rate coupled with the need to withstand high pressures and temperatures. The use of pressure to enhance the process began over 100 years ago but it was Reid at the University of Florida in the 1950s who is credited with the development of the first modern membrane, using cellulose acetate, (Channabasappa, 1982). Permeability rates were at first low but subsequent research at the University of California, Los Angeles, produced much better fluxes by the 1960s (Dresner and Johnson, 1980). The development of membranes has been charted by Hoornaert (1984) leading to lower pressure requirements and hence lower energy inputs such that the reported comparative energy costs are:

Flash distillation	37.0 kW/h m^3 of pure water
Vapour compression	24.0
Electrodialysis	10.5
Reverse osmosis	5.2

By 1979 the world's reverse osmosis capacity was 1.5 MCM per day with plants such as 208,000 m^3 per day at Riyadh (Dresner and Johnson, 1980) and 25,000 m^3 per day in Malta (Wade, 1985). Reverse osmosis has become the major competitor to distillation because it requires less energy, it is possible to employ alternative energy sources, i.e. solar and wind, it is scale flexible and the capital costs are lower. Up to 30,000 m^3 per day reverse osmosis is more economical for seawater treatment (Hoorneart, 1984). For the desalination of brackish water the main competitor for reverse osmosis is electrodialysis.

Schaffer and Mintz (1980) describe the different designs using electrodialysis. Conventionally electrodialysis consists of alternating cation and anion permeable membranes but transport depletion technology incorporates an array of cation to neutral membranes instead. The disadvantage of the latter is that more electric current is required but this may be offset by the lower costs of neutral membranes and they are less prone to fouling. While electrodialysis can compete with distillation and reverse osmosis in terms of power and capital costs, electrodialysis plants tend to have much smaller production rates and can only tolerate brackish water with salinities in the range 1–5,000 mg/l (Heathcote, 1983). At the present time reverse osmosis dominates the use of membranes for desalination.

FREEZING

Like most modern desalination technologies, freezing has also been practised since ancient times. Frost can be collected and melted while Russian fishermen traditionally collected sea ice and stored it for use during summer months. Desalination by (indirect) freezing is based on the formation of ice crystals when the temperature of a saline solution is reduced. These ice crystals are separated from the brine which is used to cool incoming seawater while the ice is melted by heat released through the condensation of the refrigerant. Tleimat (1980) lists the following advantages for freezing: a reduction in energy costs as the latent heat of fusion is less than the latent heat of vaporisation, low temperatures produce less scaling and less corrosion while inexpensive plastics can be used. Drawbacks include the need for good thermal insulation, a high vacuum for some designs and problems with the effective separation of ice crystals and brine. Water may be used as the refrigerant in a design called 'vacuum freezing vapour compression' but this requires a mechanical compressor which has presented some difficulties due to ice accumulation. The energy requirements are however, lower than for indirect freezing. Heathcote (1983) notes that freezing can handle high salinities and its energy and capital costs are comparable to distillation but it has proved difficult to construct large-capacity plants making this a specialist option.

SUMMARY

Silver (1978) and Heathcote (1983) note that each 1 ton of industrial product requires 200 to 500 tons of water while 5 to 10 times as much is needed by agriculture. Agriculture therefore tends to rely on cheap water supplies and can tolerate water of a much lower quality than industry. Not surprisingly the majority of desalination plants have been constructed for municipal and industrial supplies where high demands and revenues can offset the high costs of desalinated water. Nebbia (1968) concurs and suggests that the use of desalination for agriculture will be limited to a few specific cases where high-value crops are grown such as horticulture in Guernsey where a 2,000 m^3 per day distillation plant was constructed in 1960. Table 7.1 reveals that, during the 1970s, while the number of desalination plants doubled from 700 to over 1,500 the capacity trebled from 1 MCM per day to 3 MCM. This shows a period of growth in the size and number of desalination plants. Yet in 1985, 80 per cent of capacity was only to be found in the Middle East, North Africa and the USA, regions which either through the availability of cheap fossil fuels and/or a sufficiently high enough demand and capital were able to support large-scale desalination production. This polarisation in the location of desalination reveals its major handicap – the costs of construction and operation.

The annual rate of growth of desalination between 1961 and 1968 was 23 per cent (59.8 to 247.2 mgd or 0.272 to 1.125 MCM per day) with a predicted growth rate of 26 per cent (Office of Saline Water, 1970). If sustained this would have produced a worldwide capacity of 9 MCM per day by 1977 whereas only 3.7 MCM was observed (Evans, 1979) reaching 6.8 MCM per day by 1985. The reasons for this shortfall lie with increasing energy costs and environmental constraints. The reported cost of producing a cubic metre of pure water had fallen from US$0.88 to US$0.28 by the early 1960s with a predicted cost of US$0.17 by the 1970s (Thorne, 1963). Pereira (1973, p. 16) reported current costs (in the early 1970s) of still US$0.22 per cubic metre of water with US$0.15 for a plant at Tijuana, Mexico. Increases in oil prices during the 1970s contributed to a doubling of energy costs and anticipated reductions in production costs did not materialise resulting in the technology only being attractive in certain locations. In addition there is the problem of pollution created by the production of salt waste. Popkin (1969) estimated that a 10 mgd distillation plant would require 18 mgd intake of seawater, and if the salt concentration was 9,000 ppm then 400 tons of solid waste would be produced daily. Channabasappa (1982) reports that the desalination plant on the Colorado River designed to reduce the river's salt concentrations of 3,100 ppm down to 115 ppm produces 590 tonnes of calcium carbonate sludge daily; 113 tonnes of calcium oxide is reclaimed but the rest is used as land fill. Desalination is also strategically a vulnerable supply for water and security measures may have to be included in costs.

Despite these constraints desalination remains an important supply of water in some parts of the arid world and recent reports are optimistic. Ukayli and Husain (1988) estimate that desalination unit costs have fallen from US$2.7–10.8 m^3 before 1978 to US$1.7–2.4 m^3 afterwards. The major recent change in this technology has been the rise in the use of reverse osmosis which Hoornaert (1984) predicted would increase from 1.5 million m^3 per day to more than 4.0 million

Table 7.4 Relative costs of desalination (US$ millions)

	MSF(1)	*MSF(2)*	*RO*
Capital costs			
Desalination	28.9	28.9	29.5
Power plant	6.3	14.2	3.5
Total	35.2	43.1	33.0
Operating costs	14.7	22.7	8.3
Unit costs (US$/$m^3$)	1.83	1.73	1.63

Calculations based upon a fuel cost of US$185 per tonne; a plant life of 20 years; membranes lasting 5 years; and a capacity of 20,000 m^3 per day.
MSF(1) = Back pressure steam turbine for power generation
MSF(2) = Gas turbine for power generation.
(From Wade, 1985)

m^3 per day during the 1980s, although World Resources Institute (1987) report that capacity was still only 1.8 million m^3 by 1985. Multiple Stage Flash requires twice the energy but until recently had advantages of large-scale construction and the commercially unique ability to desalinate seawater. There is growing confidence, however, in the use of membranes with the construction of a 0.2 to 0.5 million m^3 per day reverse osmosis plant at Yuma, Arizona (Dresner and Johnson 1980, Hoornaert, 1984 and World Resources Institute, 1987) with an investment cost of $792 m^3 per day compared to $1,000 m^3 per day for an equivalent MSF unit in Kuwait.

Table 7.4 reveals some of the assumptions that have to be made in order to calculate desalination costs. This is a complicated procedure, open to error and misinterpretation. The figures quoted above by Pereira and Ukayli and Husain differ by a factor of 10. Nevertheless, Wade (1985) shows in table 7.4 the comparative advantage of reverse osmosis; which will be attractive in high fossil fuel cost areas given the lower operating costs, providing membranes are durable. In the same article it is suggested that the problems of scale and corrosion associated with MSF are now largely solved although there remains the need for annual shutdowns and maintenance. Reverse osmosis (RO) requires less staff than MSF but they must be highly skilled and there is a particular need to pay attention to the quality of water intakes.

Cost, however, remains the major constraint for all types of desalination and Thomas (1988) reports that the relative costs of water produced for irrigation in Qatar (1980) were US$6.05 m^3 for desalination, 0.94 for treated effluent and 0.22 for groundwater, although in the same article the cost of transporting water by pipeline to Libya's coastal plain was US$10.75 per m^3 compared to US$1.8 for desalination. Fierce competition will continue between desalination technologies, although Davies (1987) suggests the future lies with reverse osmosis. With advances in technology and lower costs, desalination may well become viable outside of the oil-rich Middle East countries but mainly for potable supplies.

8

PRECIPITATION ENHANCEMENT AND CLOUD SEEDING

Precipitation can be enhanced either by stimulating more precipitation than would have otherwise reached the ground by cloud seeding; by more effective use of that actually received; or by harvesting dew and mist. This chapter is concerned with the application of cloud seeding in arid lands, but let us first consider more effective use of rainfall and water harvesting.

The term 'effective rainfall' is the proportion of the total precipitation that is available for a specific purpose (Agnew 1985, p. 150). This is problematic as it can be defined in a variety of different ways, each depending upon the use to which the water is being put (Dastane, 1974). Chapter 1 introduced a number of strategies developed by arid land inhabitants to combat the vagaries of rainfall, which demonstrates that the effective use of rainfall is both a cultural and technical issue. Barrow (1987) reviews a number of examples of runoff harvesting and rainfall collection from the use of mulches through to the construction of earth walls, terraces and artificial catchments. McPherson and Gould (1985) discuss rainwater catchment systems in Kenya and Botswana and conclude that roof systems have distinct advantages of supplying uncontaminated water directly to the home but suffer from the cost of installation. Waller (1989) reviews worldwide examples of rain harvesting and notes that 1 million people in Australia rely upon rainwater as their primary source of supply. A full discussion of these measures is beyond the scope of this text but the ingenuity of arid realm dwellers should not be underestimated.

Although dew, mist and fog (occult precipitation) do not in general make a large contribution to precipitation totals (Sumner, 1988, p. 134) they can be important to the flora and fauna in arid environments with specific adaptations often evident. The factors influencing dewfall have been listed by Monteith (1957) as low windspeeds, clear skies and high relative humidities. Agnew and Anderson (1988) found that the duration of saturated conditions was also significant in explaining the amount of dewfall recorded but the variability in results obtained pointed to the operation of a complex process. Dewfall is not only explained by atmospheric condensation following radiation cooling but there is also soil distillation, both of which are influenced by the size, shape and colour of the surface upon which dew is forming. Anderson (1988d) has shown the difficulty there is in reliably measuring this phenomenon. As a consequence this element of the hydrological cycle has not received the attention it deserves, especially in arid lands where it can make a small but significant contribution to available water.

Mist and fog require similar conditions as the production of dewfall but involve either more intense cooling of the atmosphere near the ground or some movement of moist air over a cold surface. The environmental permutations are numerous and Matveev (1984) and McIntosh and Thom (1981) list radiation, advection and evaporation fogs while Goudie (1985) adds upslope, frontal and steam fogs. The latter is caused by cold air moving over a warmer water surface. Jiusto (1981) lists three distinct processes:

1 Cooling air through radiation losses;
2 Addition of water vapour producing frontal fogs;
3 Vertical mixing of moist air parcels producing advection fogs.

In fact this list is somewhat misleading as more than one process can take place (Rodhe, 1962); nevertheless Roach *et al.* (1976) describe radiation fogs as shallow with low windspeeds whilst advection fogs are usually substantially thicker and are associated with marine conditions. Advection fogs can be found in the coastal regions of many arid lands such as Namibia, Oman or Peru where there is a cold offshore ocean current. Onshore breezes arrive laden with water droplets having crossed this cold surface and create high humidities locally. Optical sensors have been developed to monitor fog and mist (Puffett, 1979 and 1980) but as with dewfall it has proved difficult to quantify the effective rainfall of such conditions.

The mining of water from mist- and fog-laden winds has tended to be considered an ecological process in arid lands. Pereira (1973) reports that pine trees in Hawaii were capable of extracting an extra 760 mm from heavy cloud while coastal forest in Kenya was characteristic of a region receiving 2,000 mm annually yet little more than half actually fell. There have been attempts to mine the atmosphere for water by collecting dewfall and fog. Barrow (1987) reports 50 m^3 per day was collected in traps of 6,000 m^2 area in Peru; while polythene sheets in the Negev yielded 3,631 mm per m^2 per month. Barrow suggests that in the right conditions significant amounts could be collected, a view endorsed by Stanley-Price *et al.* (1986). At Dhofar (Oman) they found that using a tower with four corrugated iron sheets at 4.2, 3.1, 2.0 and 0.9 m heights the upper sheet collected 34–35 litres per m^2 per day with 9–13 litres from the lower sheet. An overall total of 50 litres per m^2 per day was obtained suggesting such devices could be used to enhance water supplies at particular times of the year.

An alternative to trapping fog, mist or dewfall or harvesting rainfall is to stimulate more rain to fall through cloud seeding.

CLOUD SEEDING: A BRIEF HISTORY

The ancient Greeks expressed interest in rainfall enhancement with Hippocrates considering health and climate, Aristotle's publication of *Meteorologica* in 350 BC described major rainfall types, Theophrastus suggested methods for predicting rainfall in his 'book of signs' and Plutarch observed that a relationship existed between great battles and subsequent rainfall (Thomas, 1978). Despite this early empirical work and interest, activities involving rainfall enhancement tended to concern ritual dances, chants and superstition until scientific instrumentation led to better understanding of atmospheric processes. The thermometer was devised by Galileo in 1593, the barometer by Torricelli in 1643, the anemometer by Hooke in 1661, the hair hygrometer by Saussure in 1780, whilst the raingauge has been invented independently by a number of cultures although Shaw (1983) suggests the first may have originated in Korea in the 15th century. Scientific discoveries paralleled these developments in instrumentation during the 17th and 18th centuries with Pascal investigating the relationships between pressure and altitude, Halley mapping wind systems and Boyle establishing the behaviour of gases. In the 19th century discoveries included Dalton's, Aitken's and Wilson's work on condensation, Kelvin's examination of anticyclones and Coriolis's study of air motions leading to the establishment of weather forecasting, for example the US Weather Bureau in 1870 (Bosart, 1985). By the end of the 19th century the mechanisms leading to rainfall had been established sufficiently to encourage a number of attempts to enhance precipitation artificially. But before embarking upon a discussion of such attempts there follows a brief explanation of rainfall processes as this is necessary to understand the technique of cloud seeding. More detailed descriptions can be found in Barry and Chorley (1978), McIlveen (1986) and Sumner (1988).

The amount of water vapour in the atmosphere can be measured in a variety of ways including the pressure it exerts (vapour pressure – mb) through to the mass of water per kilogram of dry air (humidity mixing ratio – g/kg). Through evaporation, water vapour is added to the atmosphere and this may continue until the air becomes saturated when the relative humidity reaches 100 per cent. The amount of water contained by the atmosphere at saturation is dependent upon a curvilinear relationship with temperature such that at 0 °C 6.11 mb or 3.84 g/kg are required while at 10 °C a higher moisture content is necessary with 12.27 mb or 7.76 g/kg. It follows that cooling an air mass can also induce saturated conditions which can be achieved by adiabatic expansion, radiation cooling, contact cooling or mixing of air masses. Wilson demonstrated in 1897 that condensation of water vapour in a particle-free atmosphere required a moisture content seven times the saturation expected for that temperature. Coulier in 1875 and Aitken in 1880 demonstrated that condensation could occur at much lower vapour pressures on airborne particles (Mason, 1962). In fact hygroscopic nuclei, i.e. those having an affinity for water such as sodium chloride ($NaCl$), could initiate condensation at vapour pressures well below saturation. Hence the significance of condensation nucleii became realised and many early cloud seeding experiments involved firing sand and salt grains into clouds. However, owing to their smaller radius and the action of surface tension, small water droplets can only continue to grow if supersaturated conditions prevail and it was clear that condensation alone could not produce rainfall-sized water droplets.

It is believed that growth occurs through collision and coalescence as droplets fall through clouds. In addition many clouds contain supercooled droplets at temperatures down to −30 °C Below about −10 °C ice crystals start to form through spontaneous, heterogeneous and contact nucleation. As the saturation vapour pressure over an ice surface is lower than that over a water surface, lower atmospheric moisture contents are required for condensation and growth. Ice crystals therefore act as very effective condensation nuclei. Hence the ingredients for the initiation of precipitation can include high atmospheric humidities, the presence of airborne particles, the presence of ice crystals and low temperatures.

Byers (1974) has created a useful list of the different 19th and early 20th century attempts at cloud seeding:

1839 James Espy proposed that large fires should be built to generate updrafts to stimulate convective rainfall. (Theoretical proof of such updrafts, being significant, followed in 1938 through work by the Hungarian G. Rovo and later attempts in Europe by the Frenchman Dessens in the 1950s and 1960s actually produced a small tornado!)

1871 Edward Powers described how, during the US civil war, rainfall tended to follow battles. The notion of a relationship between noise, fire and rainfall was mentioned by the ancient Greeks (Plutarch) and Congress allocated $10,000 for experiments with rockets, balloons and explosives. But the frequency of rain in the USA is every 3–5 days and battles are normally prepared in good weather; hence the association between rainfall and battles is probably due to the weather pattern. Funding stopped in 1893 but by then there were a plague of con-men offering to initiate rain throughout the USA.

1924 Dr E. Chaffe (Harvard) fired charged sand grains into a cloud 5–10,000 ft high, resulting in the cloud's dissipation.

1930 A. Veraart a Dutchman, used (by chance it appears) dry ice (CO_2) as a seeding material but his claims of success were largely ignored.

1930s MIT undertook a series of experiments aimed at mist/fog dissipation with calcium chloride.

1932 The USSR established an Institute of Artificial Rain.

The growing interest and scientific discoveries culminated in 1946 with research at General Electric Laboratories, Schenectady, New York. The work was led by Irving Langmuir and was concerned with aircraft icing and radio static when flying in storms. Freezers were being used to create conditions characteristic of supercooled clouds and Dennis (1980) describes how one of Langmuir's associates, Vincent Schaefer, dropped a pellet of dry ice (carbon dioxide) into a freezer in order to speed up the reduction in temperatures. The dry ice left a trail of ice crystals which Schaefer recognised as a possible method to glaciate supercooled clouds and hence produce precipitation. Braham (1986, p. 1) recounts:

> On 13 November 1946, Vincent Schaefer scattered a little dry ice into the top of a supercooled stratified cloud over some low mountains east of Schenectady, New York. Within minutes the seeded portion of the cloud was transformed into a mass of snow crystals.

Byers (1974) reports 3 lb of dry ice was used and the rain fell 2,000 feet before evaporating and did not reach the ground. Nevertheless the significance of this discovery was that ice crystals could be produced artificially. Another General Electric researcher (Vonnegut, 1947) suggested that silver iodide (AgI) could be used as an alternative to dry ice to seed clouds as it had a similar crystalline structure. Vonnegut found that as silver iodide was more efficient than natural nucleii it could produce precipitation at temperatures of only −4 °C. But if too much was used the result was dissipation of the cloud through supernucleation as the water vapour content of the atmosphere is reduced by condensation. As a result cloud seeding has also been widely used to suppress hail rather than to enhance rainfall. These two seeding agents form the basis of much modern precipitation enhancement which has been called glaciogenic seeding because of its emphasis upon ice crystals and supercooled clouds.

Sewell (1973, p. 7–19) characterises subsequent development of cloud seeding in the USA as:

> **1946–1952 Period of exploration** This involved experimentation, much publicity and controversy with the rapid acceptance by commercial firms contrasted by scepticism from scientists and government agencies.
> **1952–58 Evaluation** This period coincides with the retirement of Langmuir and the cessation of droughts in the south-west of the USA resulting in a reduction of the tempo and more time to evaluate claims. Controversy and debate still continue with conflicting evidence over the impacts of cloud seeding.
> **1958–66 Scientific re-evaluation** Congress provided funds for universities and the National Centre for Atmospheric Research to investigate cloud seeding more rigorously. At the end of this period the National Academy of Sciences Panel on Weather Modification reported that in some circumstances cloud seeding was effective and could produce increases of 20 per cent. This view was endorsed by other influential groups such as the National Science Foundation and later by the Committee on Atmospheric Sciences of the National Research Council.
> **1966–72 Accelerating the effort** Qualified endorsement of cloud seeding resulted in more funds being allocated for weather modification but also identified gaps in knowledge where research was needed.

The enthusiasm with which these discoveries was met was astonishing. Dennis (1980) suggested that every 1 mm of added precipitation in N. Dakota enhances yields of hay by 9 kg/ha and wheat by 7 kg/ha. A 10 per cent increase in rainfall would increase production in the state by $400,000,000 a year. CAS (1980) report that by the early 1950s possibly as much as 10 per cent of the USA's land area was under commercial seeding operations at an annual cost of several million dollars. The President's Advisory Committee on Weather Control announced in 1957 that in the mountainous areas of the Western USA there had been a 10–15 per cent increase in rainfall due to cloud seeding. Sewell (1973) notes that the Federal Government allocated $2.7 million in 1963 for weather modification rising to $10 million in 1967 with $25 million requested in 1973 and this does not include commercial operations.

Yet there has been much criticism of cloud seeding with Dennis (1980) reporting it had been called the 'crime of the century' and outlawed in Pennsylvania. Breuer (1980) cites an example from Colorado when a bomb damaged the trailer of a firm of weather consultants as part of a protest by farmers opposed to cloud seeding. Much of the controversy surrounding cloud seeding results from the difficulty in proving its effectiveness when precipitation is such a variable phenomenon and the processes involved are still only partly understood. Before investigating the social and legal obstacles to cloud seeding we will first consider the technique itself and the scientific proof.

PHYSICAL METHODS

From the above brief introduction to rainfall mechanisms it is hopefully apparent that condensation takes place through the introduction of condensation nuclei and atmospheric cooling. There are two main types of cloud seeding: glaciogenic in subzero conditions where the aim is to enhance ice crystal growth; and warm cloud seeding where buoyancy processes dominate raindrop growth mainly through coalescence and collision.

Warm cloud seeding

Langmuir suggested the use of water sprays in warm clouds in the 1940s to enhance the rate of precipitation formation. Braham *et al.* (1957) provided some evidence of this effect while the University of Chicago working in the Caribbean showed precipitation within a cloud could be increased by introducing water droplets although the precipitation did not reach the ground (Dennis, 1980). It is also possible that this method will reduce precipitation through the disintegration of droplets colliding with each other plus the release of latent heat during condensation. In addition the transporting of water in aircraft increases the costs substantially for this type of operation making it of questionable usefulness. Alternatively hygroscopic nuclei (such as sodium chloride) can be sprayed into clouds to promote the growth of large droplets. Dennis (1980) reports work in the 1960s off Hawaii introducing salt powder into warm moist air with relative humidities of 80–90 per cent producing visible cloud but no rainfall and no increase in buoyancy. Cotton (1986) suggests that the results of a reported 42 per cent increase in rainfall in India through this method are now being questioned and that with so much natural atmospheric salt it is necessary to introduce very large amounts artificially to have any effect. In addition salt is corrosive and it is difficult to obtain specific grain sizes. There are alternative substances, e.g. microencapsulated urea, but cost is a major obstacle. Warm clouds tend to be shallow clouds (Sumner 1988) and therefore the opportunity for much rainfall enhancement is restricted whereas cold clouds offer the opportunity to seed in subzero conditions where ice crystals are present.

Cold cloud seeding

Cold clouds typically consist of both supercooled water droplets and ice crystals especially in convective clouds where the vertical extent takes the cloud top beyond the isotherm (0 °C). Braham (1986, p. 2, 3) suggests seeding can be effective in cold clouds where

1 rainfall efficiency is limited by a shortage of ice nuclei,
2 increasing the buoyancy of such clouds will invigorate the rainfall process.

Hence there are two fundamental approaches to seeding cold clouds:

1 static seeding to augment the presence of ice crystal nuclei,
2 dynamic seeding which aims to enhance vertical motion and uplift.

Silverman (1986) notes that static seeding is the most commonly used approach but points to a continuing debate concerning this technique. Up to the 1960s it was commonly believed that ice crystal growth by water vapour diffusion was the dominant process while it now appears that freezing of coalescence-grown drops may be significant in some cases. This illustrates the continuing uncertainty over the actual processes involved in cloud seeding owing to the dynamic and complex reactions within clouds.

Work on dynamic seeding of clouds received attention in the 1960s in campaigns in Arizona and the Caribbean. It essentially involves the introduction of a seeding agent into clouds around −10 °C where supercooled water is present, resulting in conversion of supercooled water droplets into ice and thus releasing latent heat which adds to the buoyancy of the air giving rapid vertical motion leading to precipitation. Orville (1986) summarised Simpson's (1980) work on the key stages in dynamic seeding of initial vertical growth for 10–20 minutes, followed by horizontal expansion with low-level convergence of cloud systems leading to increased areal rainfall.

Seeding agents

Although a variety of substances from sand to water have been employed to seed clouds, the two most common substances are silver iodide (AgI) and dry ice

(CO_2). There has been some concern over possible pollution through these emissions. Dry ice vaporises into carbon dioxide which can be leached out of the atmosphere by rainfall, although there is obviously concern over releases of any greenhouse gas. Silver iodide is insoluble and therefore persists in soils and water courses. Although no negative effects have been detected there is still debate over possible contamination (Breuer, 1980).

Although substances such as cupric sulphide (CuS) and lead iodide (PbI_2) have been employed, silver iodide remains the most commonly used seeding agent. Vonnegut (1949 and 1974) promoted the use of silver iodide over the use of dry ice because of the possibility of releasing the former from ground-based generators as opposed to expensive airborne campaigns. The principle of groundbased emissions is to spray a solution of acetone and silver iodide into a flame leading to the silver iodide being vaporised and rising through the atmosphere where it cools to form fine particles. The problem with ground-based generators is that the plume emitted from the burners may be dispersed by winds and therefore cannot easily be directed towards specific clouds. Aircraft have therefore been widely used for greater precision in impregnating particular clouds at specific altitudes. Pyrotechnic devices which burn slowly and emit silver iodide can be mounted on aircraft or dropped into clouds but costs are greater than ground-based generators. The latter has the advantage that they can operate continuously for longer periods but diffusion of the plume is a serious problem. Rockets and shells have also been used but they pose a hazard to aircraft and persons on the ground beneath the seeded area.

Problems over the dispersion of silver iodide has led to a renewed interest in dry ice. This has coincided with a move to cloud top seeding and the realisation that the number of ice crystals produced is almost independent of temperature below −2 °C (Silverman, 1986). The amount of ice crystals produced has increased from 10^8 crystals per kilogram to 10^{16} per kilogram with a mean value of 10^{14} or 10^{15} per kilogram suggested by Dennis (1980) but the exact number depends upon the sizes of the pellets dropped. Silver iodide behaves in a more complicated manner and it has proved difficult to predict the production of ice crystals although they appear to start around −5 °C and increase until around −16°C.

SCIENTIFIC PROOF

CAS (1980) provide a lengthy list of cloud seeding experiments of the 1950s and 1960s using a range of ground and airborne generators and pyrotechnic devices. The Kings River (California) study showed an increase in runoff of 6 per cent; the Sierra Cumulus seeding of 45 paired clouds yielded precipitation from 42 that were seeded compared with only 3 that were not seeded; while the Skagit River Project (Washington State) resulted in a significant increase in stream flow. All of these studies have been criticised for their collection of data, in particular the lack of randomisation in the selection of seeding opportunities. This reiterates the conclusions reached in an earlier report (CAS, 1966) that from 41 seasons of operational orographic seeding and 14 cumulus cloud seeding projects, it was evident precipitation could be modestly increased although the evidence was ambiguous. A study in the early 1960s of all experimental cloud seeding studies to date concluded that only 23 were statistically rigorous and of these only 6 increased precipitation, 7 cases were ambiguous and in 10 cases precipitation had reduced (Breuer, 1980).

It is very difficult to prove that seeding actually caused enhancement of precipitation because of atmospheric variability and poor observational networks. Decker (1978) cites an example where a 30–40 per cent increase in rainfall was not statistically significant. Breuer (1980, p. 9) quotes Dr. Joannne Simpson's remark that:

> The most striking lesson these writers have learned from 30 years of cloud study is that a cumulus cloud can do virtually anything all by itself without any interference by man.

Braham (1986) summarises the problems as being:

1. the physical mechanisms of rainfall are complex,
2. results from seeding clouds can be both positive and negative,
3. the variability of clouds has not been appreciated,
4. raingauge networks and monitoring are rarely adequate and
5. experiments must run for a long period of time and are therefore expensive.

In particular Cotton (1986) criticises the 'black box' studies of seeding followed by observed precipitation and argues that more comprehensive monitoring is required including cloud droplet distribution and cloud chemistry. In this way it is suggested that the full impact of cloud seeding upon rainfall processes can be evaluated. In addition clouds seeded should be selected randomly from a suitable larger group to remove bias. Griffith (1984) lists three main types of analysis:

1. **Percentage of normal analysis**, based on precipitation, snow pack or stream flow to establish normal conditions from which change is measured. This method is beset with random variability and requires several analyses of this type before any confidence can be placed in the results.
2. **Target control analysis**, when an adjacent site is selected as a control to assess the impact of seeding. There are problems over selection of an appropriate area with the same precipitation systems and environmental conditions while again long-term monitoring is necessary.
3. **Predictor analysis**, when actual precipitation is compared to that predicted through meteorological observation, but this is notoriously difficult to achieve reliably.

There are major practical problems reviewed by Smith (1986, p. 153), including the delivery of the seeding agent to particular clouds and the timing of the operation, who concluded

> work is underway to develop better techniques ... for evaluating operational weather modification projects ... much remains to be done in this important area.

The problem of verifying the impact of cloud seeding has beset the technique and Byers (1974) recounts the controversy surrounding Langmuir's early attempts to seed a hurricane in 1947. It was 350 miles east of Jacksonville (Florida) travelling north-eastward, but shortly after seeding it changed direction to a westerly path and consequently struck South Carolina and Georgia, thus raising the question of whether seeding caused the change in direction and hence was responsible for the damage. Langmuir claimed this was possible but the US Weather Bureau, it appears, were not as easily convinced. Their own studies on cloud seeding found little evidence of rainfall enhancement. Some 40 years later there is perhaps more careful optimism. CAS (1980) concluded their review of projects with generally positive effects for cold winter orographic clouds, but with mixed results for convective clouds revealing the complexity of the processes involved. For non-orographic and non-convective storms no firm conclusions were reached and it was recommended that studies using randomised experiments continue. Given this scientific debate and controversy surrounding the impact of cloud seeding it is not surprising to find that political and legal issues are equally confusing.

SOCIAL AND LEGAL ISSUES

As populations increase in arid lands and the demand for water rises so there will be more attention paid to cloud seeding particularly in times of water crisis, i.e. drought. Yet if rainfall is enhanced is it of benefit to all? Sewell (1973) notes that up to the 1970s most efforts and funds had been directed towards establishing whether cloud seeding 'could be done', i.e. scientific evaluation with little attention being paid to 'should it be done?', i.e. social and legal evaluation. Haas (1974) cites some controversial examples of cloud seeding. In Florida (1971) beef producers and citrus growers requested cloud seeding following low rainfalls; however, melons are damaged by heavy rainfall, tomato growers find their crop needs careful regulation of water supplies and damage can result from rain at the wrong time, while dairy farmers have little to gain from increased spring rains. There was therefore a range of opinion from those welcoming more rainfall by cloud seeding through to those opposed to it. After the seeding campaign, 61 per cent of tomato growers felt that the experiment had cost them financially, but there was no legal action. In another example, barley growers in the San Luis Valley (Colorado) hired a commercial firm to produce additional rainfall during the early growing season, to suppress hail in midsummer and to suppress rain in weeks before harvesting. Cattle owners and market gardeners opposed this programme claiming it had caused reduced rainfall and public hearings were organised by the state with a view to preparing new weather modification legislation. Dennis (1980) reports on protests in South Dakota against the weather modification programme following droughts in 1973 and 1974. It was argued that raindrop size had reduced resulting in more evaporation and less effective rainfall. As a consequence from 1976 it was noted that no state funds were released for cloud seeding. A major problem in all of these disputes is proving that cloud seeding actually had any effect and confusion surrounding the ownership of clouds. R.J. Davis (1974) provides the following three examples which demonstrate different interpretations over the vexed question of 'cloud ownership'.

In 1950 New York City embarked upon cloud seeding to combat a drought; but the Catskills Holiday Resort sought an injunction because an increase in rainfall would threaten business and destroy the environment through soil erosion and flood. The court ruled there was no evidence that there would be harmful effects and dismissed the notion of vested rights in clouds or rainfall. That is, the city had the right to seed clouds and there were no individual cloud property rights.

In 1958 South West Texas cereal farmers attempted to reduce hail damage by cloud seeding, while cattle ranchers claimed that seeding of clouds over their lands should stop as they owned the rain that would otherwise fall. The farmers argued that nobody owned the weather but the ranchers' view was accepted by the court upholding the belief that clouds do belong to the landowners immediately beneath.

In 1968 a Pennsylvania judge argued that clouds and rain do not belong to the landowner but are common property; however, if seeding reduces these rights of access to rain then it must be illegal.

That is, three cases with three different rulings on cloud ownership and the rights to seed clouds. Given the dispute over the effectiveness of cloud seeding it is not surprising to note that one of the major obstacles to challenging seeding is the proof of damage occurring due to seeding or potential damage likely to occur. Another major area of concern is the effect of cloud seeding upon weather that passes over state and national boundaries, which can lead to claims of poaching. Hence cloud seeding remains a minefield of potential litigation and claims for damage compensation.

SUMMARY

Much of the above discussion has concentrated upon examples from the USA yet parallel developments have taken place in other arid environments. CAS (1980) summarised worldwide studies with contributions from Australia, Canada, India, Israel and the USA. Federov (1974) has reviewed early Soviet research from the 1930s while Australia began flights in 1947 (Smith, 1974) and Israel in 1948 (Gagin and Neumann, 1974). Israel has obtained some of the most impressive published rainfall enhancement results with significant increases around Lake Tiberias (Breuer, 1980) and from coastal campaigns (Goldreich, 1988). Warner (1974), however, concludes that Australian studies using silver iodide from aircraft have proved cloud seeding unconvincing with ambiguous results.

Cloud seeding is a complicated technology. If too much or too little seeding agent is introduced at the wrong time and location the results may be ineffective or, worse, may even reduce precipitation with dissipation of the cloud. In addition not all clouds are suitable for seeding and moisture must be present in the air near saturation for condensation to commence. Controversy still surrounds the scientific evidence while the costs of airflights may prove inhibitive. Breuer (1980, p. 68) states

> comparing the production cost of a unit volume of water, cloud seeding is considerably cheaper than desalination of sea water,

but this assumes it will work!

Cloud seeding is often used in desperation as a last resort to combat water shortages and it is likely this will continue, but the method is still not fully proven. As a consequence cloud seeding will continue to be used in arid lands but until more is known on precipitation mechanisms and there are improvements in instrumentation it will continue to be a 'chancy' technology. As Braham (1986, p. 2) concluded:

> Cloud seeding is promising, unproven and worth pursuing.

9

WATER STORAGE

This chapter has been divided into three sections dealing with surface storage, interbasin transfers and finally groundwater storage and recycling. This reflects the growing opposition to large-scale water development projects embodied in dam construction and interregional water transfers. Their environmental impacts and huge capital costs have promoted the search for alternative means of storing water or greater efficient use of that available. Groundwater storage is being looked upon as being more effective than surface storage while recycling and purification of water is often achieved by artificial recharge of aquifers. Hence these two strategies are dealt with together.

SURFACE STORAGE

Dams are constructed for a variety of purposes including the regulation and augmentation of surface flows to improve navigation, to prevent floods, to provide water for irrigation, to conserve water resources, to preserve ecological systems, for aesthetic reasons possibly promoting tourism and for power production. Between the 1970s and the end of this century, the river discharges regulated by reservoirs worldwide will increase from one-tenth to two-thirds (Canter, 1985). Balek (1983) reports that the potential for hydroelectric power is still enormous. In Australia 45×10^3 MW is available for HEP while in Africa it is estimated that a staggering 780×10^3 MW is available, but in both cases only 2 per cent is actually developed. Given all these possible benefits it is not surprising to find dam building throughout the arid world.

Overman (1976) noted that King Menes of Egypt was perhaps the first to build a significant barrier across the Nile some 5,000 years ago, to be followed by the first Aswan Dam in 1902 rising 27 m above normal water levels. This structure is dwarfed by present-day dams. Goldsmith and Hilyard (1984) predicted that by 1990 at least 113 dams would be over 150 m high of which 49 will have been constructed in the last 10 years. They note that the Volta Dam in Ghana is large enough to impound a reservoir the size of Lebanon while the Sanxia Dam on the Yangtze River will generate 40 per cent of China's electricity output. By the start of the 1980s 1,220 dams had been completed or were under construction in India, with two-thirds of the total water storage coming only from 26 major dams including the Bhakra Dam (226 m high) and the Tehri Dam (260 m high) (Dogra, 1986).

Beaumont (1989) reports that dam construction in North Africa and the Middle East began in earnest in the 1950s in the wetter highlands of Algeria, Tunisia, Turkey, Iran and Iraq, peaking in the early 1970s. The aim was generally to capture water that was otherwise 'lost' but often discharges were overestimated and the effects downstream ignored; for example, the Latian Dam on the River Jai collecting water for Tehran which results in less water being available for irrigation and recharge (Beaumont, 1989). Twelve dams constructed in Saudi Arabia have a storage capacity greater than 3 MCM, with the Al-Mudhiq Dam retaining 86 MCM alone and rising 62 m in height. The Anatolian Project (GAP), in south-eastern Turkey, is a major irrigation scheme partly completed and partly under construction. It will provide some 50 per cent of the irrigated area and 50 per cent of the power generation of the country. Eventually it will include 22 dams and 19 hydroelectric plants, constructed on the Euphrates, the Tigris and their tributaries. The major element is the Ataturk Dam (see figure 10.3 and plate 19) the lake behind which commenced in January 1990. This reduced the flow of the Euphrates to a trickle for 1 month causing political stirrings in both Iraq and Syria, who feared that this demonstrated the Turkish potential for further water geopolitics (see chapter 10). The total area to be irrigated is estimated at about 700,000 ha requiring 10,000 MCM of water per year. Indeed, it has been suggested that, should the project be fully implemented, Syria's share of the Euphrates would be cut by up to 40 per cent and Iraq's by 80 per cent (*The Economist* 12 May 1990).

A major problem experienced by reservoirs is the loss of capacity through siltation. Time and again pre-construction predictions of the amounts of material transported by the fluvial system underestimate post-construction yields. The Tahri Dam in India is now expected to silt up in 40 years rather than the 100 years predicted (Dogra, 1986). In the same article further evidence is provided of this problem in India, as indicated by table 9.1.

From the 1970s there has been growing criticism of the effects of large-scale water projects which Canter (1985), in the case of dams, considers to be due to construction; water impoundment; and downstream changes. A comprehensive review of the impacts of large-scale dam construction has been provided in the

Table 9.1 Annual rate of siltation for selected Indian dams (in acre feet, after Dogra, 1986)

Reservoir	*Expected*	*Observed*	*% Increase*
Tungabhadra	9,796	41,058	419
Bhakra	23,000	33,475	146
Ukai	7,448	21,758	292
Pandut	1,982	9,533	481

Plate 19 500 cumecs discharging through the Ataturk Dam, Turkey (this equals the agreed flow into Syria).

Table 9.2 Social disruption associated with large dams (after Goldsmith and Hildyard, 1984)

Dam	*Displaced People*	*Area (ha)*
Aswan High Dam, Egypt	120,000	400,000
Volta Dam, Ghana	78,000	848,200
Kariba Dam, Zimbabwe	50,000	510,000
Kainji Dam, Nigeria	42,000	–
Keban Dam, Turkey	30,000	–

publication by Goldsmith and Hildyard (1984). These impacts can be grouped under the following headings.

Resettlement and social disruption

In Africa, during the 1970s, some 350,000 people were displaced due to the construction of five large dams (Baillon, 1987). The figures for resettlement listed in table 9.2 are dwarfed, however, by the Three Gorges Dam in China which is expected to displace 1.4 million people. Such massive movements and relocation of population result in social disruption which can threaten the success of water development projects. De Wilde (1967) describes how the Office du Mali scheme in Mali relied upon the migration of farmers from Burkina Faso. The disruption of village life and breakdown of social networks resulted in disenchantment with the project and the return of many farmers to Burkina Faso. It is apparent that local populations are largely disregarded when such schemes are designed; there is little local participation and a lack of compensation with often inferior conditions being offered to people forced to leave their homes and lands.

Environmental and ecological degradation

There is a direct loss of land through impoundment of waters, including fertile valley floors and wooded hillslopes. The impact upon the ecosystem of large dams is significant. It is often assumed that wildlife will relocate naturally but this view is directed at large herbivores and carnivores and disregards interactions and dependency within the ecosystem and the land available elsewhere. The damming of the Nile is an example of environmental degradation where the loss of silt has resulted in more use of artificial fertilisers and the retreat of the delta at the mouth of the Nile. A further downstream effect of the reduction of flow in the Nile has been a complete ecological change in the eastern Mediterranean. In former times, the outflow of freshwater checked the ingress of any Red Sea flora or fauna through Suez. Since the Aswan High Dam came into operation, the reduction in freshwater flowing from the delta has resulted in major ecological changes. Red Sea species have colonised large areas of the eastern Mediterranean and their effect is felt as far westwards as Malta. Lavergne (1986) lists seven impacts concerning the Aswan High Dam including erosion of the river bed; retreat of the delta; the reduction in silt carried by the river; the consequent increase in fertiliser use; higher evapotranspiration rates from aquatic weeds; a rising water table in places; and declining soil fertilities.

Inland fisheries on the other hand can benefit from dam construction through the release of nutrients from rotting of submerged vegetation. Fish stocks in both Lake Volta and Lake Kariba soared after impoundment but then reduced as the supply of nutrients inevitably fell. This is a familiar pattern which can be used to exaggerate the positive effects of dam construction when it is in fact a short-lived phenomenon, favours lacustrine species and takes no account of the reduced flows and supplies of nutrients downstream. Canter (1985) suggests that thermal stratification within the impounded waters may well be a significant factor in water quality changes, resulting in a general lowering of water quality.

Increased water losses

The average natural flow of the Nile at Aswan is 84 BCM (billion cubic metres of water), of which 17 BCM reaches the sea and 13 BCM is lost through evaporation from the major reservoirs in Egypt and the Sudan (Chesworth, 1990). Evaporation losses can be reduced by covering the water surface with chemicals or artificial skins but the costs are normally prohibitive for large-scale water schemes. Although water bodies can influence their immediate environment through the establishment of a vapour blanket where humidities are high and consequently evaporation is lower, such atmospheric effects require a large expanse of water to be significant. Most artificial lakes in arid lands will suffer from high evaporation losses and this can be exacerbated by the presence of aquatic weeds which can increase water losses beyond potential rates (Agnew, 1984a, Crundwell, 1986 and Holm *et al.*, 1971). Evaporation losses from Lake Nasser are currently estimated to be some 10 BCM per year. With the burgeoning populations of Egypt and, to a lesser extent, Sudan, these losses are seen increasingly as catastrophic. If the states of the Nile Basin could develop a cooperative programme, it has been suggested that the main storage unit in the system could be moved from Lake Nasser to Lake Tana in the Ethiopian Highlands. This would result in a vast decrease in evaporation losses and would make more water available for irrigation schemes in the headwaters' riparian states.

Increased prevalence of pests and disease

Bodies of stagnant water provide excellent breeding grounds for the vectors of harmful diseases and water resource projects are often accompanied by an increase in schistosomiasis and malaria (see chapter 6 on irrigation for more details). Yacoob *et al.* (1990) suggest primary health care programmes have avoided tackling disease through improved water resources owing to the relative costs cited (US$200–250 per child death avoided by strategies over breast feeding, tetanus, child immunisation, anti-malaria and oral rehydration; compared to US$3,600–4,300 for water supplies and sanitation). They challenge these figures and argue for a focus back onto potable water supplies while making a strong case for more attention to be paid to the debilitating impacts of the guinea worm (dracunculiasis).

Increased threat of dam failures

Because of their great size, 20th century dams pose a serious threat of damage and loss of life. May and Williams (1986) note that between 1970 and 1980 in the USA dam failures caused 500 deaths and $2 billion of damage. In 1972 in Buffalo Creek, West Virginia, 125 people died while in 1976 there were 11 deaths due to failure of the Teton Dam and $1 billion of damage. The Committee on Safety Criteria for Dams (1985) believe that there are 10 significant dam failures each decade. The reasons lie in faulty design, inadequate maintenance or mismanagement, unanticipated floods or earthquakes and fewer suitable sites being available. It is now recognised that the weight of impounded water can result in earthquakes and Goldsmith and Hildyard (1984) note increased seismic activity has been associated with the Hoover Dam (USA), Kariba Dam (Zimbabwe) and Konya Dam (Israel).

Myth of flood control

It would appear that floods are becoming more frequent and destructive. One cause is upland deforestation resulting in increased surface runoff and erosion but the prevalence of major floods worldwide reveals the inability of dams to prevent flooding. The Committee on Safety Criteria for Dams (1985) note that dams generally do not protect property and people downstream. Extreme flood events can overwhelm the storage capacity of a dam possibly leading to its breaching unless the water is allowed to run off downstream. The need to maintain reasonably high levels of water behind a dam for water supplies and water quality regulation result in little capacity to deal with extreme flood events which are inevitably allowed to pass through. Dogra (1986) reports that in 1978 in India, waters rose in the Sagar, Kalagadh and Begulbandh reservoirs requiring the release of flows resulting in serious flooding in Uttar Pradesh. The author cites other examples from Pakistan where large reservoirs were unable to prevent serious flooding downstream.

Poor management and maintenance

Barnett (1977) blamed poor, inflexible management for the disappointing performance of Sudan's Gezira scheme and the same comments have been applied to other water resource projects (Biswas, 1978 and Houston, 1978). Partly this is a problem of communication difficulties associated with large-scale schemes but is also due to inadequate training at all levels and too much emphasis upon design and construction with inadequate attention being paid to management and maintenance. Bureaucracy and complex administration also contribute to this problem. Vohra (1975) described the management of India's water before reorganisation when land and groundwater was the responsibility of the Ministry of Food and Agriculture; the Ministry of Irrigation was responsible for surface flows; and prospecting for groundwater was the concern of the Ministry of Mines and Metals. Such overlaps create confusion and anomalies which can be tackled by water projects having a multidisciplinary organisational structure that transcends Ministerial boundaries – which is not only a 'mouthful' to say but also difficult in practice to achieve.

When the dams of the Anatolian Project are complete, Turkey has pledged the release of an average 500 cumecs of Euphrates water at the Syrian border. There would be an even flow throughout the year, rather than the erratic discharge of the present regime. While this may seem to be a major management advantage, it will undoubtedly cause great problems in Iraq. For millennia, irrigated agriculture in Iraq has been developed in phase with the flows of the Euphrates. Thus, during the low-flow period, the farmers have little need for water. As a result, regular flow may well militate against the efficient use of water, given the present Iraqi agricultural practice.

Water quality

Industrial and urban development associated with the provision of water and cheap HEP can lead to serious pollution. Goldsmith and Hildyard (1984) suggest that hazardous industries are finding it easier to locate in Third World countries where environmental controls are often less exacting. Arid lands with their perceived 'open spaces' and remoteness are particularly susceptible to this kind of pressure. Balek (1983), writing on the tropics, believes that artificial pollution is limited to cities and industrial centres, but with intensification of agriculture and population growth this will become a more severe problem for arid lands. World Resources Institute (1987) report that nitrate pollution from fertilisers is becoming a worldwide problem and in 1984, 6 per cent of wells sampled in the USA exceeded safe drinking water standards for nitrates. Water quality is also altered by changes to the hydrological system of an area by land degradation through salinization (see chapter 6) and deforestation (Grainger, 1990) leading to erosion and high sediment yields. The latter reduces the capacity and operation of a dam without periodic cleaning but also accelerates eutrophication which can lead to algal blooms toxic to humans and livestock.

The FAO wrote in 1975 of their concerns over water mismanagement with the title 'Water: available but mismanaged', and in the same publication Vohra (1975) voiced opposition to large-scale water projects with the headline 'No more gigantism'. This view has recently been reiterated by World Resource Institute (1989) who noted an increase in small dam construction as large-scale projects become unpopular and it is recognised that many large-scale water projects have failed to reach their original objectives (Matlock, 1988). Carruthers and Clark (1983, p. 2) are more emphatic:

> The era of the big dam is, with one or two notable exceptions, now over.

The use of small dams and barriers to check surface flows can be traced back to antiquity but they are evident in many arid areas today. That such techniques have survived requires that those interested in water resources consider small-scale strategies for water resources development. Barrow (1987) provides a useful review of methods of small-scale water diversion and storage and notes that out of 34.5 million ha irrigated in India, 4 million ha were watered from small reservoirs called tanks with each capable of supplying 9 ha. World Resources Institute (1989) report that India has recently ignored such water tanks; the area irrigated by which, reached a peak of 4.8 million ha in 1958–9 and has then declined. The same publication contrasts India's experience with that of china where 6 million small reservoirs have been constructed between 1950 and 1980 and now 40 per cent of rural townships receive their power from small HEP dams, although Cummings (1990) suggests that there is a limited opportunity to use small-scale, rural HEP plants to supply urban industrial centres. Frasier (1988) classifies small-scale water storage into soil storage systems or tank/pond systems for use in water harvesting. He concludes that there is no universal best design with each site having unique requirements.

Small reservoirs suffer from many of the same problems as large reservoirs, e.g. siltation, water quality and evaporation, but their size can enable cleaning and repair to be undertaken using local labour supplies and technology. The advantages of small-scale surface storage are that it has a reduced impact on landuse and ecosystems, the threat of dam failure is reduced, it can be constructed using local knowledge and small dams can be erected more rapidly. The disadvantages are that benefits accrue to a smaller number of people and that multi-purpose projects are unlikely, e.g. for navigation and low-flow maintenance, large-scale irrigation, fisheries and the development of tourism, although technological developments may change this with HEP for instance now produced from small, local dams. A further advantage of small-scale storage is that their construction and maintenance can promote community participation. Frost (1987, p. 75) argues that, when developing water resources in arid areas

> systems must be cheap, small scale and decentralised ... capable of being locally erected, controlled and maintained by the people themselves.

This view is supported by Donalson (1987, p. 86):

> Experience has shown that a key element in ensuring long term success is that of involving the user in every level and phase of the effort.

Golladay (1983) raises a note of caution before embracing local community participation by arguing that such activities require sacrifices of time, resources and individual freedom and that failure may jeopardise the whole group rather than the individual household. It is also noted that community participation is not spontaneous but requires organisation and perceived advantages. Nevertheless there is now a strong move towards small-scale, community-based water resource projects which are environmentally sound. Goldsmith and Hildyard (1984) conclude their study on the social impacts of large-scale dams with:

> We thereby call upon those organisations, herewith, to cut off funds from all large scale water development schemes that they may plan to finance, or are involved in financing, regardless of how advanced those schemes might be.

It is worth noting finally the point made by Matlock (1988, p. 939):

> Small scale water management systems represent a modern technology; they need not be construed as being second class.

Small dams can then be an opportunity to combine traditional practices and experiences with modern technological developments.

INTERREGIONAL WATER TRANSFERS

The opportunities for large-scale water storage projects are diminishing as suitable sites for huge dams are used up. Some areas benefit from 'natural' interregional water transfers (IWT), for example the Nile for Africa, the Rio Grande for America and the Euphrates for Asia. There have been many suggestions to emulate nature by the artificial construction of canals, rerouting rivers, etc. although the proposal to tow icebergs is one of the more fantastic. Al-Faisal (1982) reports that at the turn of the century, small icebergs were being towed by ships and that the Antarctic provides the best source as the ice shelves are more extensive and accessible. Calculations indicate that icebergs would melt before crossing the equator unless protected and there are obvious difficulties in harvesting the water once the ice has arrived at its destination. Hult (1982) estimated that the cost of a pilot scheme to bring 100 MCM of iceberg to the Californian coast in the 1980s was US$20 million or US$0.2 m^3.

Water transfers by canal can be traced back to at least 4,000 years ago in Egypt (Lake Qarun) while the Romans built aqueducts in 200–300 BC up to 90 km in length delivering more than 1,000 l per capita per day to Rome (Overman, 1976). In Africa, plans have been drawn to direct the Congo River into Lake Chad to produce a Congo sea which would be linked to the North African coast by canal and used to irrigate the Sahara (Balek, 1983). Groundwater was discovered in the Libyan Desert in the 1960s and a scheme devised to transport 2 MCM a day northwards along a 1,900 km pipeline (Alam, 1990). In China work commenced in 1987 to direct 5 per cent of the flow of the Yangtze River by a 1,000 km long canal to the northern semi-arid region (World Resources Institute, 1989); while in Australia the Snowy Mountain scheme transfers water from the Snowy River across and through the Great Dividing Range to augment flows of the Murray River in the drier interior. In 1967 the arid Negev region was linked by pipeline with Lake Tiberias in Northern Israel (Overman, 1976) while in Africa the unfinished Jonglei canal is designed to divert upper Nile flows in order to bypass the Sudd marsh, thus reducing evaporation losses, and is expected to yield 2.13 BCM for Sudan (Pemberton, 1987). The second phase includes diversions that it is hoped will provide an extra 6.87 BCM a year. Perhaps one of the best known IWT proposals is from the USSR to divert the northern-flowing rivers, the Ob and the Yenisei, southwards into the dry steppe region to supplement irrigation and to tackle the falling levels of the Aral Sea (Goldsmith and Hildyard, 1984 and Vorapaev, 1979); while Micklin (1971) reviews similar plans for increasing flows in the rivers Volga and Kama draining into the Caspian Sea. It was planned that some 20 BCM would be diverted southwards by the year 2000 with construction started on a further 27 BCM of transfers (Micklin, 1986). Opposition to such water transfers, based on ecological impacts, has become more public in the USSR since 1983 and in 1986 work on the Aral Sea project was stopped for further investigation (World Resources Institute, 1989).

Among international transfers of water are the following:

1 A 40 cm pipeline to supply drinking water to Kuwait from the Euphrates has been opened (*Financial Times* 25 September 1990).

2 The 'Peace Pipeline' project was announced by Turkey in 1986. The proposal is that surplus potable water should be piped from the catchments of the Seyahn and Ceyhan rivers in two pipelines. The west pipeline would link Turkey with Syria, the West Bank, Jordan and Saudi Arabia; and the east pipeline would go through the Gulf States to Oman. The proposed routes have changed at least twice since the original announcement, but the major contractors, Brown and Root, have now completed initial feasibility studies. It appears that physical obstacles can be overcome relatively easily, as can threats from terrorist action, but political objections are likely to be far more substantial. The vulnerability in obtaining even part of the national drinking water supply from an outside source might well prove unacceptable.
3 The proposed Nile–Saudi Arabia pipeline, originally suggested by President Sadat, seems unlikely to be built owing to objections from Ethiopia.
4 Feasibility studies are just being completed on the possibility of exporting water from Turkey to Israel. It would be transported in large plastic bags, which could be carried on container ships. The proposal, which seems likely to go ahead, is code named 'Medusa'.

There are then a number of IWT projects planned or underway worldwide. Golubev and Vasiliev (1979) estimate that at the turn of the century IWTs totalled 500 to 1,000 million m^3 a year which had risen to 10,000 million m^3 a year by the 1970s and a tenfold increase was expected during the next 20 to 30 years. Yet in the same publication Golubev and Biswas (1979) predict only steady growth and they state (p. 60):

> At the present time it is hard to foresee the construction of any new major IWT in Canada and the U.S. before the end of this century.

Biswas (1979, p. 87) quotes Howe and Easter (1971) saying:

> Large scale transfers of water are likely to cost more than they are worth to a nation except in certain resource operation cases.

This is a response to the growing criticism of the environmental effects and high costs of major IWT projects; for example, the California State water project diverting water from northern parts of the state to serve the southern region has been criticised for the effects upon water quality in the San Francisco Bay area (Ortolano, 1979). Apart from environmental consequences involving seepage losses, possible climatic changes and alterations of water quality there are immense political and legal obstacles. Balek (1983) reports on the legal constraints for water resource schemes including the difficulties in establishing rights to and boundaries of water resources which can result in armed conflict. The ill-fated Jonglei Canal (Collins, 1990) provides an example of political and military opposition to water transfers. Although there were technical difficulties initially with the canal excavation in the heavy soils of Sudan using the 'Bucketwheel' device, these were overcome once work began in June 1978. Local political opposition, however, developed into an armed struggle with canal construction a focus for dissent. Hostages were taken and in 1984 operations ceased. Collins (1988, p. 53) writes:

> And so the Bucketwheel rests silent and forlorn, quietly rusting away at km 267 . . . behind stretches the big ditch.

There are legal obligations on both upstream and downstream states when water is diverted and the difficulty in obtaining international agreements concerning ownership and access to water severely constrains such developments. It is then not surprising to find an interest in small-scale water development and groundwater storage. As World Resources Institute (1989, p. 133) state:

> In the United States the era of building huge federal dams and long aqueducts and canals appears to have ended.

GROUNDWATER STORAGE AND RECYCLING

Arid lands with their high evaporation rates, large amounts of sediment transported in flood conditions and flash floods pose particular environmental hazards for dams. Often valleys are ill-developed resulting in comparatively few suitable sites yet the demand for water storage is high. The opportunities provided by storing water in aquifers beneath the ground are therefore increasingly attractive. The obvious advantages of groundwater storage are reduced evaporation rates (providing the water table is not too close to the surface) and reduced construction costs although wells and infiltration ponds may have to be installed. Other benefits include less loss of land although wells and recharge dams need to be considered; reduced threats of ecological damage, land degradation and sedimentation; and water quality is less variable. Water purification is also possible as water percolates to and is stored in an aquifer while with careful management groundwater flows can be used to redistribute water without the construction of canals and pipelines.

Groundwater storage is not a recent idea and has long been practised through barrages and rainfall harvesting, and Jones (1981) suggests aquifers can be considered as subsurface dams. Stoner (1990) estimates that while Lake Nasser has a capacity of 130 BCM, the groundwater storage capacity in Egypt is around 400 BCM and much is recharged naturally by Nile flows. Control of groundwater storage can also improve water quality and prevent seawater intrusions. Mandel (1977) reports that in 1963 the Tel Aviv to Beersheba area (4,000 km^2) contained two-thirds of Israel's population and at least half of the irrigated lands. In 1948 most water was supplied from the coastal aquifer satisfying the demand of 110 MCM a year but by 1954 demand had risen to 360 MCM a year. Seawater intrusion became a problem in 1958 because of overextraction of groundwater and was tackled by pumping surface water into boreholes supplied by a pipeline linked to Lake Kinneret in the north. By 1969 it is reported that the situation had been stabilised by controlling groundwater conditions through artificial recharge.

In Morocco 30 per cent of demand is satisfied by groundwater. Margat and Saad (1984), writing on 'Deeplying aquifers: water mines under the desert', observed that there are extensive sedimentary basins concealing deep aquifers in all continents and that it is only in the last few decades that intensive use is being made of this resource. They mention projects in Algeria and Tunisia planning to withdraw 30–40 BCM, 70 BCM in Egypt and 11 BCM in Libya. Simpson (1985) reports that in Saudi Arabia, 70 per cent of all water is from groundwater with 6 per cent from surface supplies and the rest from desalination. World Resources Institute (1990) note that 90 per cent of water consumed in Saudi Arabia is from non-renewable, fossil groundwater which could well be exhausted by the year 2007. Such huge withdrawals can result in declining yields, reduced water quality and often subsidence. Concern of overexploitation of groundwater has then given rise to greater awareness of recharge and the opportunity of using aquifers to store surface waters.

Techniques for computing natural recharge are discussed in chapter 5 based on studies by Bear (1979) and Chidley (1981). Bear demonstrates the potential for groundwater storage by considering an aquifer with a storavity of 15 per cent (storavity is the volume of water released from storage per unit horizontal area and per unit decline or rise of piezometric head). Such an aquifer measuring 10 km by 10 km could store 15 MCM of water with a 1 m rise in water table. However, water percolating into an aquifer will spread out and flow towards the boundaries where it will seep out. Aquifers therefore only provide a temporary store but given the generally slow rate of movement this can be quite satisfactory for flood, seasonal and even annual storage. Bear (1979, p. 44) states:

> In many instances, especially in arid and semi arid regions, storage in (phreatic) aquifers has been proven to be more economical than surface storage.

Simpson (1985) notes that arid land groundwater is recharged naturally in three ways:

1 By exogenous rivers, e.g. Nile, Euphrates.
2 By springs and seasonal wadi flows.
3 By runoff in the foothills.

Huisman and Olsthoorn (1983) suggest that the safe yield from aquifers is commonly 20 to 80 per cent of natural recharge which can be substantially enhanced artificially. They compute that a city of 100,000 with a consumption of 200 l per capita per day would need an aquifer with a catchment area of 30 to 120 km^2 in a moderately wet region; but this area can be reduced to only 0.15 km^2 with artificial recharge where this is defined as (p. 11):

> The planned activity of man whereby surface water from streams and lakes is made to infiltrate the ground, commonly at rates and in quantities many times in excess of natural recharge.

Artificial recharge can be undertaken by the following.

Indirect artificial recharge

Wells are located as close as possible to areas of natural surface discharge (rivers, wadis or lake). Extraction of water from the wells draws down the water table which induces a flow into the aquifer from surface sources. The advantage of this approach is that the surface water is filtered and cleaned before extraction at the wells. Bear (1979) suggests a retention time of 2–3 months is typical for purification. Thomas (1987) mentions how this technique was used for refugee water supplies in Somalia in saturated sands 1 to 3 m deep. This technique is best where stream beds are highly permeable and aquifers are thick (Huisman and Olsthoorn, 1983). A major constraint is the deposit of debris on the bed of the stream reducing its permeability and hence recharge rates. In the past flood flows could have scoured such material away but with increased regulation of discharges such natural cleaning may no longer be effective. Indirect recharge is more traditional but is now giving way to direct artificial recharge because of the better control afforded by the latter.

Direct artificial recharge

Infiltration enhancement

Surface runoff is arrested by small barrages or terraces to enhance infiltration. This is feasible where the land surface is permeable but there is a tendency for sediment carried by flood runoff to reduce infiltration. In addition only a relatively small area of recharge is obtained without large structures.

Surface spreading

Water is directed to ponds and basins although peak flood water is avoided because it contains silt and sediment. Unfortunately in arid lands flood flows are characterised by high sediment contents hence reducing the effectiveness of this technique. Settling basins can be used before infiltration ponds to remove suspended load although this does increase the amount of land required. Cleaning of the ponds is inevitable as infiltration rates decrease over time.

Wells

Aquifers are recharged from water pumped into wells. This is useful for confined aquifers where an aquifuge beneath the surface makes infiltration ponds ineffective. Clogging of wells is a problem particularly as fine material will be deposited some distance from the well where velocities decrease. Hence cleaning is difficult and pre-treatment of the recharge water is essential. Roosma and Stakelbeek (1990) argue that recharge by infiltration lagoons and canals can have undesirable aesthetic impacts as well as encouraging vegetation growth; while deep well infiltration can overcome these criticisms and they review plans for such a scheme in the North-Holland dune area.

Care is needed to maintain groundwater free from pollution, especially unconfined aquifers, and Huisman and Kop (1986) suggest careful regulation of extraction rights to maintain water quality. Without sufficient regard to the balance between recharge and extraction it is clear that aquifer storage can be overexploited with a subsequent decline in water quality and even land subsidence. Larson and Pewe (1986) report that in Arizona water levels have declined by 90 m with associated subsidence of 1.05 m. Land subsidence up to 9 m (due to aquifer depletion) has been reported in California and Mexico City (World Resources Institute, 1987). Groundwater extraction from the Ogallala aquifer in the Western Great Plains (USA), which stretches from South Dakota to Texas, began in the 1880s using wind power but major increases in groundwater exploitation in response to drought in the 1930s and 1950s and increased irrigation areas meant that by 1978 170,000 wells had been installed irrigating 13 million acres (Taylor *et al.*, 1988). Speth (1988) reports that withdrawals have been averaging 7 to 8 million acre feet a year but with a recharge rate of only 150,000 acre feet a year leading to serious depletion of this resource. Speth continues to note that in South East Arizona, half of the available groundwater of 283 million acre feet has been removed in the last 80 years.

Huisman and Kop (1986) note that purification of water along with storage is a major objective of artificial recharge and they suggest a detention time of a few days is sufficient to remove suspended sediment from surface water; for pathogen-free water a detention time of 2 to 6 months is required; while to maintain water of a constant quality a detention time of 0.5 to 2 years may be necessary. Sand filtration acts as a three-way filter: physical, chemical (nitrates and phosphates, salts, toxic chemicals and heavy metals) and biological (pathogenic bacteria and viruses, parasite eggs, worms and helminths) using micro-organisms and bacteria to decompose organic matter. Sewage effluent has long been used to supplement water supplies for agriculture (Biswas, 1988) but interest in water purification has increased since the 1980s owing to the greater emphasis upon water conservation and recognition of the possible role of aquifers in the process. Montgomery (1988, p. 38) states:

> In the context of Jordan and other semi arid and arid countries recharge is potentially a simple, safe, and inexpensive method for purifying and subsequently storing effluent for reuse when required.

Israel reused 35 per cent of its municipal waste water in 1986 and plans to recycle

80 per cent by the end of this century primarily for irrigation (M. Buerk, BBC Nature, 3 May 1990). Montague Keen (*Financial Times*, 18 April 1990) notes that 15 per cent of Israel's water comes from sewage and this will increase to 25 per cent by the year 2,000 but a cholera outbreak in 1970 traced back to using untreated waste water points to the need for caution when recycling water. There are now numerous waste water recycling plants operating in the Middle East listed by Touqan (1988) including the Tubli plant at Bahrain for 500 ha of irrigated crops. Kuwait has used effluent for agriculture since the 1960s and produces 27 MCM a year of recycled waste water at present with plans to increase output to 125 MCM a year. Similar projects can be found throughout the Middle East although there are environmental constraints. For example, Nasser (1988) reports that surface water supplies in Jordan supply 880 MCM and groundwater 220 MCM a year for a present demand of 420 MCM year which will increase to 1,100 MCM a year by the end of the century. Hence the need to consider recycling and water conservation. A waste water treatment plant was opened at Amman in 1985 with an anticipated output of 128,000 m^3 a year by 1995 but Kilani (1988) notes that the hydrogeological conditions are complex and poorly understood resulting in uncertainty over the possibilities of groundwater recharge. At Aqaba, there are problems of salt water intrusion and a shallow groundwater table (Hegazin, 1988) requiring water to be pumped via recharge wells. In Saudi Arabia, out of a total demand of 7 BCM a year, which is expected to increase to 17 BCM by the year 2000, only 100 MCM is supplied by treated waste water recycling. This technology will receive more attention with a new 11,400 m^3 per day plant at Riyadh (Kabbara, 1987).

There is then a shift of emphasis away from the large-scale water projects involving huge dams, interbasin transfers of water, massive irrigation developments and HEP production. It is suggested, however, in a recent publication of the World Resources Institute (1990) that water storage in large basins such as Lake Victoria still offers enormous capacity worldwide and that construction may well continue because of the growing demand for HEP. Despite nearly all 'super dams' built in India over the last 50 years suffering from severe siltation problems and failing to provide expected benefits, Lewis (1991) reports that 'super dams' are enjoying a boom in the developing world. Nevertheless the environmental costs, capital costs and the poor operation of past projects coupled with the lack of suitable sites have led to a greater emphasis upon small-scale storage, groundwater storage and water conservation and recycling. Cummings (1990) reports that, in 1989, the US$500 million World Bank grant for large dam construction in the Amazon basin was shelved and instead talks were initiated on grants for environmental projects. This reflects the growing awareness of the need to avoid the adverse environmental impacts associated with large dams. Water is no longer regarded as infinite and naturally replenished but a resource that needs to be conserved with water efficiency a high priority.

Part IV

MANAGEMENT OF WATER RESOURCES

Part III has commented upon the unwarranted environmental impacts of many attempts to enhance water supplies from the salinization of soils due to irrigation to the increase in diseases associated with large dams. In addition cultural constraints, legal obstacles and economic problems pervaded much of the discussion, especially when large-scale schemes were considered. Failure to predict or to tackle these problems was often blamed on 'poor management'. This has become a ready excuse for all bungled or disappointing attempts to develop water resources such that inefficient management almost appears inevitable for large-scale operations. But is this true or inevitable? Do the problems of environmental hostility, variable water supplies, poor data bases and accelerating demand that beset arid lands create a situation where efficient planning, design, implementation and operation are impossible? Part IV tries to answer these questions by comparing the ways in which water management has been tackled in arid lands. But given the heterogeneous nature of arid environments, those expecting a universal management strategy as the conclusion of this discussion will be disappointed. The approach we use is to compare the different solutions employed by different arid lands to emphasise that there is no single universal management solution.

10

THE GEOPOLITICS OF WATER

NEW WORLD ORDER

The early 1990s have been dominated by two changes with profound political implications. From late 1989, Soviet control over Eastern Europe and, indeed, many of the constituent states of the USSR itself has crumbled. The underlying weakness of the Soviet economic system has been exposed and the state appears to be in danger of complete breakdown. With such internal problems, little attention can be paid to foreign policy or the projection of power, still possible through the vast Soviet military arsenal. As a result, there is effectively only one superpower, the United States, and the North Atlantic Treaty Organization is desperately searching for a new policy. The diffusion of the Cold War East–West tension has brought into focus the potential for North–South conflict. The North is taken to represent those countries which are economically developed, technologically advanced and, relatively, militarily strong. Whereas the South comprises, essentially, the Third and Fourth World. Trade from the South to the North is composed, basically, of raw materials and therefore the most likely potential conflict would concern the North's access to or denial of resources.

The disparity between the North and the South is highlighted by the other major change, the development of increasingly closer bonds, amounting virtually to unification, between the countries of Western Europe. Following EC '92, Europe, with a population of some 320 million, will become the most powerful economic bloc in the world. Thus, the Mediterranean will represent the major geopolitical fissure, separating the 'have' from the 'have not' nations, the wealthiest from the poorest. As the gap between the two increases, so dispute becomes more likely, much of it focusing upon resources. Therefore, this would seem to be increasingly the era of resource geopolitics, a concept fore-shadowed, if not explicitly, in the earliest writings on geopolitics.

GEOPOLITICS

The term 'geopolitics' was first used by Rudolph Kjellen in 1901 to combat the view, then prevalent in political science, that the state was merely a legal concept.

In his book, Kjellen (1916) envisaged the state as a unit of power, this power being derived from five key variables, one of which was Geopolitik or the geographical characteristics of the state. In Kjellen's terminology, 'geography ...' is effectively limited to physical geography, and Geopolitik thus subsumed three elements: *Topopolitik*, *Morphopolitik* and *Physiopolitik*. The term *Physiopolitik* was used to denote the content, and particularly the physical resources, of the state.

However, the subtleties of these initial distinctions rapidly became blurred as the general term 'geopolitics' impinged upon the awareness of statesmen. The key components increasingly became physical geography – the size, shape and general relief of a state, its position with regard to the global sea and its distance from other states – and military power. In the writings of Mackinder (1904, 1914 and 1943) in particular, geopolitics is seen as the interrelationship between physical geography and international politics. This accent upon global location as the major contribution of geography can be traced through the writings of Mahan (1890), Spykman (1944), de Seversky (1950) and Walters (1974).

In describing geopolitics as one of the 'grand theories' of International Relations, Sloan (1988) suggests that geopolitical theory has a close affinity with the Realist approach to International Relations. This posits that international relations can only be understood in terms of the development and projection of national power. As already indicated, geographical variables can be seen as major contributors to the power of the state. This idea is encapsulated succinctly in the definition of geopolitics adopted by Cohen (1973):

> The relation of international political power to the geographical setting.

Cohen is more explicit than previous writers on the components of 'political power', but it can be argued that the epithet 'international' is unnecessarily restrictive. Geography embraces the total global environment, and thus factors at a very local level may influence international decisions and, conversely, global politics may exercise a strong influence at the local level (Anderson, 1988b). Therefore, while the accent will probably remain upon the effects of geography at the international level, there can be interaction throughout the environmental system. The question of scale is also introduced by Zoppo and Zorgbibe (1985) in the context of low politics in geopolitics. If low or local politics can be accepted as the generating point for ripples of geopolitical consequence, it would seem illogical to exclude low or local geography.

In his updating of the term 'geopolitical', Parker (1985) illustrates developments occurring within the subject, often on a local scale, and, in so doing, illuminates the original distinction made by Kjellen:

> Mention of the term (geopolitics) nowadays is calculated to conjure up visions of oil supplies, strategic minerals, agricultural potential, dangerous sea routes, vulnerable frontiers and, possibly, dwindling natural resources.

This view of the increasingly broad perspective of the subject is reinforced by that of Anderson (1984) who, in his consideration of the contribution of geography to geopolitics, states that:

> It should be understood that geography properly includes consideration not only of locational and environmental circumstances but also of demographic, social, and economic conditions.

In tracing the relationship between the predominant geopolitical world view and technology, Child (1985) distinguishes five geopolitical perspectives including the resource perspective. The development of this perspective Child attributes to the post-war period which witnessed the intensification of competition for scarce resources such as petroleum and strategic metals. The implementation of such competition can be seen in policies of 'access' and 'denial'. If a state lacks a particularly critical material, it may attempt to obtain supplies by non-economic means. In contrast, a country possessing the particular mineral may, in pursuing its foreign policy, deny access to another country. The position was well stated by the British Foreign Secretary, Sir Samuel Hoare, during a speech at Geneva (ll September 1935) when he stated that:

> Abundant supplies of raw materials appear to give peculiar advantage to the countries possessing them. It is easy to exaggerate the decisive character of such an advantage, for there are countries which, having little or no natural abundance, have yet made themselves prosperous and powerful by industry and trade. Yet the fact remains that some countries, either in their native soil or in their colonial territories, do possess what appear to be preponderant advantages; and that others less favoured, view the situation with anxiety.

The major elements of resource geopolitics are designated strategic and are generally considered to be petroleum and strategic materials (e.g. Jordan and Kilmarx, 1979, Conant, 1982, Maull, 1984 and Westing, 1986.) However, in certain parts of the world, no resource could be more strategic than water. Indeed, water even more than petroleum can be related directly to conflict.

WATER AS A STRATEGIC RESOURCE

If, since it is both vital and scarce, water provides a legitimate target for resource geopolitics, it can, like petroleum and certain minerals, be considered strategic. Strategic minerals are mostly required in small quantities and are of high value, petroleum is needed in significantly larger quantities and is of medium value, while the demand for water is high, but the commodity is of low value. Thus, the strategic nature of each of the three is different. In the last resort, since it is a basic for life, water is the most strategic commodity. Nonetheless, there are sufficient elements in common that it is valid to discuss the counter-measures normally advanced to combat shortages of strategic minerals of petroleum in the context of water. The main options are:

1 stockpiling or storage;
2 research and development in relation to substitution, recycling and conservation; and
3 alternative sources.

Precipitation is, of course, naturally stored in aquifers which, unlike the similar stockpiles of oil, occur virtually worldwide. In the Middle East, in areas of ephemeral flow, a major emphasis of water management is upon the enhancement of this storage through recharge projects. However, in contrast to surface reservoirs for which direct measurements can determine water budgets, the relationship between inputs and outputs in the case of aquifers is much more difficult to ascertain. In the Middle East, no detailed experiments have been carried out to allow water yields, before and after emplacement of a recharge dam, to be compared.

On the other hand, while a surface reservoir represents, instrumentally, a simpler system, there are still notable losses through seepage and evaporation. In the Middle East, impounded lakes are restricted, almost exclusively, to areas of surface flow and, in particular, the Nile and the Tigris–Euphrates drainage basins. They, together with the few naturally occurring lakes, such as Lake Kinneret, provide what are likely to be the only viable examples of surface storage. Water loss by evaporation from lakes established elsewhere is likely to be prohibitively high. Similarly, recharge is liable to be limited to flood water which would otherwise be lost. The potential for loss would seem to be too high for any procedure which involved transporting water to recharge sites. Finally, it must be stressed that to be of any real significance, in national water management, volumes of water recovered would need to be large and therefore, given the losses, quantities stored would have to be considerably larger. Thus, other than recharge or reservoirs in the basins of perennial rivers, the option of stockpiling to alleviate water scarcity is likely to be confined to emergency storage only.

Water is, of course, a unique resource and there is no question of substitution, particularly not in the quantities required. However, it is possible to consider the substitution of sources. The origin of freshwater, both on the surface and underground, is precipitation, but it is also possible to produce freshwater from brackish water or seawater. With surface sources, scarce as they are in the region as a whole, approaching capacity utilization, and with groundwater in many areas declining towards exhaustion, there is great urgency throughout the Middle East attached to the search for alternative sources. Chief among these, by far, has been desalination. More money has been spent on such installations in this region than in any other part of the world, and some 60 per cent of all plants currently in operation are located in the Arabian Peninsula. Most are multiple stage flash (MSF) plants, but other methods such as electrodialysis (ED), vapour compression desalination (VCD) and particularly reverse osmosis (RO) are being introduced. Reverse osmosis plants can be small and portable and therefore are of particular use in emergencies. According to the method, whether or not the plant is single purpose and whether the water is saline or merely brackish, the production price varies from approximately $1.7 to perhaps as low as $0.7 per m^3. However, these are running costs and capital costs add significantly to them. Therefore, at present, the process is too costly to produce anything other than potable water.

Other innovations include the introduction of hybrid plants, using a mixture of techniques. The first of these was established at Jubail (Saudi Arabia) and produces 15 million gallons per day. Research into effective desalination procedures is paralleled by that into the use of solar energy. For many Middle Eastern countries, of course, energy costs, at present, are of little concern, but, in the longer term, such developments as solar power technology will allow small units to be viable. This would eliminate the strategic risk, at present evident in the Gulf, where several countries rely for their main supplies of drinking water upon one or two major installations (Keen, 1985), see figure 10.1. Thus, with this particular definition, substitution is likely, for a long time ahead, to offer only a very limited solution to the problem of water scarcity.

Recycling can be interpreted in a variety of ways. Already in many irrigation systems in the region there is recycling in that water is reused downslope in one irrigated area after another. A basic problem is that the quality of water declines during each cycle as the concentration of dissolved matter increases. Cloud seeding may also be seen as an aspect of recycling, although for most of the time throughout most of the Middle East the relative humidities are too low for it to be an effective technique.

However, a closer parallel with the case of strategic minerals is provided by the recycling of waste waters, chiefly those from sewage disposal. While potable water could be produced, it seems likely that the use of recycled water will be limited to irrigation, particularly of amenities. However, this will release other water for social uses, including drinking. At present, few of the large cities in the region have adequate sewage treatment systems, but increasing attention is being paid to recycling. Perhaps one of the best known and certainly one of the earliest schemes is the Dan Wastewater Reclamation Project, which was built to treat the sewage from Tel Aviv (Shuval, 1980). In 1987 a very large plant was completed at Al Khobar and further developments are planned in Saudi Arabia. At present, the countries making most use of recycled water are Jordan, Qatar, Kuwait and Saudi Arabia. Although it will always remain limited in scale and possibly in application, recycled water will be an increasingly important source in the Middle East.

Conservation is obviously a major subject in its own right and has been discussed in some detail in the context of water management. It ranges from the conservation of aquifers through licensed and therefore regulated pumping to the elimination of wastage in the infrastructure and during water use. In the Capital Area of Oman, it is estimated that some 30 per cent of the supply is lost through leakages in the infrastructure. Furthermore, as mentioned earlier, the urban population, removed physically and spiritually from concern with daily fluctu-

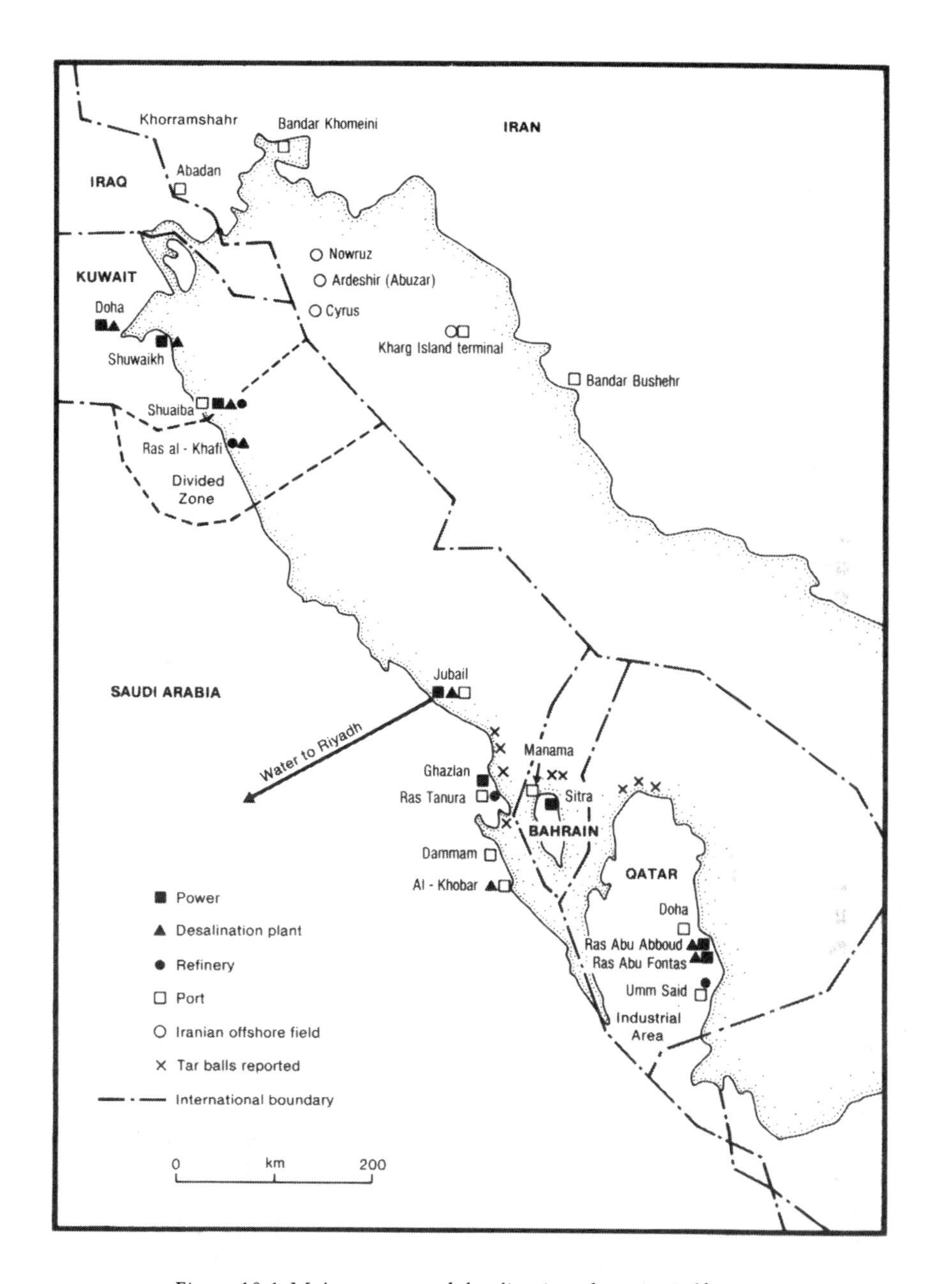

Figure 10.1 Major power and desalination plants in Gulf region.

ations in well levels, tends to be profligate with water. Therefore, it is essential to enact laws which maintain stringent control over urban water use. However, by far the greatest part of the water budget is used in agriculture and great savings can be made through the implementation of improved irrigation practices. These may vary from the covering of storage tanks and the lining of channels to the adoption of trickle rather than inundation techniques. All this requires firm legislation and a high level of public awareness. As yet, throughout most of the Middle East these are lacking.

The possibility of finding alternative sources on any reasonable scale internally is, for most of the Middle Eastern countries, unlikely. The development of hard rock aquifers presents some opportunities and it may be possible to obtain more water from deep aquifers. In contrast to strategic minerals exploration, which is often constrained, as in the United States, by environmental legislation, test drilling for water can take place virtually anywhere in the Middle East. Thus, most countries in the region maintain exploratory drilling programmes. The other internal source which has been seriously examined is dewfall. Although the actual volume of water derived from this source can never be large, it may be sufficient to maintain plant growth.

As a commodity of low value which is required in bulk, water, other than for drinking, could not be imported economically over large distances. Therefore, since the countries in the region with a water surplus are all peripheral, the international transport of water on any scale is likely to be very restricted. There is also the strategic consideration, resulting upon the dependence on a foreign source for such a vital commodity. There would seem to be a strong preference for self-sufficiency at whatever cost. Although pipelines have been mooted from the Nile to Saudi Arabia and from the Euphrates to Kuwait, bulk imports of water by any Middle Eastern country remain small. The only significant trade is in bottled mineral water. The latest project, and by far the most significant, hydrologically and geopolitically, is that of the 'Peace Pipeline', promulgated by Turkey in 1987. The plan achieved great publicity, not only because of its nature and scope, but, in particular, because it was said to have been conceived by the Turkish Prime Minister, Turgut Ozal. It is envisaged that some 6 million m^3 per day would be diverted from the Seyhan and Ceyhan rivers in Anatolia by two routes into the Arabian Peninsula. The western pipeline would supply 3.5 million m^3 per day to Syria, Jordan, Saudi Arabia and local areas of Turkey, while the eastern pipeline would provide 2.5 million m^3 per day to Kuwait, Saudi Arabia, Bahrain, Qatar, the United Arab Emirates and, if required, Oman. The basic idea would be to provide drinking water at a price less than that for desalinated water.

With improvements in desalination technology, it seems doubtful whether the final price of 'Peace Pipeline' water would be competitive. However, a far greater objection concerns the geopolitical question. Apart from securing the agreement of all the states concerned through which the pipeline would pass, there is the strategic vulnerability which would ensue. Even though it would seem to be in the interests of all to maintain the pipelines, in the volatile atmosphere of the

Middle East, dependence on an outside source for such a crucial item as drinking water is virtually unthinkable.

A further possibility discussed has been the import of water in cleaned products tankers. Water could, by this means, be imported at a competitive price, but the question of vulnerability remains. Furthermore, with an average products tanker-load of some 7,000 m^3, a shuttle service would be required and this would certainly cause logistical problems and probably affect the reliability of supply. There are precedents for imports by tanker, but these are on a small scale and have, in most cases, been in response to particular shortages. Finally, as with the 'Peace Pipeline', since only potable water would be concerned, special reception facilities and a separate infrastructure would be required.

The possibility of obtaining water from icebergs has also been discussed. It is considered feasible for icebergs towed to the Persian–Arabian Gulf to provide sufficient drinking water for the project to be viable. However, the question of controlled melting and the necessary links to the drinking water system have not been seriously addressed. Therefore, chiefly as a result of the strategic vulnerability, the use of imported water on any scale would seem to be restricted to emergencies only.

From this analysis, it can be concluded that water, like certain minerals and petroleum, must be considered strategic. Throughout most of the Middle East it is scarce and, with the quickening pace of development, becoming scarcer. As with strategic minerals, research and development can provide some alleviation, but any more general solution is unlikely to be found until well into the future. The most promising form of storage procedure is likely to be through aquifer recharge, but many hydrological aspects of this within the region remain conjectural. Import is of course a possibility, but as with strategic minerals and petroleum, this results in strategic vulnerability. Taking all this into account, there are unlikely to be any large-scale favourable changes in the water position in the Middle East in the foreseeable future. Water will remain a scarce commodity and the water system, both natural and artificial, will remain fragile. Given this situation, the potential for resource geopolitics remains high. Indeed, the position is so volatile that small-scale developments become critical and even the effects of climatic fluctuations can be wrongly attributed. Furthermore, it has been suggested that the effects of global warming in the Middle East will produce temperature rises of around 4 °C (Rind, 1984) and the effects of this can only exacerbate tensions (see chapter 3).

Despite the fact that in the Middle East it is the areas of ephemeral flow which predominate and have the greatest water supply problems, it is the river basins which have provided the focus for geopolitical activity. As with strategic minerals and petroleum, actual cases of proven resource denial are few, but foreign policy essentially consists of many strands, one of which, and perhaps not the most obvious, may comprise resource geopolitics.

THE POTENTIAL FOR WATER GEOPOLITICS

Water in the form of rivers or swamps can obviously be of great importance in boundary disputes, but it is as a resource, with the potential for access or denial, that water is most significant geopolitically. The potential for water geopolitics is also indicated by Chaliand and Ragaeu in their strategic atlas (1985). In a world map for 1980 they identify the major world region with the greatest potable water deficiency as the Arabian Peninsula and Egypt. By 2000, the Levant coastal area, the Maghreb and Pakistan are also included. It is only where supplies are scarce that water can be considered a strategic resource and therefore the discussion on water geopolitics or hydropolitics will focus on the Middle East. Furthermore, the Middle East is the one arid region with internationally shared drainage basins in which hydropolitics has been not only threatened, but actually demonstrated. Even during the Iraq–Kuwait crisis of 1990–1, media focus was far more on water than on the other more obvious geopolitical liquid, oil. Not only have the fundamental problems of water deficiency been addressed, but also the possibilities of deliberate denial (Lawson, 1990).

While instances in which the abundance of water in the form of controlled flooding may have been used geopolitically (Anderson, 1991a), the use of the 'resource weapon' and resource geopolitics generally has been taken to be possible as a result of scarcity. The degree of vulnerability is clearly related to the availability of supplies and since the ultimate determinant is precipitation, regions with an arid climate are those most at risk. Most vulnerable among these, since it has not only very low rainfall over much of its area, but also a burgeoning population, is the Middle East. In particular, countries in which the population increase results from oil revenues and is unrelated to the potential for food production are likely to encounter the greatest problems. These countries are mainly located in the Arabian Peninsula and its immediate vicinity, reinforcing the judgement of Chaliand and Ragaeu (1985).

Throughout the Middle East, other than in a few of the higher mountain districts in which low temperature is important, water availability imposes the major constraint upon agricultural development. Annual precipitation totals range from well below 100 mm in the Sahara and Arabian Deserts to 200 mm around the desert fringes and up to 500 mm in the higher areas. The 400 mm isohyet is generally taken as that limiting cultivation, while areas with less than 250 mm are considered only suitable for rough grazing. Thus, only in limited peripheral areas is water not a scarce resource. Furthermore, total annual rainfall is not the best indicator of water availability. Since in hot climates, a proportion of the precipitation is lost through evapotranspiration, the surplus which remains, the sum of the monthly differences between precipitation and potential evapotranspiration, provides a more accurate prediction for water availability (Beaumont *et al.*, 1988). Areas of water surplus are confined almost exclusively to the northern highlands of Turkey and Iran, so that when effective rainfall is considered, the area of water scarcity is even greater than that indicated by absolute rainfall totals. The situation is exacerbated if the annual distribution is taken into account. Water surplus normally occurs within a very limited period of the year, although, as a result of recharge, its effective period will be somewhat extended. As the annual total falls, so the period of surplus declines. However, in most parts of the Middle East, particularly towards the south, the distribution itself is irregular, so that the most influential characteristic of rainfall is its annual variability. The lower the total, the less reliable the rainfall becomes. Thus, several years of drought may be followed by the incidence of widespread flooding (Anderson, 1986). This point can be illustrated by a transect, approximately north–south, across the region (Blake *et al.* 1987).

Table 10.1 illustrates clearly the southward decline of water availability. Not only is there a marked diminution in the mean annual total, but also a sharp increase in variability (see chapter 2).

In a region where, with regard to water availability, much of the cultivated area can be regarded as marginal, rainfall variability exercises a profound influence. The position is further complicated by the fact that annual variability appears to be superimposed upon cyclical oscillations. For example, the advance and retreat of agriculture around the scarp foot villages of the Northern Oman Mountains can be related directly to the occurrence of wet and dry periods over the past 50 years. Thus, not only is there scarcity throughout most of the Middle East, but the rainfall total may range disastrously from year to year or over a period of years, threatening cultivation and the very existence of settlements. One solution is to develop a reliable water supply by digging deeper wells locally or by importing water. The reliance upon precipitation is thereby replaced by that upon either deep aquifers or transport links. Both involve a range of possible difficulties. Furthermore, with reliance upon irrigation, not only is water consumption likely to rise, but supply interruptions can be catastrophic.

It can be concluded that precipitation, the ultimate source of water, is throughout most of the region both low in annual total and unpredictable. As a result, recharge rates to shallow aquifers are modest. Nonetheless, until this century and, in particular, the period since the Second World War, a basic balance was achieved between water, agriculture and population. The agricultural system which had evolved appeared able, even in dry periods, to produce sufficient food to safeguard the population against extreme shortages and famine. Other than for irrigation, the demand for water was, at this time, very limited.

Table 10.1 Annual precipitation (mm)

Station	*Mean annual total*	*Maximum annual total*	*Minimum annual total*
Rice	2,440	4,045	1,758
Konya	316	501	144
Baghdad	151	336	72
Cairo	22	64	1.5

However, the situation has been transformed by the extremely high rates of population growth throughout most of the Middle East. From a figure estimated to have been between 40 and 60 million at the turn of the century, the population had risen to 150 million by 1950, 270 million by 1980. The present population is estimated to be 327 million (World Population Data Sheet, 1988) and for the years 2000 and 2020, the predictions are respectively 450 million and 700 million. In the 12 years remaining of this century therefore, it is considered that the population of the Middle East will more than double. Such increases must impose intolerable pressures upon food and water supplies.

Furthermore, in many of the countries, the oil boom has resulted in accelerated industrialization and the rapid development of elaborate services infrastructures. This in turn has facilitated urbanization at an unprecedented rate, so that in certain countries the urbanized population now exceeds 75 per cent of the total. The result of all this has been increased demand for water, particularly with the creation of an urban middle class. Whereas domestic water consumption in a village is probably less than 20 litres per capita, in the cities, the average consumption is between 10 and 12 times as great. A further factor affecting water supplies is pollution. In urban areas, the population increase has commonly exceeded by far the capacity to dispose of waste of all types and, with no properly developed treatment systems, pollution of aquifers is always a danger. Also, in an effort to satisfy the demand for food, more intensive agricultural techniques have been introduced with the result that excess fertilisers and sprays constitute a further source of aquifer pollution.

It can be seen that with the accelerating use of resources, water is not only naturally scarce, but becoming markedly scarcer. The complete water system, both natural and artificial, is, throughout most of the Middle East, extremely fragile. While, globally, water management is seen as an increasingly significant element of national planning, in much of the Middle East it is becoming vital. One aspect of particular relevance is likely to be water allocation. There must be a case for transferring water from comparatively low-value uses such as agriculture to high-value industrial uses. However, the obvious concomitant of this is an increasing reliance upon offshore sources for food supplies. Thus, for most of the Middle East, the provision of water supplies must impinge upon most aspects of government planning. Throughout the Middle East, all governments are now giving a high priority to water policy, investment in water exploration, the construction of barrages and the development of alternative supplies (Anderson, 1988b and c).

Whatever is achieved through national water management programmes, water will remain for most countries of the Middle East a scarce resource. Further, for those countries relying upon deep aquifers, it is likely to be also a declining asset. Therefore, for most of the region, water geopolitics, at any scale from local to national, will always remain a possibility. More specific geopolitical potential can be identified if the various sources of water are examined.

WATER RESOURCES AND GEOPOLITICS

Throughout a large part of the Middle East and effectively all of the Arabian Peninsula, there is no permanent surface flow. Runoff, when it occurs, tends to be either very localised or of extremely limited duration. Indeed, with the exception of the Nile and its tributaries, all of the perennial rivers of the Middle East are to the north of latitude 30°N. Even beyond that line, in such countries as Algeria, Tunisia, Libya, Jordan, Syria, Iraq and Iran, considerable areas only experience ephemeral flow. Furthermore, the Nile itself receives no surface additions in Northern Sudan or Egypt and is largely dependent for its discharge upon precipitation which falls outside the region.

Within the region, but north of 30°N, surface flow predominates throughout Turkey, in most of Iran and on the seaward side of the Atlas Mountains. Additionally, the Lebanon, as the source of the Litani and the Orontes, is comparatively well watered. However, by far the most hydrologically significant system is that of the Tigris–Euphrates, while the most strategically important is that of the Jordan. Therefore, in the Middle East as a whole, apart from the Nile and the restricted flows of the Levant, surface water is peripheral. The possibility of making good shortages by imports from neighbouring countries is therefore very restricted. With comparatively few permanent rivers of any size within the region, opportunities for barrage construction are very restricted. In practice, the majority of the perennial rivers have one dam and multiple damming is increasingly common. In the case of international rivers, such constructions not only introduce competition for water, but present opportunities for resource geopolitics. All dams, but in particular the large multi-purpose barrages, exercise an influence on development downstream, since they affect both the quantity and quality of the water available. For example, in the south–east Anatolian project, the largest of its kind in the world, water will be required to irrigate some 727,000 hectares (Mitchell, 1986). The completion of the Ataturk Dam in 1991 which will facilitate this and particularly the period of lake filling will exercise a profound influence on the various irrigation projects in both Syria and Iraq. Altogether at present there are at least 35 major dam schemes projected for the permanent rivers of the region.

For the greater part of the region in which ephemeral flow predominates, the key problem is that of flood water utilization. This is being most commonly approached through the construction of dams on ephemeral rivers, in a few places to form reservoirs, but in most to check water loss and enhance recharge. For example, the Saudi Arabian programme included the provision of some 60 such dams, of varying sizes, by 1990. In Oman, two major recharge dams have been constructed on the Batinah and feasibility studies are being made for several installations on Salalah Plain. However, for all this construction, a major handicap is the general lack of data, particularly on discharge.

Dam construction on permanent rivers can provide the opportunity for water denial and can therefore lead to direct geopolitical action. In contrast, the problems imposed by rainfall variability and lack of discharge data mean that water management must be difficult and imprecise. Programmes throughout the region to install recharge dams have so far done little to alleviate water scarcity. Consequently, geopolitical activity is more likely to arise directly, as a result of this scarcity, rather than through any aspect of water denial.

The accent upon recharge projects throughout large parts of the Middle East illustrates the importance which has, for thousands of years, been attached to groundwater. Two major categories of aquifer, each with its own particular problems, have been distinguished (Beaumont, 1981): the shallow and the deep. Shallow alluvial aquifers are normally fairly limited in extent, but are found throughout the region. They are located in river and wadi bed sediments and beneath alluvial fans and plains. Other than when such features as calcretes interpose, they are generally unconfined and therefore the height of the water table within them responds rapidly to changing meteorological conditions.

Shallow aquifers provide by far the more important groundwater source throughout the region, since water may be obtained, in many cases, from hand-dug wells, or directly from springs or seepages. For deeper wells, drilling may be required, together with a variety of lifting devices, ranging from the most primitive to modern diesel pumps.

A good example of a problem of geopolitical significance on a local scale arises when wells are sunk to tap the same aquifer as that which supplies a spring or qanat. Competition leads to depletion, with the result that the well may be dug deeper, leaving the spring or qanat completely dry. However, there is a greater issue at stake in that two lifestyles are brought in direct opposition. The qanat is owned or controlled by a complete settlement and requires cooperative activity of a high order for the equitable distribution of the water. In contrast, a well is likely to be owned by one family and water can be drawn as required. One lifestyle fosters cooperation, the other competition. The meeting of the two can produce not only economic, but also social and political disruption. The potential for resource geopolitics is, of course, more evident on a larger scale. Aquifers are not confined by national boundaries and, especially in the case of shallow aquifers with a limited capacity, extraction on one side of the boundary can affect water availability on the other. In general, however, knowledge of aquifers and particularly recharge mechanisms is so limited that a direct cause and effect would be difficult to establish.

The major deep rock aquifers specific to the Arabian Peninsula and North Africa are, as yet, little understood. They are on a totally different scale from the shallow aquifers, extending over very large areas, although the actual outcrops themselves may be very restricted. If there is only a limited or no clearly related outcrop, rates of recharge may vary from negligible to non-existent. As a result, the aquifers may include water which can, in places, be many thousands of years old. They may also, due to the environment, be brackish, requiring some measure of desalting.

The major hydrological problem concerning deep aquifers, and one critical to water management, is whether or not the source is renewable. Abstraction, particularly of the older, deeper water, can be designated mining, since there seems little chance of replenishment. The potential geopolitical problem arises from the fact that aquifer boundaries, such as they are known, rarely coincide with political boundaries. If the volume of available water is finite, extraction by one country depletes the resource base of others which share the aquifer.

A spectacular example of both these problems is provided by the Great Man-Made River Project in Libya, which is intended to irrigate some 18,000 hectares, at an annual cost of over $500 million (figure 10.2). When complete, the scheme will transport groundwater from Kufra in the southern Fezzan to the coast and will facilitate the development, both agricultural and industrial, around most of the Gulf of Sirte (Odone, 1984). If, as seems virtually certain, there is no recharge

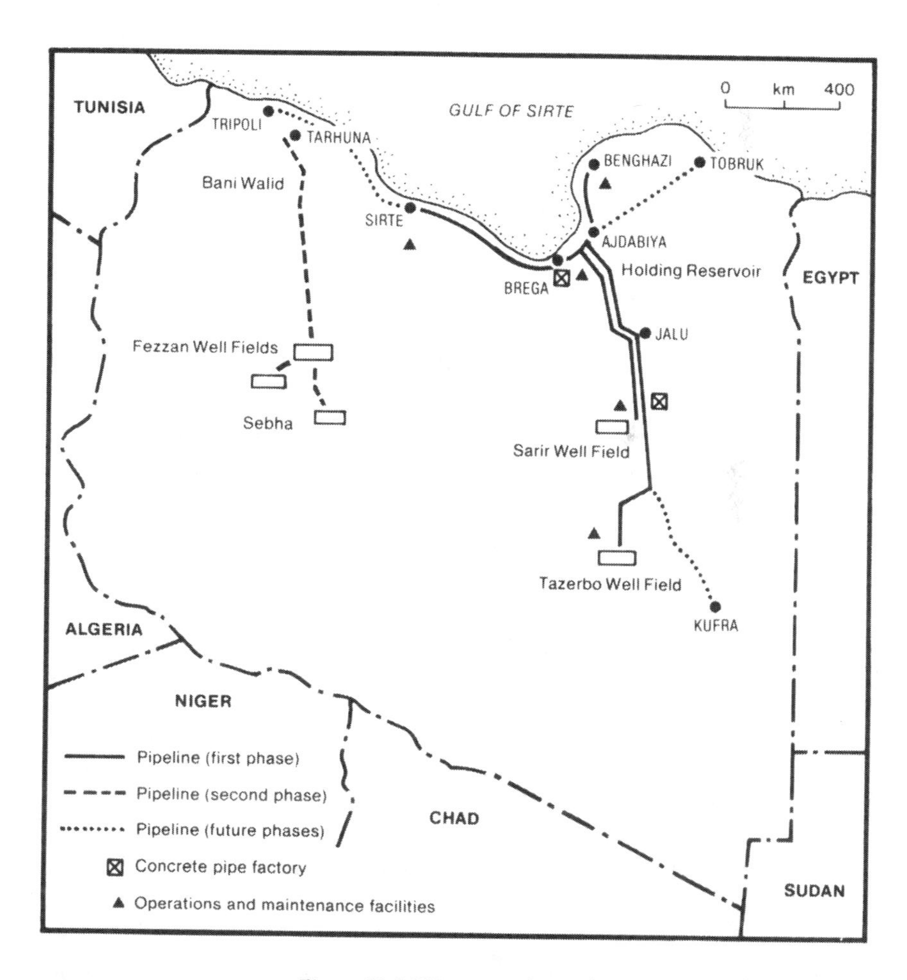

Figure 10.2 Water transfers, Libya.

to the aquifer, the life of the project must be considered medium term only. Furthermore, the aquifer itself extends well under Egyptian territory and, as that state struggles to meet increasing water requirements, there is an obvious conflict of interests.

As deep aquifers may only allow a medium-term solution and shallow aquifers are everywhere in the Middle East under threat of depletion or even exhaustion, attention has focused upon hard rock aquifers. Although, potentially, these suffer from the same problems, hydrological and geopolitical, as deep aquifers, they do offer possibilities for future development (Hoag and Ingari, 1985).

At present, the greatest potential for a border incident over groundwater resources occurs in Oman, where aquifers to the east of Buraimi are particularly high yielding. During the 1980s, overpumping in Al Ayn (United Arab Emirates) has resulted in a rapid decline of the water table, by as much as 5 m per year beneath Buraimi. Since Buraimi is the centre of an Omani development area, this occurrence has caused obvious concern. Should Israel be forced to withdraw from the West Bank, hydropolitical activity in the context of aquifers could become a reality.

INTERNATIONAL RIVER BASIN: LEGAL ASPECTS

Law, with regard to international river basins, is enacted in the light of four alternative theories, which govern the rights and obligations of co-riparian nations.

1 **The Theory of Absolute Territorial Sovereignty** is the oldest, having been defined in 1895, in connection with the Mexico/United States dispute over the utilization of waters from the Rio Grande. The application of this theory works entirely in the interests of upstream states:

> states in possession of the upper waters of a river have not recognized any general obligation to refrain from diverting its waters and thereby denying to the states in possession of the lower waters the benefits of its full flow.
>
> (Berber, 1959)

2 **The Theory of Absolute Territorial Integrity** has, in practice, the opposite effect:

> a state is not only forbidden to stop or divert the flow of a river which runs from its own to a neighbouring state, but otherwise to make such use of the water of the river as neither causes danger to the neighbouring state or prevents it from making proper use of the flow of the river on its part.
>
> (Utton, 1973)

3 **The Theory of Community or Integrated Drainage Basin Development,**

adopted by the International Law Association in Dubrovnik (1956) and New York (1958), is regarded as an agreed principle of international law pertaining to the utilization of international river basins (Laylin and Bianchi, 1959):

> Riparian states should join with each other to make full utilization of the waters of a river ... so as to assure the greatest benefit to all.

This theory thus seems more genuinely equitable as the rights of all riparians are taken into account.

4 **The Theory of Limited Territorial Sovereignty or Equitable Utilization.** This theory, which stresses that a state may make use of international waters to the extent that it does not interfere with their reasonable use by other riparians is the most commonly and widely advocated by the international community. It was developed by the International Law Association more broadly into the so-called 'Helsinki Rules' (1966), which are considered the definitive expression of international law, pertaining to the management of international river basins (Utton, 1973).

When examined chronologically, it can be seen that international law has been developed in favour of the community and equitable utilization theories. However, in an attempt to produce a more widely acceptable and applicable set of principles, the United Nations International Law Commission, in accordance with General Assembly Resolution 2669 (XXV), began a study which is still in progress on the law with regard to the non-navigational uses of international water courses.

However, there is at present no generally accepted principle and agreements must depend upon the mutual goodwill of co-riparians in any particular drainage basin. On the other hand, it is clear from a perusal of the theories that the likelihood of conflict or cooperation depends very much upon a number of geopolitical factors. Naff and Matson (1984) distinguish three key variables:

1 the relative positions of the co-riparians within the drainage basin;
2 the degree of national interest in the problem; and
3 the power available externally and internally to pursue policies.

Thus, the upstream state is in the strongest position, provided it has sufficient interest in the problem and power to enforce its wishes. For example, in the case of Turkey in the Tigris–Euphrates basin, it has the key position and also the military power, but, with many water sources available, may lack total commitment. In contrast, Iraq has the worst position, but is reasonably powerful and is desperately interested in the outcome of any waters agreement. In the case of the Nile, at present power and interest appear to increase downstream.

A more comprehensive approach, taking into account possible cooperation as well as conflict potential, was designed by Nizim in 1968 and later broadened by Al-Himyari (1985). In their model, factors including hydrological, climatic,

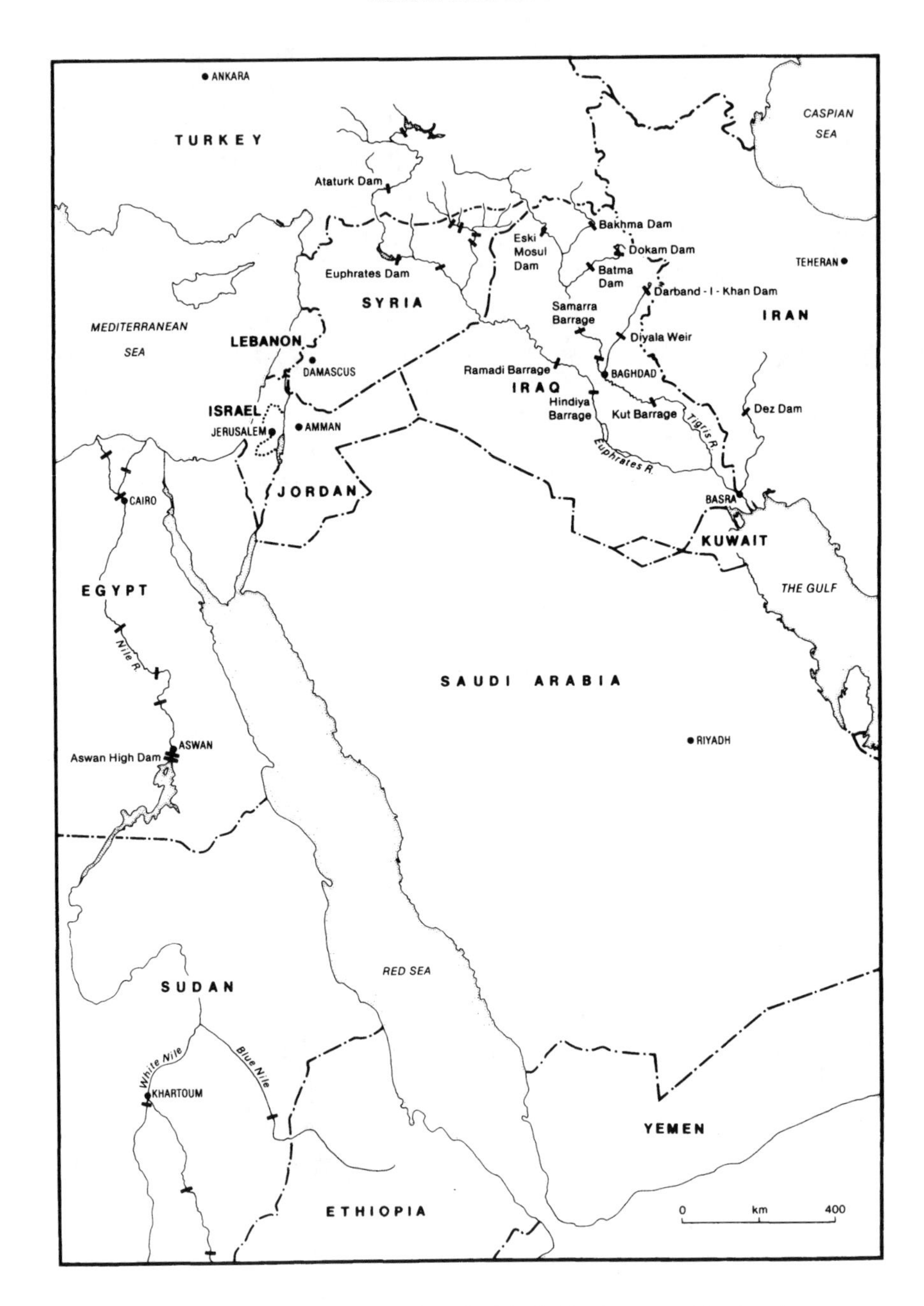

Figure 10.3 Catchments of the Nile, Tigris–Euphrates and Jordan rivers.

developmental, demographic, cultural, political and ideological were rated on scales according to their conflict potential and their possibilities for cooperation.

Peaceful cooperation has been achieved through international agreement in several international river basins (United Nations Legislative Series, undated). A good example in an arid context is that between the United States and Mexico over the utilization of the Rio Grande. There have been two treaties, the first signed in 1906 and the existing in 1944. This treaty, operated by the International Boundary and Water Commission, oversees the allocations of the average annual flow equally between the two states. In the Middle East, the Soviet Union and Turkey signed a convention on the use of frontier waters and a protocol on the Araxe River in particular, in Kars, on 8 January 1927. These, both of which came into force on 26 June 1928, provide that:

> the two Contracting Parties shall have the use of one half of the water from the rivers, streams and springs which coincide with the frontier line between the Turkish Republic and the Union of Soviet Socialist Republics.

THREE KEY CATCHMENTS

In the more arid part of the Middle East there are three shared basins of significance, each of which exhibits potential for conflict over water (figure 10.3). The Nile catchment is shared by a number of countries, but only Ethiopia, Sudan and Egypt are in positions of influence. Of the three Egypt is the country with the most obvious water crisis. Less stable in political relationships, since there has been no agreement, is the basin of the Tigris–Euphrates. The co-riparians, Turkey, Syria and Iraq, are all undergoing rapid development, but Iraq is the most at risk from water shortages. The third basin, that of the Jordan, is by far the smallest, but also the one in which resource geopolitics has already been most significant. The basin includes not only the boundaries between Israel, Jordan, Lebanon and Syria, but also the major frontier between the Arab world and Israel.

The Nile basin

In draining 10 per cent of Africa, the Nile (figure 10.4) passes through more different climatic regions and therefore has a more complex hydrological regime than any other river. It consists essentially of the White Nile and two major tributaries, the Blue Nile and the Atbara, each with different hydrological characteristics. From its confluence with the Atbara to the Mediterranean, some 1,800 km, there are no perennial tributaries and therefore Egypt contributes virtually nothing to the flow.

Egypt, with a fast growing population of, at present, some 60 million, virtually all settled in the Nile valley, is witnessing increasingly heavy demands for water. Water requirements underlie every facet of life and therefore there are

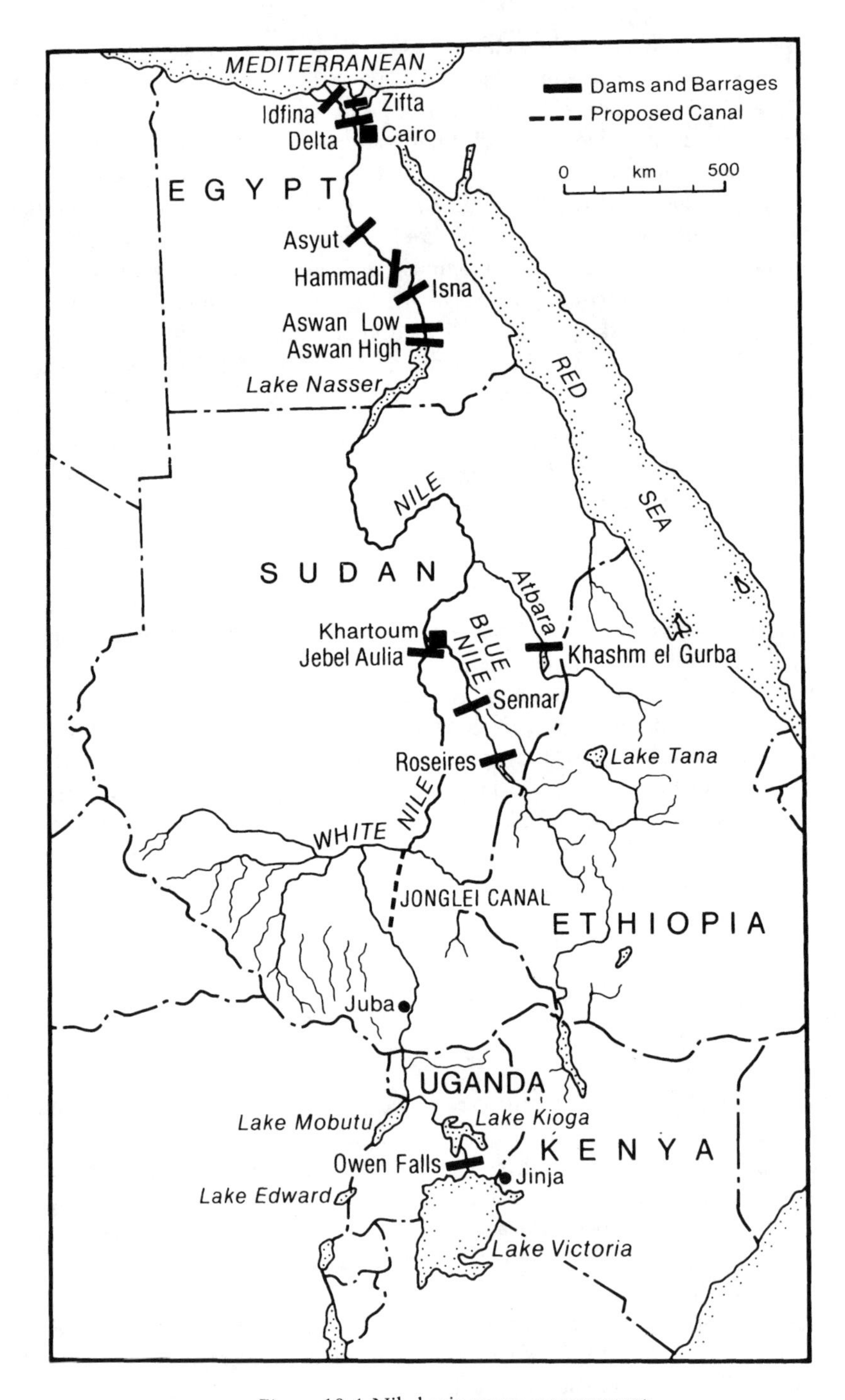

Figure 10.4 Nile basin water management.

obvious potential geopolitical implications. Thus, it is hardly surprising that estimates for water supply and demand vary wildly. Depending upon the extent of agricultural expansion, the development of conservation projects and the increase in recycling, estimates of the total water balance vary from a very small surplus to a deficit of 4,000 million m^3 annually by 1990 (Waterbury, 1979).

Reliable statistics are in even shorter supply for the Sudan and deficits as high as 14,000 million m^3 per year have been forecast for the 1990s. However, these result from a number of grandiose agricultural schemes, few of which are likely to come to fruition. Thus, in the comparatively near future, it seems reasonable to suppose that both major riparians of the Nile will be experiencing water shortage.

The first extensive Nile Water Agreement between the two was signed in 1929, with Egypt receiving 48,000 million m^3 and Sudan 4,000 million m^3, the remaining third of the discharge passing, unused, to the Mediterranean. As a result of the Aswan High Dam project, tension increased between the two countries in the 1950s, until a position of military confrontation was reached in 1958. One result of this was that Sudan, in disregard of the 1929 agreement, raised the height of the Sennar Dam. However, the Agreement for the Full Utilization of the Nile Waters was signed in 1959 and remains in operation. The 1929 apportionment was ratified but equity was restored by the distribution of water from the Aswan Dam. Of the 22,000 million m^3 net storage, 7,500 was allocated to Egypt and 14,500 to Sudan. It was further agreed that any increase in available water as from for example the Jonglei Canal Scheme (4,700 million m^3) should be shared equally and the two states established a permanent Joint Technical Committee. Finally, and of increasing significance, Egypt and Sudan decided to adopt a unified view on claims from other riparian states (Kirshna, 1986). With developments in the other riparian states, the geopolitics of Nile water may alter considerably and Egypt's long-term foreign policy interests have been directed accordingly towards Ethiopia, Uganda and Zaire.

Recent geopolitical concerns have centred on the relationship between Egypt and the upstream states, in particular, Ethiopia (Anderson, 1991b). In 1979, after President Sadat of Egypt had stated his intention to supply water from the Nile to Southern Israel, President Mengistu of Ethiopia retaliated, announcing that he would develop a number of separate projects on the Blue Nile, reducing its flow by perhaps as much as 4,000 million m^3 per year. The response from Sadat was that any such unilateral action would be opposed by force.

Subsequently, relations between the two countries improved, but, as the main source of Nile discharge, Ethiopia undoubtedly holds a key position in the catchment. Indeed, it has used this advantage to negotiate with certain Arab nations it has accused of supporting the Eritrean breakaway movement. There is, however, some evidence that Israeli engineers have begun preparations for building three dams in the basin of the River Abay, a tributary of the Nile.

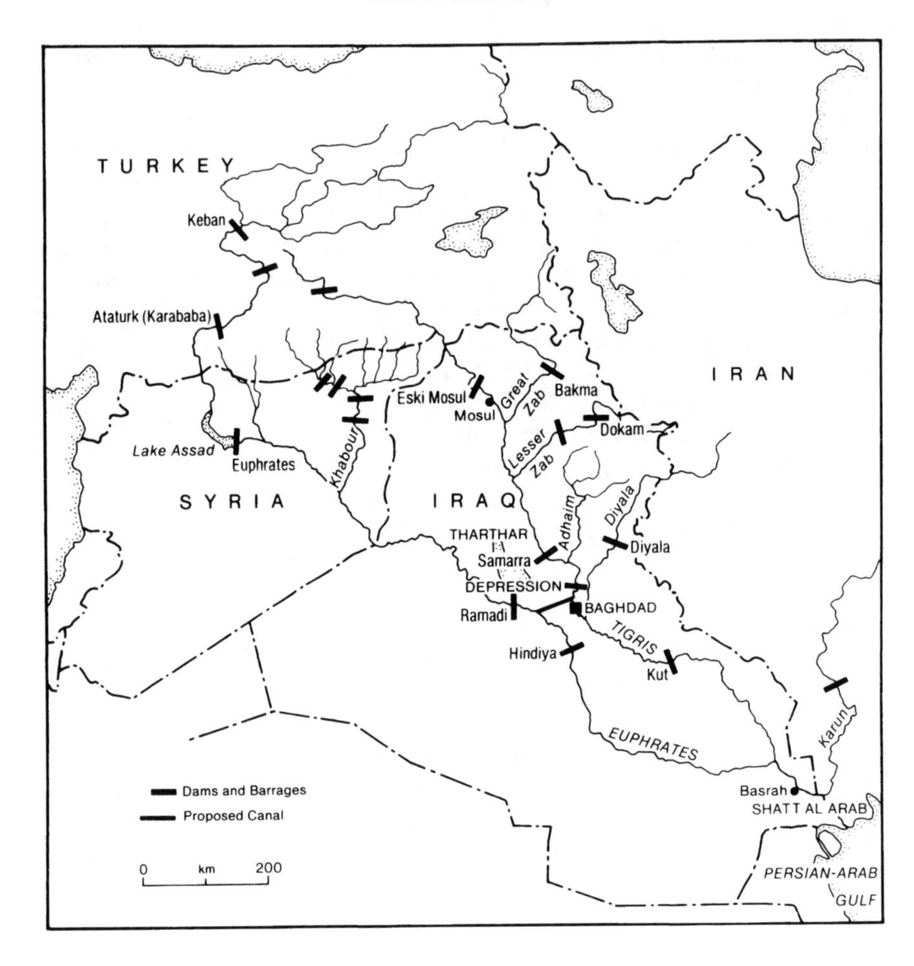

Figure 10.5 Tigris–Euphrates water management.

The Tigris–Euphrates basin

This is the only basin (figure 10.5) among the three with a marked surplus of water, but, bearing in mind the scale of current developments, Iraq may experience shortages during the 1990s. The Euphrates is divided between Turkey, Syria and Iraq, with a minor contribution from Saudi Arabia. The mean annual flow is effectively generated within Turkey (89 per cent) and Syria (11 per cent). The Tigris is smaller in area, has been, as yet, less utilized and receives several important Left Bank tributaries from Iran. Of the annual water potential, Turkey provides 51 per cent, Iraq 39 per cent and Iran 10 per cent. As a result, while the annual discharge of the Euphrates is approximately 35,000 million m^3, that of the Tigris is 49,000 million m^3.

The area of modern Iraq has been concerned with irrigation since antiquity, and it was in this region that the first modern dam, the Hindiya Barrage, was completed in 1913. Today, Iraq is considered to be the one Middle Eastern country self-sufficient in agriculture, based on irrigation (Allan, 1985). Syria is also a developing economy, dependent to a large extent on agriculture. Its chief rivers are the Euphrates and the Orontes and on the former, in particular, there are several major projects, either under construction or planned. Turkey began large-scale irrigation comparatively recently but its plans for the Anatolian Project (GAP) are the most ambitious in the Middle East. Geopolitical problems are likely to arise not only from a decline in water quality and quantity once the schemes are all operational, but also during the late filling phase. Indeed, the only documented crisis over water in the basin occurred between Syria and Iraq in 1974 during just such a phase.

However, on 13 January 1990, the diversion channel beneath the Ataturk Dam was closed allowing the lake above the dam to commence filling. The flow of the Euphrates was greatly reduced as a result and despite the fact that the governments of Syria and Iraq had both been warned in advance, and the discharge had been enhanced in compensation, prior tothe cut-off, immediate alarm was expressed over water security. Although this event was important chiefly as an example of geopolitical symbolism, it was influential in strategic thinking on resource denial, during the later Iraq–Kuwait crisis (Anderson, 1991a).

There are, as yet, only a few limited agreements between the co-riparians. The Protocol of Constantinople (1913), between Turkey and Iran, provided for an equal sharing of the frontier streams; it has since been violated several times by Iran. The Franco-British Convention (1920) dealt with water allocation between their respective mandates, Syria and Iraq. The Treaty of Lausanne (1932) was the first to address the total Tigris–Euphrates system as an international river, shared between Turkey, Syria and Iraq. However, the most important settlement so far was the Friendship Treaty (1946) between Turkey and Iraq, the key section of which states that:

> Turkey shall keep Iraq informed of her plans for the construction of conservation works on either of the two rivers or their tributaries, in order that these works may as far as possible be adapted, by common agreement to the interests of both Iraq and Turkey.
>
> (Berber, 1959)

These various agreements, therefore, were all comparatively imprecise and none specified actual quantities of water. However, in 1984, Syria called for the establishment of a multinational Euphrates River Authority and for a joint meeting to discuss riparian rights (Allan, 1985). With the various tensions including two wars in the region in the last decade, it seems more than ever imperative that some agreement between the co-riparians in the Tigris–Euphrates is reached.

The Jordan basin

The Jordan basin (figure 10.6) comprises only 11,500 km^2 and the proportion of the area of each riparian state within the basin is: Jordan 54 per cent, Syria 29.5 per cent, Israel 10.5 per cent and Lebanon 6 per cent. However, only 3 per cent of the area of the Jordan basin lies within the pre-1967 boundaries of Israel. The Jordan has three headwaters: the Hasbani, rising in Syria and Lebanon; the Banyas, with springs in Syria; and the Dan, wholly within Israel. Downstream, the Jordan receives virtually no perennial inflows other than the Yarmouk until it reaches the Dead Sea with a flow equivalent to about 2 per cent of the annual flow of the Nile. Thus, water in the basin is extremely scarce and three of the co-riparians, Israel, Jordan and Syria, are all facing water shortages. In the case of Israel and Jordan, consumption is virtually equal to potential supply. It is therefore a zero-sum position in which gains by one side must result in equal losses by the other. Furthermore, by the end of the century, Israel's population will have increased from about 4 million to 5 million, Jordan's from 2.7 million to 4 million and that of the West Bank from 0.8 million to 1 million (World Population Data Sheet, 1988).

For such a small basin as that of the Jordan, the geopolitical situation is complicated. The major tributary, the Yarmouk, provides the boundary successively between Jordan and Syria and Jordan and Israel. The River Jordan itself forms the boundary between Israel and Jordan and, to the south, the West Bank and Jordan. Thus, in the Jordan valley, Syria and Lebanon are upstream of Israel and Israel is upstream of Jordan. However, with the southern third of Lebanon in Israeli hands, the position is, at present, distorted. Along the Yarmouk, Syria is upstream to Jordan and Jordan to Israel. The two most critically interested states, from the point of view of water supply, are Israel and Jordan and Israel is without doubt the major power in the region.

The history of interstate relations within the Jordan basin reflects the importance of water, although whether the so-called Israeli hydraulic imperative can be substantiated is largely a matter of opinion. In most of the conflicts, major and minor, there were a number of objectives, one of which could well have been the securing of water supplies. Since 1951 and the draining of the Huleh Marshes there has been a continuing series of water-related or pseudo-water-related incidents. Eleven such incidents have been documented in the period 1951 to 1967, three of them of a particularly serious nature. In 1953, Israel began constructions to divert water from the Jordan above Lake Kinneret at Jisr Banat Yaqub. There were major objections from Syria, supported by the United Nations. Following pressure from the United States, work was halted and the National Water Carrier extraction point was moved to Eshed Kinrot on the lake itself. Then followed the Bunger Plan, comprising the construction of the Maqarin and Addisyah dams and the East Ghor Canal. Israel objected strongly to the Maqarin dam, insisting that its rights on the Yarmouk were at stake.

Finally, potentially the most dangerous of all the incidents, and one which was

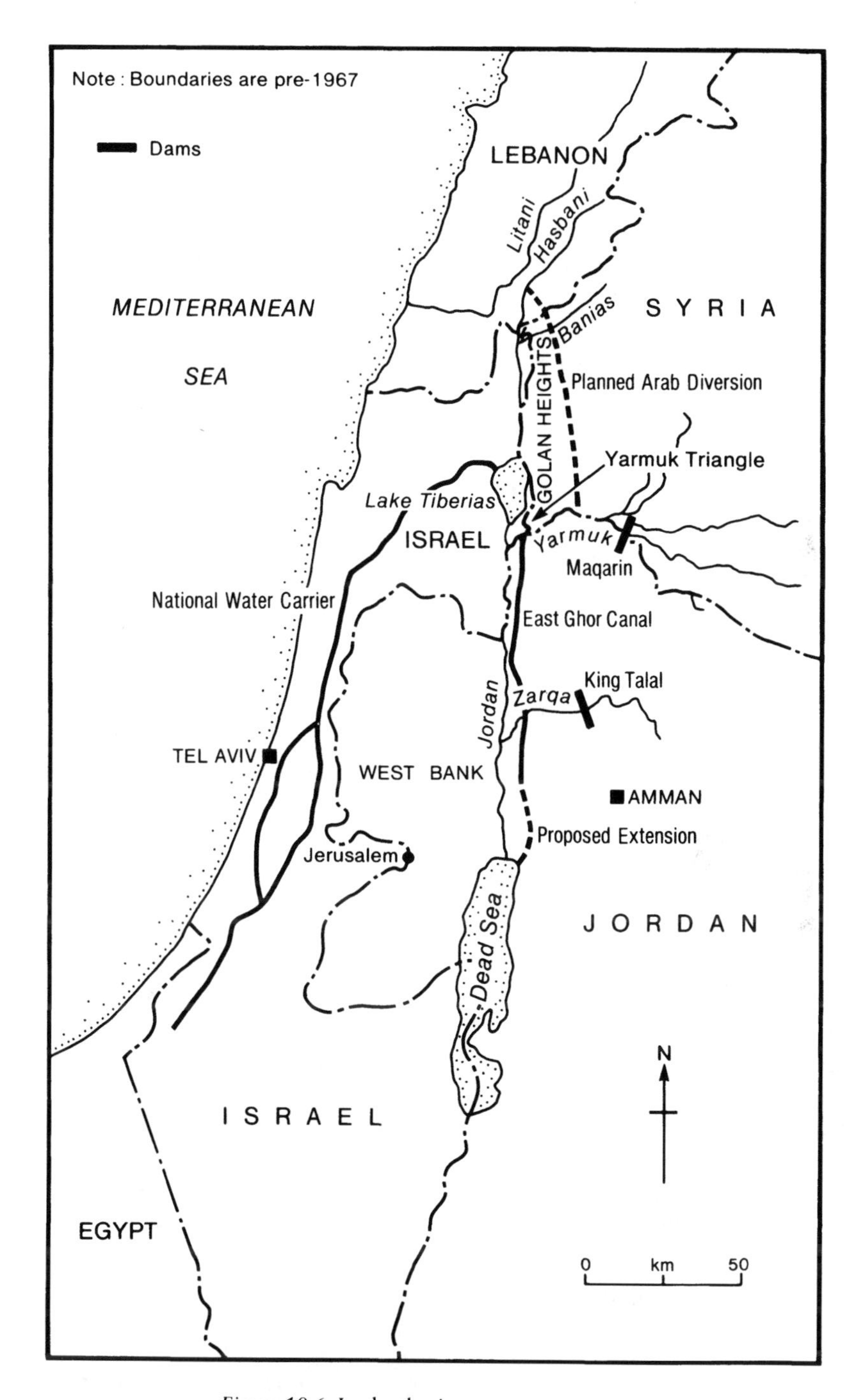

Figure 10.6 Jordan basin water management.

undeniably geopolitical, resulted from the resolution of the 1964 Arab Summit, to divert the headwaters of the Jordan. The scheme was either to divert the Hasbani to the Litani, or the Banyas to the Yarmouk, or both to the Yarmouk, the preferred scheme. The cost was estimated to be between $190 to $200 million, which was underwritten by Saudi Arabia and Egypt, and work started in 1965. Israel calculated that this would remove approximately 50 per cent of the intake from the National Water Carrier and would reduce the national budget by over 10 per cent. A series of military strikes followed and these continued until the 1967 war.

Following the war, the Palestine Liberation Organization made a series of attacks on Jordan pumping stations in 1968 and 1969 and, in response, in the summer of 1969, Israel attacked the East Ghor Canal. However, relations improved in 1970, particularly following the expulsion of the Palestine Liberation Organization from Jordan.

Nonetheless, the area remains tense, particularly in view of the Palestinian issue, and the water situation can only deteriorate. Furthermore, speculation on possible pipelines from the Nile, the Euphrates and Turkey, has interconnected all the surface water problems of the region. Also, the symbolic nature of water in countries such as Israel and Jordan, in which water is considered such a 'primary need', must be taken into account since in its defence, drastic action is justified. The Zionist ideology has always involved close association with the land and 'making the desert bloom'. Jordanian perceptions are similar, but piquancy is added by the fact that Israel, as a more urban state, has a far higher water consumption per capita and there is a strong Jordanian distrust of Israeli water statistics.

A further key factor in the resource geopolitics of the basin is that while Israel obtains only 40 per cent of its water from surface sources, the figure for Jordan is 75 per cent. For Israel, the West Bank aquifers are vital as the source for some 80 per cent of its groundwater. There are three major aquifers in this region and prior to 1967 Israel was exploiting two of them almost to the maximum by pumping from within its own borders. After 1967, control of the West Bank allowed access to the eastern aquifer, with an estimated yield of 66 million m^3 annually. However, there is a major disagreement over West Bank water statistics. Israel officially considers the area to be self-sufficient, while Jordan has calculated that there is a large surplus for use within Israel itself. Undoubtedly, the growth of Jewish settlements has increased water requirements and Arab agricultural development has been, as a result, constrained. In some cases deep drilling has resulted in the desiccation of Arab springs and wells and this has again been a source of tension. Although the data cannot be accurately checked, it has been stated by some authorities that the increase in water consumption by Israel since 1967 has only been possible because of its territorial expansion.

Three areas of key geopolitical significance can be identified. Occupation of the Golan Heights has not only allowed Israel to construct six new reservoirs, but also facilitated protection of the Jordan headwaters and control of half the

Yarmouk. The occupation of the West Bank has provided additional water sources, which are now crucial to the water budget of Israel. The virtual annexation of southern Lebanon has not only brought the Hasbani totally under Israeli control, and with it additional water, but has reopened the possibility of diverting the Litani. However, to obtain economical amounts of water from the Litani it would be necessary to disrupt upstream installations funded by the United States. Thus, although surveys appear to have been completed, political problems will probably militate against such action.

The Jordan basin is clearly well suited to integrated development, but all the schemes proposed so far have failed, as a result of the extreme enmity and distrust between the Arabs and Israel. Since the Ionides Survey (1939) at least 17 major plans have been put forward, but the most significant was that proposed by Eric Johnston, appointed as Special Ambassador by President Eisenhower (1955). The Johnston or Unified Plan was generally accepted, although not ultimately ratified by the Arab League Council. Nonetheless, the proposed percentage allocation of water (Jordan 52, Israel 36, Syria 9 and Lebanon 3) has been retained by all states as a guideline.

Thus, while there has been some very broad agreement, the lack of data has made verification impossible, with the result that there is very little trust among the co-riparians. In such a basin, with limited and fixed supplies, but increasing demand, there is great scope for resource geopolitics. Indeed, the Jordan basin remains the most likely flashpoint in the Middle East.

CONCLUSION

While the number of incidents directly attributable to water concerns is small and restricted to areas of surface flow, particularly the basin of the Jordan, there is potential for denial elsewhere in the Middle East. In areas of ephemeral flow, with increasing dependence upon desalination and a fragile infrastructure, the destruction of facilities would bring life to a halt in a very short time. All key facilities, the sources, the treatment plants, the storage facilities, the main structures and the power systems are vulnerable. There are, in many countries, vital points such as desalination plants or pumping stations on which the whole system depends. The danger of supply disruption was illustrated in 1983 when a threat to installations was posed by a major oil slick (figure 10.1). Saudi Arabia, Kuwait, Bahrain and Qatar all erected booms or similar structures to protect the intake pipes of their desalination plants. In Kuwait, it was estimated that should the Shuaiba plant be damaged, there would be a loss of 44 million gallons of water per day, making rationing a necessity.

Throughout most of the Middle East, water is scarce, demand is increasing rapidly and the possibility of obtaining additional supplies is very limited. As water must therefore be considered strategic any planning to safeguard resources against geopolitical activity must parallel that adopted in the cases of strategic minerals and petroleum (Anderson, 1991a, b and c).

11

THE MANAGEMENT OF WATER RESOURCES

THE HYDROLOGICAL SYSTEM

The distorted nature of the hydrological cycle in the arid zone presents unique problems for water management. In humid latitudes the key element is runoff or surface water, whereas in hyper-arid areas, an equally vital component is groundwater. The management system is therefore often focused on the aquifer rather than the river.

In hot, truly arid environments, the operation of the hydrological cycle is incomplete and intermittent in both time and place (figure 11.1). Precipitation, the motive factor in humid climates, is scarce and irregular. Thus, except as a result of the occasional storm or aquifer leakage, there is little runoff. Permanent surface storage is also extremely limited in extent and, in many significant cases, has been retained as a result of civil engineering rather than any natural features. Infiltration rates are influenced not only by the low antecedent moisture conditions, but also by the texture of the surficial material. Even in the case of coarse gravels, the pore spaces are frequently packed with the pervasive desert dust. Therefore, when flash floods occur, the eventual groundwater enhancement may be comparatively small, a high proportion of the runoff being lost to the sea. Given this fact and the generally discontinuous nature of any covering which could be classified as soil, interflow is very limited. Thus, with the exception of unconfined open-textured shallow aquifers, recharge rates tend to be slow. In the case of most deep aquifers, it is doubtful whether any contemporary recharge occurs. Semi-arid lands typically have more surface flow but are still characterised by variable atmospheric supplies and an increasing awareness of the importance of groundwater resources.

The depletion of aquifers results predominantly from deliberate abstraction through pumping, the use of various lifting devices or the construction of qanats. Natural freshwater springs do occur on land and offshore, but such losses are comparatively minimal. Since throughout most of the arid region the dimensions of the aquifers themselves are unknown, subterranean leakages are largely speculative. Other water losses occur near the surface through evaporation and evapotranspiration. However, there is evidence that evaporation is a surficial

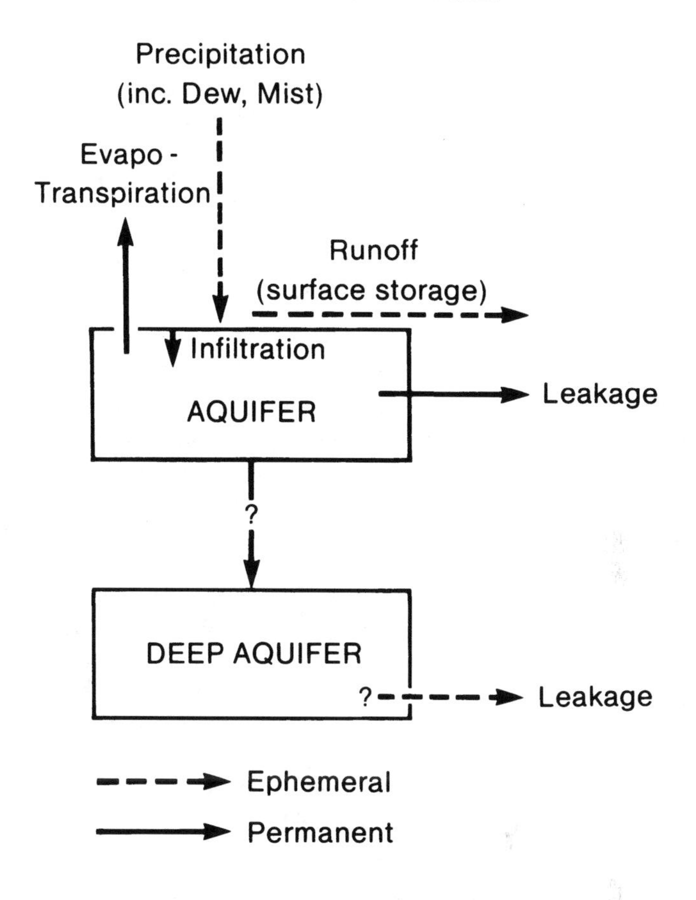

Figure 11.1 The hydrological system.

process and water below a relatively shallow covering is protected by the drier layers above (Agnew, 1988a). Aquifer depletion by evapotranspiration depends upon the density of the vegetation cover, which is everywhere sparse and in the truly arid areas much reduced in size and restricted almost exclusively to ephemeral water courses. Moreover, there is evidence that transpired water may not originate exclusively from aquifers. Certain genera, notably *Prosopis*, appear able to take up moisture through their leaves (Anderson, 1988d). Indeed, in some parts of the arid zone, dew and mist, or occult precipitation, may provide sufficient moisture for recharge. This is surmised in particular in the case of the seaward slopes of the mountains in the Dhofar Province of Oman. Measurements made on the Jebel Qara indicate that mists condensed by vegetation yield an estimated precipitation of 500 to 650 mm per annum (Halcrow, 1975).

The hydrological cycle in the arid zone (figure 11.1) is therefore a system characterised by irregular inputs, but, in general, regular outputs. Precipitation and occult precipitation are not only scarce, but also highly variable, whereas evaporation, evapotranspiration, aquifer leakages and especially human abstraction cause regular depletion. Moreover, as a result of forecast climatic change, precipitation could possibly diminish even further and increase in variability, while with population growth throughout the region, abstraction can only increase. This is the crux of the management problem.

Models and management

In the classification of Chorley and Kennedy (1971), the key elements of the hydrological system have been described in the form of a cascading model, but one in which the cascades are both irregular and, for much of the time, incomplete. At one level higher in the classification, the construction of a process–response model is hindered by the time lags in the system. Also, in the settled areas, human influence frequently distorts the effect of process on the landscape. For example, wadi profile adjustment to runoff is clearly distorted by road building.

Already, throughout much of the area, the most significant factor in aquifer water loss is human abstraction. With the proliferation of recharge dams, increasingly the inputs into the aquifer will be artificially directed. Most surface storage on any scale results from hydraulic engineering, while increases in agriculture must lead to enhanced losses through evaporation and, particularly, through evapotranspiration. With improved techniques of exploration and exploitation, the key aquifer variables will become better understood and an increasingly more systematic aquifer control will become possible.

Management aspects

Thus the hydrological cycle in the arid zone is best envisaged as a control model in which the human influence is significant and increasing. To facilitate management, this model can be extended to a management model which includes all relevant aspects of the water system (figure 11.2). Water from the rivers and aquifers, the major natural and local sources, is collected in the infrastructure where it is supplemented from a number of other sources. Chief among these are desalination, storage and recycling. The major proliferation of desalination plants has occurred on coastal sites in the driest and wealthiest countries, most notably in Saudi Arabia and adjacent areas of the Arabian Peninsula. In several of these countries, production from this source predominates in the water system. This is the case in Kuwait, Bahrain, Qatar and certain of the Emirates. In contrast, in countries with large-scale surface water sources such as Iran and Iraq, desalination, in some cases of even brackish water, will eventually provide some input into the water systems of many countries in the truly arid zone. Recycling is also of growing concern throughout the area and is, in many countries, the major component of the conservation programme. From the management viewpoint,

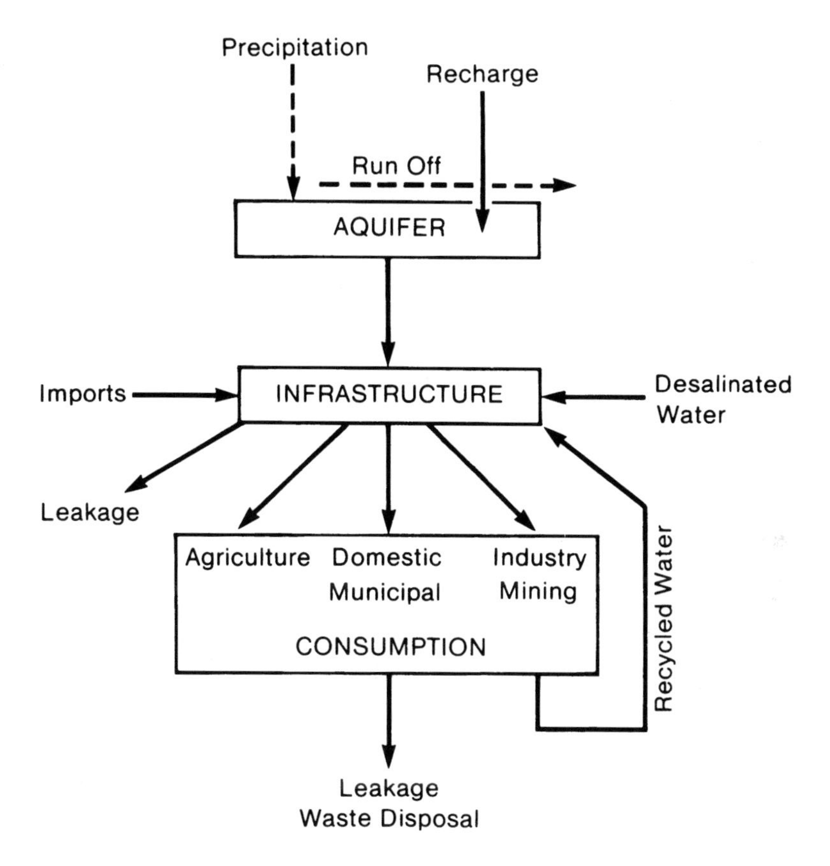

Figure 11.2 The hydrological system and water management.

there is a major distinction between the two types of water. Theoretically, either could be produced to any required purity, but in practice, desalinated water, being considerably more costly to produce, is reserved for human consumption and commercial activities with a high return, while recycled water is used for the irrigation of public amenity areas and, in some cases, crops. Far less costly, but applicable only under special circumstances, is irrigation by seawater. A number of trials have been conducted and the results appear impressive (Hodges *et al.*, 1987).

The other main supplement to the water system is imported water. In a few cases this is on the scale of a pipeline, but given the regional climatic constraints and the political vulnerability involved, this is not a preferred method. Tankers may be used in an emergency, or where the requirement is very limited, but the most common method of transport is by bottle. Many countries have developed their own sources of spring water, but international trade appears still to be growing. Such water does not of course enter the physical infrastructure.

Management of the infrastructure therefore entails controlling water from a number of different sources, a varying proportion of which, depending upon the country involved, is restricted to specific uses. Like the aquifer, the infrastructure also suffers losses through waste and leakages. In the Capital Area of Oman, this is estimated at approximately 30 per cent, while the water budget for Malta gives a figure of almost 50 per cent.

Water is delivered through the infrastructure for the various categories of use, which may be broadly distinguished into: domestic and municipal, agriculture and industry. Although the water requirements of these three categories differ, they can be grouped within the consumption unit of the model, since, increasingly, allocation between them will become an important component of water management. Throughout the arid zone, agriculture, through the requirements for irrigation, is almost everywhere by far the largest consumer of water. However, in the Middle East, for example, the burgeoning population and the development of industry are modifying this pattern. Right through the consumption unit there are leakages, but there is an increasing emphasis on recycling water after industrial and domestic use.

Conservation

Recycling is a key facet of conservation, a management concern of growing importance throughout the entire water system. In the area of physical hydrology, the emphasis is upon artificial recharge of aquifers together with the collection and storage of surface runoff, but no element of the hydrological cycle is neglected. The reduction of evaporation losses, for instance, has a high priority. The reduction of losses throughout the entire infrastructure and consumption unit is of great concern, but as throughout the entire system, effective programmes depend upon provision of accurate data. Therefore conservation is closely linked to the exploration and monitoring functions of management.

Conservation is, moreover, concerned not only with quantity, but also with quality and the monitoring and control of pollution throughout the system is a further vital role of management. Pollutants can enter at many stages during the normal working of the system, but there are a number of particularly vulnerable points. With coastal aquifers, abstraction in excess of recharge can lead to saline encroachment. Particularly striking examples of this occur along the Batinah coastline of Oman. In many arid zone countries, the infrastructure is either incomplete or poorly developed so that there is always potential for pollution from sewage or other waste matter. However, it is with the consumption unit that most pollution problems are likely to occur. The serial use of water in irrigation naturally results in an increase of dissolved solids, but the danger of serious pollution is frequently compounded by the lack throughout much of the rural area of any domestic waste disposal system. Indeed, throughout most of the urban areas of developing countries, rates of material consumption have far outstripped the capacity for waste disposal. Pollutants of course enter the water system through industry, both manufacturing and extractive. This may occur because the cost of installing cleaning equipment is considered prohibitive or, more commonly, because the legislation requiring such installation has not been enacted.

Thus, it can be seen that besides the integration of the system, possible allocation and the development of a water budget, management involves conservation, pollution control, monitoring and water exploration.

River basin planning

The river basin is often considered to offer the most suitable geographical framework within which data collection, analysis, forecasting and planning can take place. The advantages of using the watershed to demarcate hydrological management boundaries are:

1. disparate data sets can be compiled and examined,
2. water supplies can be more effectively regulated thus reducing the threats of both drought and flooding,
3. conflicts between demand from different parts of the catchment can be more easily resolved, and
4. water management is more effective through integration.

Rodda (1976) charts the scientific interest in river basin studies which perhaps commenced in the 17th century with studies of rainfall and runoff in River Seine, France. Efforts accelerated in the 20th century through investigations of the impacts of landuse changes, in particular deforestation, upon catchment water and sediment balances. These studies revealed the need to consider the hydrological cycle within a single geographical framework. Management practices followed these scientific studies with the UK for example reorganising its water planning units several times during the last 50 years from Catchment Boards in 1930; to River Boards in 1948; to just 29 River Authorities in 1963 and 9 Regional Water Authorities in 1974. The scale of operation increased at each stage so that each Water Authority now covers a sufficiently large catchment (or catchments) to enable integrated water management.

Perhaps the best known example of river basin planning is the TVA in North America which was initiated in the 1930s with the professed aims of power production, flood control, soil conservation and economic development (Saha, 1981). The same book (Saha and Barrow, 1981) includes a number of other schemes from throughout the world showing the widespread acceptance of this unit for water management. In 1987, the 13th International Congress on Irrigation and Drainage emphasised the need for integrated management of water resources with the river basin forming the basic management framework (World Resources Institute, 1989).

However, the river basin approach has not met with universal approval and there are some disadvantages. At a meeting at the Institute of Hydrology, Pereira (1973) commented upon the problems for scientific studies including the great expense and lengthy periods required for such studies with often inadequate networks of observations. Problems specific to water management are:

- many basins are too small for integration,
- some basins are too large and require international agreements and regulation, and
- resource inventories are often inadequate and basins often 'leak'.

In addition, river basins in arid lands are often only poorly demarcated where groundwater, not surface water, is the more obvious geographical unit for management. Unfortunately there is little reliable information on the extent of arid land groundwater systems. Whitlach and de Velle (1990) also point to the dangers of a regional, integrated approach where planning is inevitably more complex, political constraints are greater and the environmental consequences of failure are significant. Nevertheless integrated river basin management is still widely accepted as both a goal and a framework, but within this complete management system, each element such as recharge programmes, irrigation and recycling requires its own management procedures (figure 11.2). Furthermore, within the common framework illustrated, management strategies must vary from country to country, according to the national development context. Therefore, in discussing each key element of the management system, practical application can only be illustrated through detailed case studies, which follow. These are taken largely from areas such as the Middle East where the most extreme problems occur.

Oman will be used to illustrate a developing country which is in the early stages of water management. Malta provides an example of a more developed country, with a longer history of water management, but still facing many problems. Israel, in many ways the most developed country in the Middle East, will be used to furnish illustrations of more advanced water management techniques.

CASE STUDIES OF WATER MANAGEMENT

The Sultanate of Oman

Oman (Yiassin, 1985, Anderson, 1986 and Johnson, 1986) is included as an example of an arid zone country in the early stages of development. Therefore statistics on water supply and demand are scarce, exacerbating the problems of water management. However, Oman has less immediate water problems than all the other Gulf Cooperation Council states, except Saudi Arabia. Groundwater is a major source and is still capable of further development, while the effects, as yet, of saline encroachment from the coast are comparatively limited. Nevertheless, locally there are potential problems, particularly within the Greater Capital Area.

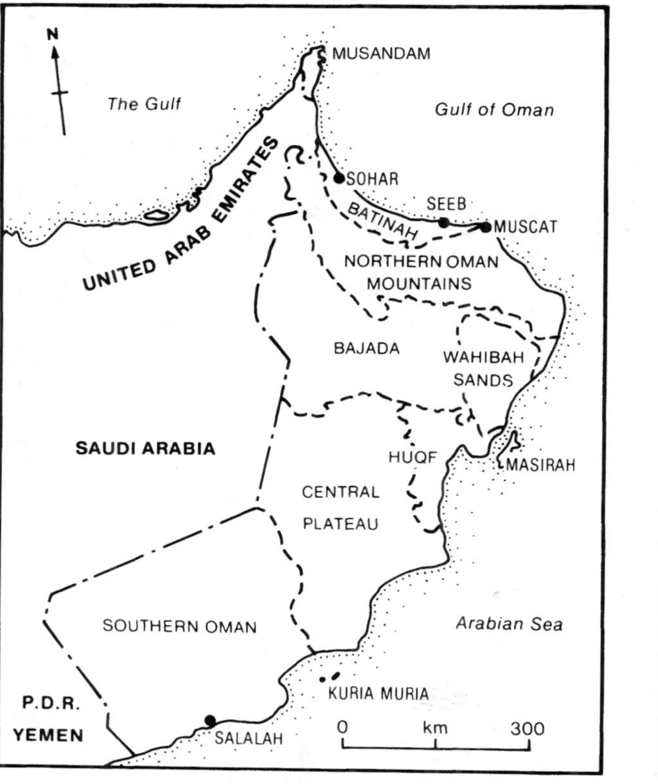

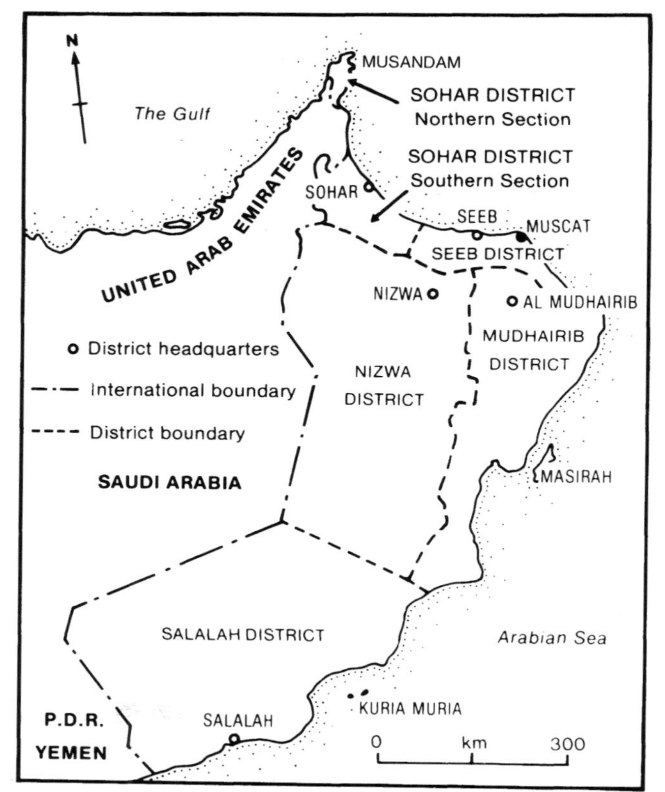

Figure 11.3 Water management regions, Sultanate of Oman.

Owing to its size and variety of climate and terrain, together with the scattered nature of the population distribution, water management problems in Oman produce different challenges in different regions. The effect upon the climate of the mountains, in both the north and the south, is profound, so that a high proportion of the population, together with most of the agricultural development, occurs along the Batinah in the north and in the plain of Salalah in the south. However, in the interior, there are important centres, concerned with agriculture and particularly petroleum, by far the major source of national income.

The Public Authority for Water Resources was established by decree of the Sultan in 1979, to develop a water data base, to conduct hydrological studies, to propose water laws and regulations and to train local experts (PAWR, 1983). To provide accurate estimates of national water resources, a 5 year drilling programme was initiated and a network of 626 observation wells and 360 aflaj measuring locations established. With this programme, it is possible to measure the effect of wet and dry seasons, to investigate the effect of pumping on the water table and to monitor changes in water quality. In recognition of the wide physical and human variations, for data collection and monitoring, the PAWR divided the country into five regions, each with a District Office (figure 11.3). These are: the Greater Capital Area, with an office at Seeb; the remainder of the Batinah, adjacent mountains and Musandam, based on Sohar; the southern area, especially the Dhofar, centred on Salalah; and the interior divided between the Sharqiya, with an office at Mudhairib to the east, and the Interior region, based on Nizwa to the west.

Such a division has an obvious hydrological underpinning, but is also one of administrative convenience, in that none of the regions could be considered a hydrological unit. For example, the Northern Oman Mountains form an obvious hydrological unit, but for data collection are more conveniently divided into northern and southern sections, according to the orientation of the drainage basins. Thus, the result is that hydrological characteristics and attendant problems need to be described by a hydrological area (figure 11.3), whereas water use data and related difficulties can only be discussed by a PAWR region.

Hydrological Areas

The Musandam Peninsula

Separated from the remainder of Oman by some 75 km of the United Arab Emirates, this peninsula consists essentially of limestone mountains rising to some 2,100 m, deeply dissected by wadis, each with a limited catchment area. Mean annual rainfall is thought to be about 125 mm (measured over the limited period). A maximum of 431.9 mm was recorded in 1976 and a minimum of 16.4 mm in 1978. Runoff measurements, scanty at best, indicate as expected that flood duration is short. Thus, given the terrain and climate, the occasional heavy storm produces very rapid runoff and a high percentage of water loss to the sea. In an effort to overcome this, there are plans for a number of dams, the major one of which is at present under construction. It is hoped that these will not only retard flood water, so that more irrigation will become possible, but they will also facilitate increased recharge. There are indications that the main zones of recharge coincide with active wadi channels and water quality declines away from these and also towards the coast (Saines *et al.*, 1981).

The northern coastal plain of Oman

Known as the Batinah, this is the key region in the country for agriculture, industry and settlement. It extends from the face of the mountains to the coast and can be divided into three geomorphological regions. The northern front of the mountains is drained by a series of deeply incised wadis, enlarged in places into flat-floored erosion bowls. The relief and the problems are similar to those in Musandam in that occasional heavy rainfall produces large volumes of runoff and, as yet, there are no retaining dams to check the flow. Coastwards is a wide zone of dissected gravel terraces, formed by at least three levels. These are interrupted at varying distances from the mountains by outcrops of calcareous rock which produce a narrow discontinuous strip of great significance for potential enhanced recharge projects. Groundwater movement within the terrace deposits is dependent upon their thickness, the presence of buried channels and the degree of cementation. Furthermore, recent isotope studies indicate that recharge processes vary according to the age of the channel. The calcareous rocks are thought to be of only minor significance as an aquifer, but evidence for this is very thin.

Beyond the terraces is the coastal plain, consisting of alluvial deposits, scattered sebkas, a comparatively wide strip of fertile soils and coastal sand dunes. The depth of deposits varies from some 250 m to 600 m and the sequence is essentially: upper gravels, clayey gravels and cemented gravels. The most productive of these aquifers is the upper gravels, while the clayey gravels have markedly lower yields (Gibb, 1976).

Within the region there are problems of increasing salinity inland from the coast and also of water retention. Recent surveys have indicated a salinity of coastal groundwater greater than 6,000 micromhos per centimetre for virtually all of the eastern Batinah. As agriculture has spread inland, increased pumping has caused a decline in the water table and a reduction in seaward pressure, allowing the encroachment of seawater. Undoubtedly, salinity has also been increased as a result of irrigation returns and direct evaporation.

Planning is hampered by a lack of data concerning both rainfall and runoff. Statistics for the period 1976–1981 indicate a mean annual rainfall of 85 mm at Seeb and 95 mm at Sohar (JICA, 1982). However, in each case the range is very large: from 183 mm to 3.7 mm in the case of Seeb and 253 mm to 9.8 mm in Sohar. There are some instances of perennial flow in limited upper reaches of the wadis, but none regularly traverses the Batinah Plain. Surface flow over the full length of the wadis only occurs during storms, usually as short, violent floods. To lessen losses into the sea and to enhance recharge, a number of feasibility studies have been made. Two large recharge dams, each over 20 km in length, have been erected, one at Wadi al Khawd in the Greater Capital Area and the other near Sohar. However, a more efficient system could be constructed from small retaining and buffer dams.

Northern Oman Mountains

Again, the hydrological problems of the Northern Oman Mountains are similar to those of Musandam, but the implications are far more serious, in that the Northern Oman Mountains furnish the major source of water input to the Batinah to the north and the Bajada to the south. Since a significant part of the mountain chain is over 1,500 m in altitude, while the highest parts reach elevations of 3,000 m, the region presents a major barrier to northerly winds. However, rainfall, while not infrequent, is generally very localized and records are extremely scanty. The main aquifer is valley floor alluvium and water quality is good, except where irrigation reuse water or contributions from ophiolite springs predominate.

Western interior – Wahiba Sands

Consisting mainly of large dunes, surface runoff within the Wahiba Sands is extremely limited. Furthermore, while Masirah Island is reported to receive 15 mm annually and the northern mountains have a rainfall as high as 300 mm (Hunting Technical Services Limited, 1976), there is no rainfall data available for the Sands. Occasional major rainfall events have been monitored, but for most of the year evaporation rates are very high and the effects on recharge are conjectural. The presence of woodlands in certain areas, particularly along the eastern edge, indicates the probable presence of groundwater, but its quality is thought to be poor. A recent programme, sponsored by the Royal Geographical Society (December 1985 to April 1986), involved substantial work on the water system of the Sands and it is hoped that, as a result, the various physical relationships will be clarified.

The complete western area of the interior is dominated by a series of coalescing alluvial fans, known collectively as the Bajada. To the north, a rocky pediment abuts against the mountains, while stretching away to the south extends an almost level gravel plain. Interruptions to the latter are provided by the Great Sebka (Umm as Samin) and the large dunes of the Rub al Khali. Traversing the plains is a series of raised wadi channels of varying height and age. Flow patterns, past and present, are therefore complex and have so far resisted clear interpretation. Rainfall varies from about 150 mm annually along the mountain front to well under 100 mm in the Bajada proper (Water Resources

Council, 1979). Although both rainfall and runoff records are scanty, it is known that occasional floods from the mountains traverse the entire Bajada. Furthermore, borings have shown that the Najada is underlain by shallow aquifers and water quality is generally good.

The main problem in the Bajada region is the almost complete lack of data and, with a view to enhancement, the very limited area of natural recharge along the mountain front. However, southwards of the main mountain front towns, settlement is very sparse indeed and water demand is low.

The Central Plateau

This is a flat table land of carbonate sediments with occasional scattered sand dunes. The region only became of significance with the discovery of oil in the 1950s. Rainfall is highly variable since the norm is thought to be less than 100 mm, but occasional monsoonal incursions produce much higher figures. Except near the coast, the wadi systems are related exclusively to the major patterns of internal drainage. Groundwater underlies the plateau everywhere, but its quality is thought to be generally low and potable supplies are probably only local.

Little is known hydrologically about the Huqf region, although rainfall is estimated to be similar to that on Masirah Island, about 15 mm (Wilkinson, 1977). Surface drainage is extremely restricted, but water requirements are very low.

Southern Oman

This region comprises one-third of the Sultanate and can be divided conveniently into five subregions. By far the largest, stretching from the interior to the flanks of the southern mountains is the Negd, an arid landscape of basically tabular badland topography. Rainfall is thought to vary from virtually nothing to a maximum of 50 mm annually near the mountains. However, as a result of occasional depressions, predominantly the monsoon, the mountains themselves receive, in places, 800 mm and an average of over 250 mm.

To the south and south-east are three coastal plains: Salalah, Mirbat and Jaazir. After the Batinah, the Salalah Plain constitutes the second major area of agriculture and settlement in Oman. Records at Salalah itself extend over 40 years and show an average rainfall of 108 mm, but totals have ranged from 22 mm to 508 mm. There are no records for the Mirbat and Jaazir plains, but landforms and vegetation indicate strongly that precipitation is very low. Furthermore, in these cases, groundwater supplies are thought to be small and, probably in most instances, brackish.

For development therefore, hydrological investigations have concentrated on the mountains, the Jebel al Qara and the Salalah Plain. Most groundwater, from rainfall, mist and even dew, is thought to originate in the limestone of the

Table 11.1 Estimated annual water usage by PAWR districts (Sultanate of Oman)

	MCM/year
Sohar (Musandam 17.3; upland areas 125.7; Batinah coastal strip 182.6)	325.6
Seeb (Greater Capital Area 16.1; Batinah coastal strip 78.6; upland area 229.0)	323.7
Mudhairib	234.5
Nizwa	326.9
Southern Oman	49.4
Total	1,260.1

Source: PAWR, 1983

mountains and the hydraulic gradient across the Salalah Plain is such that saline intrusion appears minimal. However, the detailed mechanics of the system are, as yet, not clearly known and a detailed measurement and monitoring programme is being established. This area, more than any other, illustrates the twin problems of developing enhanced recharge and preventing flood losses. To supplement the buffer dams, protecting parts of Salalah, a series of smaller structures, further inland and within the mountain valleys themselves, is planned.

Water use

It is clear that lack of data constitutes a major problem and much of the water use pattern of Oman (table 11.1) is inferred from satellite imagery. Total usage is estimated at 1,260.1 MCM annually, over 70 per cent of this being delivered through aflaj. Of the remainder, almost 21 per cent is supplied from wells on the Batinah and 4 per cent from those on the Salalah Plain. The Greater Capital Area accounts for over 16 MCM annually and increasing demand rates indicate problems in the near future.

At present, the only area of real concern is within the Capital district, where economic growth has been extremely rapid and where an increasingly large proportion of the population of Oman has settled. There is already one major desalination plant at Ghubrah and a second, to be located at Barka, has passed the planning stage. Local wellfields are operating at virtually maximum capacity and therefore the region is likely to become increasingly dependent upon desalinated water.

Municipal water demand

Only a few areas of Oman have municipal water supply systems, based either on pipes or bowser delivery. These are: the Capital Area, Sohar, Sur, Salalah,

Table 11.2 Water consumption (Sultanate of Oman)

Region	*Consumption (litres per person per day)*
Capital Area	224
Buraimi	322
Sur	118
Salalah	40

Buraimi and Nizwa. Few statistics are available, but from various surveys during the early 1980s, consumption can be calculated (see table 11.2).

The Buraimi figure is exaggerated in that it includes all water use within the oasis, while in the case of Sur, only part of the population is actually connected to the supply. The Ministry of Electricity and Water uses a planning figure of 215 litres per person per day which, assuming a total population of some one million, would produce a requirement of 0.215 MCM per day. The growth of water use in the Capital Area has been particularly astounding, rising from 177 litres per person per day in 1984 to 224 litres per person per day in 1985. MEW has prepared a forecast of water demand for the Capital Area in Table 11.3. Thus, it was considered that over the 7 year period water demand would triple.

Since table 11.4 does not include recharge from excess irrigation, these figures overstate consumption. On the other hand, measurements from 1985 satellite imagery indicated a 50 per cent increase in the acreage of arable land between 1972 and 1985. Furthermore, much of the increase is accounted for by commercial agriculture for which water use may be particularly heavy.

Industrial water demand

There are more than 3,000 registered industrial units in Oman, varying from tiny craft workshops to major metalworking plants. Based on a sample from each main industrial sector, present demand and additional future demand can be estimated in table 11.5 (Yousif and Shajahan, 1985).

Present demand, comprising 2.21 MCM per year, together with 1.05 MCM for the Rusayl industrial estate, totals 3.26 MCM per year. Apart from the 0.82 MCM

Table 11.3 Forecast water demands (Sultanate of Oman, 1,000 m^3 per day)

	1984	*1987*	*1990*
Present Source	60.9	90.1	123.2
New Projects	1.0	31.3	48.3
New Leakage	0.1	3.2	4.5
Contingency	–	12.5	17.6
Total	62.0	137.1	193.6

Table 11.4 Agricultural water demand (Sultanate of Oman)

Region	*Consumption (1,000 m³ per year)*
Batinah	870
Oman Mountains	1,900
Desert Foreland	1,100
Wahiba Sands	N/A
Jiddet al-Harasi	N/A
Southern Region (Salalah Plain only)	53

(Based upon satellite imagery, PAWR, 1983)

listed in table 11.5, future demand includes water for new industries outside the industrial estate, predominantly fertilisers and ceramics, totally 6.85 MCM; and water for new industrial estates (1.095 MCM), giving a grand total of 8.765 MCM per year. Thus, in addition to the fast-growing municipal requirements, the demand of industry by the year 1990 will be well on the way to trebling. In addition, the requirements for the oil and gas sector, calculated separately, were estimated to be 12 MCM per year in 1986 and over 14 MCM by 1987 (Al-Baluchi, 1985).

Table 11.5 Industrial water demand (Sultanate of Oman, 1000s m³ per year)

Sector	*Present demand*	*Additional demand*
Food & Beverages		
Soft drinks & mineral water	260	68
Dairy products	115	56
Cooking oil	18	–
Flour & bakery products	15	4
Ice	40	40
Other miscellaneous food products	70	70
Chemicals & Chemical Products		
Paint	3	–
Soap & detergent	38	160
Petroleum products	22	22
Other miscellaneous chemical products	10	10
Mining & Mineral Products		
Cement	235	235
Cement products	100	25
Lime & lime products	20	–
Crusher products	92	23
Other miscellaneous products	32	32
Metals & Metal Products	1,140	55
	2,210	800

Water problems

The problems can be divided into those which are essentially technical and those which are legal. Among the former are concerns connected with seawater intrusions, artificial recharge, recycled water use and the general inefficiency of the infrastructure. Legal problems are dominated by considerations of water rights and water priorities. Water priority problems can be illustrated by the case of the Batinah. Major water demands result from the following:

1. irrigation for the traditional date gardens and agriculture along the coast;
2. expansion and modernisation of towns and villages along the coast;
3. new development along the dual carriageway, constructed a few kilometres inland;
4. proliferation of private gardens and new housing inland from the dual carriageway;
5. development of industry;
6. establishment of large irrigated commercial farms; and
7. rapid expansion in the Capital Area.

Thus, in an area of declining water resources, there is competition between and within the municipal, agricultural and industrial sectors. This highlights the necessity not only for a Water Code, but also for effective water management.

Malta

The Maltese Islands (Spiteri Staines, 1987 and Anderson and Schembri, 1989) receive an annual average rainfall of 508 mm, almost all of which falls between September and April. Since tourism is the mainstay of the Maltese economy and most visitors come in the summer, the period of least supply is the period of greatest demand. Furthermore, as in other arid zone climates, the reliability is low, droughts occur commonly and rainfall tends to be concentrated in short, sharp storms. Therefore, a high proportion of the precipitation is lost through runoff and it is estimated that, at the most, 25 per cent infiltrates to recharge the aquifers.

In contrast to Oman, Malta had developed a nationally organised approach to problems of water shortage much earlier in its history. In 1610, Grand Master de Wignacourt constructed a 13 km long aqueduct to bring spring water from Rabat to Valletta. The first well was sunk in 1851 and from 1866, subterranean collecting galleries, storage facilities and pumping stations were constructed. The first comprehensive study of the Maltese water situation was completed by Chadwick in 1883. Between 1966 and 1968, four multiple flash seawater distillation units with a maximum capacity of 20,500 m^3 per day began production. In 1971, the Gozo desalination plant, with a production facility for 20,250 m^3 per day, was inaugurated, while in 1982, the reverse osmosis plant at Ghar Lapsi, then the largest of its type in the world with a daily production of

Table 11.6 Annual water production, Malta

	G	*B*	*S*	*D*	*RO*	*O*	*Total*
1966	13.40	1.51	0.49	–	–	–	15.40 MCM
	(87)	(10)	(3)				(%)
1969	11.74	2.03	0.83	2.63	–	–	17.23
	(68)	(12)	(5)	(15)			(%)
1976	13.07	5.08	0.73	3.05	–	–	21.93
	(60)	(23)	(3)	(14)			(%)
1986	11.45	9.07	0.56	1.33	8.97	0.17	31.55
	(36)	(29)	(2)	(4)	(28)	(1)	(%)

where
G = Galleries
B = Boreholes
S = Springs
D = Distillation
RO = Reverse Osmosis
O = Other.

2,000 m^3, was installed. This was followed by further plants at Marsa, Cirkewwa and Tigne. Meanwhile, the Rizq il-Widien project to install dams in the Widien, or dry valleys, to check runoff losses, was inaugurated in 1980. In 1984, the Sewage Treatment Irrigation scheme at Sant'Antnin, with a maximum production of 9,000 m^3 daily, sufficient to irrigate 140 hectares, commenced operation. Production at the reverse osmosis plant at Tigne has now been increased to 8,500 m^3 per day and that at Ghar Lapsi to 22,000 m^3 daily. This short history illustrates the key developments in the increasingly problematical field of water management in Malta.

Apart from surface water, Malta can now call upon a full range of sources for its water planning as shown in table 11.6. Over a 20 year period, the percentage obtained from galleries has declined from 87 to 36.3 per cent, that from boreholes has increased from 9.8 to 28.8 per cent and a major element in the budget, reverse osmosis, is now responsible for 28.4 per cent of the total. During this time, annual water production has more than doubled from 15.4 MCM to 31.6 MCM. In 1966, all water came from groundwater and springs, by 1986 the contribution from those two sources was only two-thirds, while the remaining one-third was obtained by artificial means. Nonetheless, the production of groundwater from galleries and boreholes has increased from 14.9 MCM to 20.5 MCM over the 20 years, giving rise to anxiety about overpumping. Despite the heavy national dependence on tourism, water demand is dominated by the domestic sector (table 11.7).

The major change occurs between 1976 and 1979 in the government usage. The sharp decrease coincided with the departure of the British armed services from Malta in 1979.

The main problems on the hydrological side are posed by salt water intrusion into aquifers and by leakages from aquifers. The perched aquifer, which occurs in

Table 11.7 Water consumption by sector, Malta

	Domestic	*Commercial*	*Farms*	*Tourism*	*Indust.*	*Gov.*	*Total*
1967	48.5	13.0	5.5	–	13.5	19.5	100
1976	53.6	6.1	4.9	5.3	12.0	18.1	100
1979	60.8	5.8	3.6	8.6	12.0	9.2	100
1986	63.3	6.9	4.0	6.9	10.4	8.5	100 (%)

the Upper Coralline Limestone and rests upon the fissured Blue Clay, accounts for only 5 per cent extraction. The major source of groundwater is the mean sea level aquifer which rests upon seawater. The balance between the two types of water is extremely delicate and excessive abstraction can easily disturb it. A further problem concerns evapotranspiration which is thought to account for some 70 per cent of the total precipitation. Pollution, not only from seawater but from anthropogenic sources, is of growing concern in such a densely populated and small island. Limitations of space also preclude storage on any but a very modest scale.

Leakages are thought to exceed 30 per cent and a recent water budget indicates that 47 per cent of water is unaccounted for. It is essential that, in the future, a comprehensive and efficient distribution network is constructed. Malta has invested heavily in technology and non-natural resources but at a level of expense which must raise questions about the sustainability of the system.

Israel

Israel (Schwarz, 1986 and Anderson 1988c) differs from Malta and Oman not only in its level of development, but also in the fact that it can draw on surface water, particularly in the Lake Kinneret basin. The other key sources are groundwater and recycled water. Groundwater is obtained from two aquifers which occur in Israel as originally demarcated and two which are in the West Bank. The aquifers of Israel proper are the coastal aquifer, which extends from Mount Carmel to the Gaza Strip, varying in width from approximately 10 km in the north to 20 km in the south. The maximum possible potential is estimated at 280 MCM a year and at present it provides water for over 1,000 wells. It suffers from seawater incursions and pollution, particularly of salts from other sources.

The Yarkon–Taninim aquifer is the other main groundwater source of Israel proper. This extends from Mount Carmel to Beer-Sheva and from the mountain backbone in the east to the coastal plain in the west. The yield potential of this aquifer is estimated at 290 MCM of freshwater and up to 50 MCM of saline water. Great care has been taken not to overpump this aquifer, but in the dry summer of 1986, the rules were violated. In addition to these two, Israel obtains water from several small aquifers and two larger West Bank aquifers.

However, the principal national water source is Lake Kinneret from which the National Carrier transports water to the Negev. From the basin, some 400 MCM

Table 11.8 The water balance for Israel 1984–5 and projected for 2010

	1984–5	*2010*
Supply		
Groundwater	1,210	980
Lake Kinneret basin	610	580
Other surface water	90	145
Saline water	175	240
Recycled water	60	430
Total	2,145	2,375
Demand		
Domestic	436	780
Industry	100	135
Agriculture	1,445	1,250
Judea-Samaria	105	180
Losses	60	30
Total	2,146	2,375 (MCM)

('Saline water' is that used for irrigation which has a salinity of over 400 mg per litre.)

a year are transported by the National Carrier, while a further 200 MCM are drawn for local requirements from Lake Hula, the Golan Heights and directly from Lake Kinneret.

Nevertheless, the management of water within Lake Kinneret poses a number of problems, resulting from the small storage volume (500–600 MCM) in relation to the large differences in natural refill between dry years (minimum 150 MCM) and wet years (maximum 1,400 MCM). As a result, the yearly yield capacity of the National Carrier system can be realised in full only in 70 per cent of years. A key role in the utilization of the lake has been to prevent the descent of the water level beyond 212 m below sea level. At lower levels, there are risks, not only involving the penetration of saline water, but also a combination of biological, chemical and physical processes, the results of which could cause extreme pollution problems.

Other surface water is obtained from the only two perennial rivers, the Yarmuk and the Lower Jordan, and flood waters from the various wadis. This source yields some 20 MCM for seasonal reservoirs and the total utilization potential is estimated at up to 90 MCM. However, the realization of this potential involves the building of installations for impounding the flood water, either for direct use or for injection into the aquifers.

In 1985, the volume of sewage was estimated at 260 MCM and that collected through the central system, 240 MCM. Of this, 150 MCM reached the central treatment plants and 90 MCM were used for irrigation. It is intended that the annual total for use should reach 430 MCM by the year 2010 (table 11.8). However, the use of effluents for irrigation is constrained by the type of crop and also by the environment of any underlying aquifer. So far, in Israel the major technological advances have been in recycling, but to expand the supply, desalination seems likely to be needed. Indeed, a pilot and demonstration project with a capacity to produce 10 MCM a year was built at Ashdod.

The national water system is preparing for a population growth to 5.4 million in 1995 and 7 million in 2010. If an increased per capita consumption of 0.5 per cent is included, the requirement by 2000 would be 630 MCM. Within this calculation, consumption by the non-Jewish sector, estimated to be a third of that of the Israeli population, is forecast to experience a more rapid growth. Agricultural demand is related closely to the cost of water supply which, in remote arid areas, increases with the quantities provided. Supplies for agriculture have been subsidised, but resources are limited and no significant increase in agricultural water consumption is foreseen until the advent of cheap desalination. There may, however, be a reallocation of water to different regions and a replacement in some areas by recycled water. The consumption of water in the industrial sector is increasing slowly to a total of only 130 MCM by the year 2000. The only potential major increase foreseen would materialise if there were an inland power station building programme. Water requirements for cooling could completely upset current forecasts.

In addition to these sectors, water consumption in the West Bank (Judea–Samaria) is estimated separately at 105 MCM, 80 million for agriculture and 25 million for domestic use. By the year 2000, domestic consumption is expected to have risen by 70 MCM, but there will be no increase in agricultural usage. Losses, also tabulated, result from evaporation, seepage, leakages and unreported consumption. At present, the total is put at 60 MCM, but measures are in hand to reduce this figure to 35 MCM by the year 2000. A further possible source of increasing demand is the environment and, particularly with increases in tourism in mind, additional water may be required to provide amenities.

The key problems, neatly summarized by Schwarz (1986), are:

> The quantity gap, the location gap, the elevation gap, the time gap and the reliability gap.

Since the total water availability from all sources is less than the demand, there is a need for additional water to be brought into the system. The main sources being considered are recycling, storm runoff and possibly desalination. According to current plants, it will be necessary until 1990 to add a quantity of 200 MCM of water per year, but over the next 20 years, an additional 550 MCM will be needed. The main locational problems result from the fact that water is comparatively plentiful in the north, while the south is arid. However, a further problem concerns recycling, since the main producers of effluents are the large cities, which are remote from the agricultural lands on which the water will be used. The fact that the sources of water are generally at a lower altitude than the areas of major demand results in an elevation gap. At present, the pumping required amounts to some 15 per cent of the national electricity consumption. Israel's water system is therefore very energy intensive, a factor which increases considerably the management problems. There is also a time gap since precipitation occurs in the winter half-year and as a result 120 seasonal reservoirs have been necessary to even out supply over the remainder of the year. The reliability gap stems from natural causes, the fluctuations in rainfall and anthropogenic factors such as defects and failures in the system. One answer would be year-round storage to save water from abundant years for use in dry years, but the main storage area, Lake Kinneret basin, has, as already discussed, a very limited capacity. If the capacities of the coastal and Yarkon–Taninim aquifers are also added, there is a total capacity of some 2,000 MCM. However, as yet, detailed planning is lacking, despite such threats as that of global warming. Finally, the quality gap involves the necessity to separate networks for the different types of water. Freshwater of very different salinities and recycled water clearly cannot be transported in the same system. If some such separation cannot be effected, the programme for replacing expensive freshwater by effluents will be jeopardised.

STRATEGY AND PLANNING

Water is fundamental to life, particularly in the arid zone. The importance of water in the arid zone is so pervasive that its distribution reflects the basic social values of equity and justice. The influence of water can be seen in fundamental process–response systems. Resources are developed in response to environmental requirements, but the developments themselves then alter the environment. In an oasis, for example, the political and social hierarchy within the settlement governs the distribution pattern of the water, but the pattern itself determines the political and social structure. In such situations it is clear that any change in the water supply affects very much more than the irrigation of a few plants. Water management therefore must be viewed in the context of overall national strategy. Planning must be congruent with political, economic and social strategy.

An illustration of how water might be affected by national strategy is given by Creighton (1982). He constructs a matrix (figure 11.4) composed of continua from low government to high government control on one axis and from a concern for environmental quality to a concern for economic development on the other. Zone 1 is likely to support conservation or behaviourial-change solutions, zone 2 highly centralised development options, zone 3 a natural balance and zone 4 market place allocation.

National strategy is by definition implemented within the state. However, the watershed of a particular drainage basin is unlikely to coincide with political boundaries and this raises the question of the most suitable spatial unit for planning. From the hydrological viewpoint, the catchment, a complete system, is taken as the basis for study. In an international drainage basin, data is collected nationally and therefore there is a case for recognising regulatory responsibilities to accord with state jurisdiction. Certainly, an insistence on natural boundaries carries overtones of determinism (Wengert, 1976).

If the close tie between water and land is pursued further, another complication arises. Water is generally perceived as a common resource, available to all

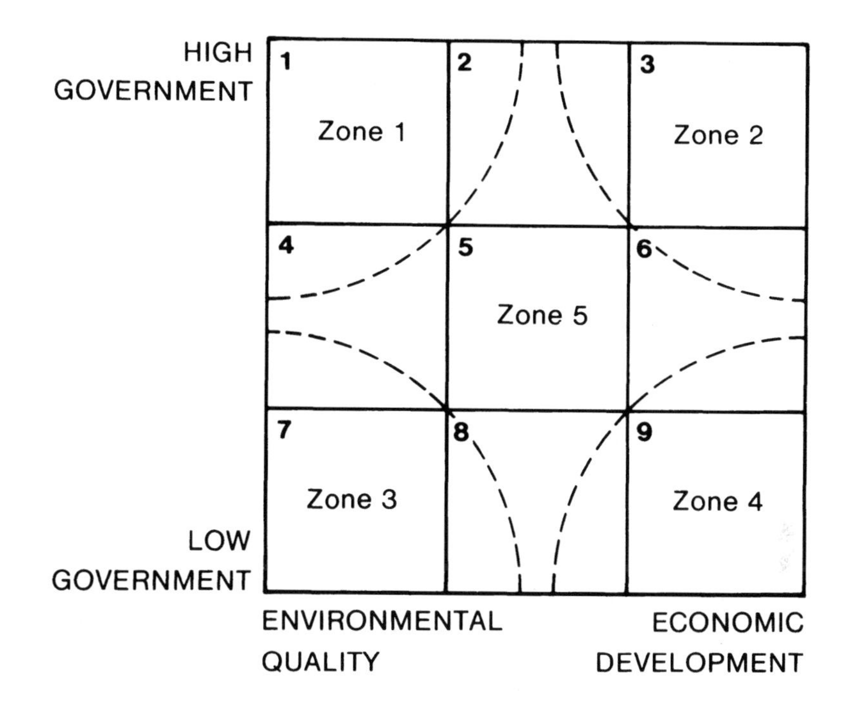

Figure 11.4 Water and national strategy and planning, after Creighton (1982).

and usually developed by public investment. Land, on the other hand, is seen as a private right and planning must take account of this. Thus, there is always the potential for conflict over the distribution of water and where there is a scarcity only exacerbates the situation. Tension may result from conflicting strategies, from boundary problems, from the public–private dispute or merely from allocation. In the arid zone, decisions allowing access to one user frequently result in denial for another. To any potential user for whom water is a salient issue, any allocation elsewhere represents relative deprivation (Priscoli, 1983).

Water strategy therefore involves some esoteric philosophical arguments, but the reality in published strategies is usually far more mundane. In *Development Guidelines for the Economic Use of Water in the ESCWA Region* (United Nations, 1986), for example, water development strategies are set out for the arid ESCWA countries. In the case of Kuwait, this consists of the following:

1 a table of existing and planned desalination plants;
2 a list of total production capacity of public wellfields;
3 a note on the distribution networks;
4 consumption rates for freshwater and brackish water;
5 the water tariff; and
6 details of recycling.

This summary of the system, while indicating in a covert fashion elements of strategy, is clearly more important as a basis for planning and management. Ideally, management plans are derived from strategic aims.

MANAGEMENT PLANNING

> No specific formula for a long range water resources plan for a country or for a hydrologic unit has ever been defined.
>
> (Dingman, 1984)

It is recognised that, to be effective, such a water resources plan must reflect the hydrology of the country or unit, while meeting the social and economic needs. Furthermore, the plan should be flexible so that it can remain dynamic and adaptable to the changing policies at all levels from national strategy to daily water requirements. The construction of such plans is therefore multidisciplinary, but, in particular, embraces elements of both the physical and the social sciences. Environmental conflicts have highlighted the point that water engineering has major social effects and objectives beyond purely technical construction and narrow economic development. Decisions may involve public against private, growth against no growth, economic against other social values, science against popularism or technical against political values.

Priscoli (1983) recognises the emergence of a new paradigm for planning which involves the confluence of public involvement, social impact assessment and futures-forecasting. The tenets include the following:

1 That planning creates as much as predicts the future. A parallel exists in theoretical physics in which the investigators find that the instrument of measurement can determine that which is measured.
2 That the validity basis of planning is found in an 'intersubjective-transfer of knowledge', not in an 'independent-observer' position. Reality is, in other words, more a shared process of creation than an independent, observable fact.
3 That planning is as much political as it is technical:

> The question is not whether planning will reflect politics but whose politics will it reflect . . .?
>
> (Long, 1980)

4 That the planner's role is to design 'win–win', rather than 'zero-sum' or 'lose–lose' alternatives.
5 That the way forecasts are made has a major impact on the type of society which results.

Thus, planning influences the environment, the changing nature of which, in turn, generates the requirement for further planning. Planning for management is therefore an iterative process. During the various phases of planning, the following tasks tend to reoccur, with varying importance at each phase:

1 Problem identification.
2 Formulation of alternatives.
3 Impact assessment.
4 Evaluation (Priscoli, 1983).

To complete these tasks, the United States Army Corps of Engineers develops in its applied social analysis training programme skills in using the following social science techniques: institutional analysis, policy profiling, values analysis, social profiling, content analysis, small group process, human cost accounting, community impact assessment, ethnographic and field analysis, questionnaire non-parametric statistical analysis, population projection, trend and cross-impact analysis.

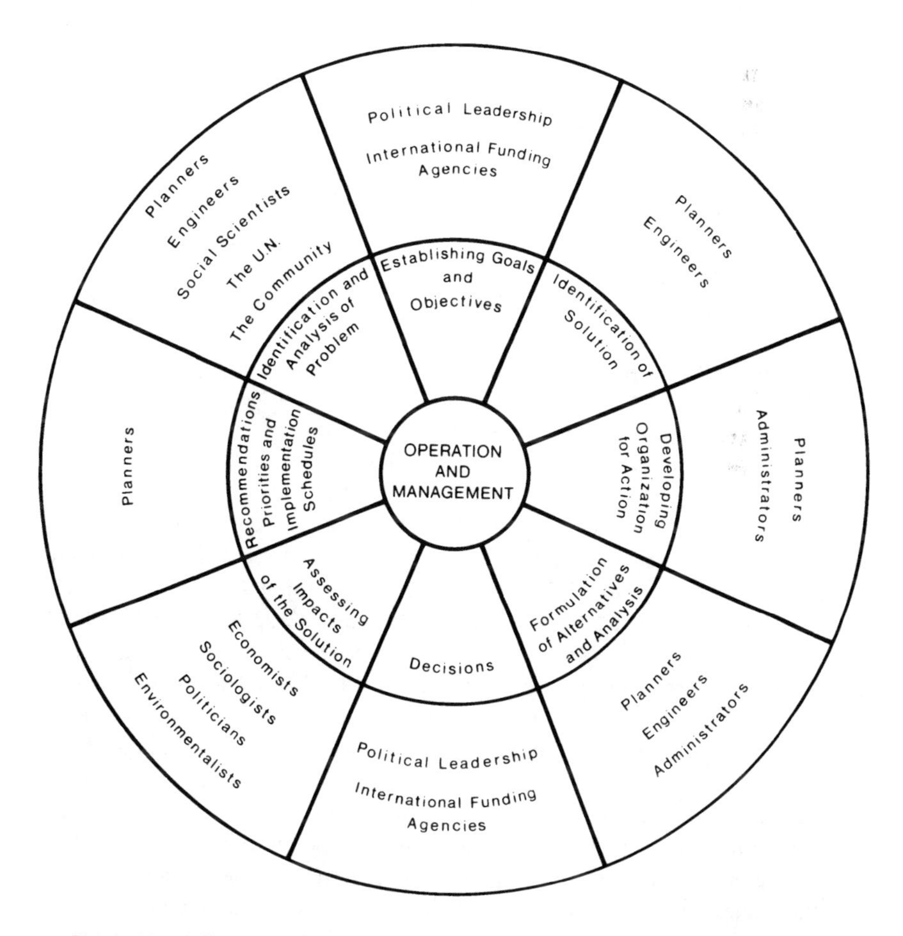

Figure 11.5 Effective water planning in developing countries, after Kadivar (1985).

Planning procedure

Water planning is a continuous process, involving a wide range of expertise. Apart from physical scientists and engineers, contributions are needed from social scientists and politicians. The procedure itself and the role of the various experts can be appreciated if the process is examined in each of its stages. Kadivar (1985) produced the outline for an effective water planning process in developing countries (figure 11.5). The stages in order are:

1 Establishment of goals and objectives (but see Agnem, 1984b).
2 Identification and analysis of the problem.
3 Identification of the solution.
4 Impact evaluation and analysis of the solution.
5 Formulation and ranking of alternatives.
6 Development of recommendations, priorities and schedules for implementation.
7 Development of an organisation for action.
8 Decision making.
9 Project operation and management.

The overall goals equate with the national strategy and the objectives, with policy at a somewhat lower level. These are therefore set by the political leadership and must necessarily be couched in somewhat general terms. However, they need to be realistic if conflict is to be avoided and this entails a good acquaintance with currently available information. This illustrates the problem already indicated that the formulation of effective strategy depends upon accurate knowledge of the water situation, but that knowledge may not be available until the strategy has been set in place. This problem can be overcome, in a measure, by using guidelines set by international experts and funding agencies. On the other hand, given the uniqueness of each country, solutions obtained elsewhere have only a limited transfer value.

Within the framework thus set, the particular problem is identified and analysed by planners and engineers, assisted by social scientists, international experts, consultancy groups and, ideally, the general public. This is a vital stage, since unless the problem can be accurately formulated, an appropriate solution cannot be developed. Hydrological aspects of the solution are suggested by engineers, while planners provide the full context. The solutions will then be scrutinised for their impacts on various aspects of society by economists, politicians, sociologists, anthropologists and environmentalists. From the various impact assessments, the possible solutions can be formulated and ranked as alternatives, before administrators, assisted by the engineers and planners, draw up a series of recommendations, priorities and implementation schedules. These, in turn, define the outline of the organization required for their implementation, but the actual structure results from a dialogue between the planners and the administrators. The success of the project depends very much upon the provision of an effective organisation. With all these stages clearly defined, the package is presented to the politicians for decision making.

This may seem a lengthy and elaborate process, but if all interested parties are to contribute effectively, it is very necessary. Indeed, while experts from abroad and consultants may be able to offer assistance, the development of a planning procedure must be completed by the country concerned. The problems of a particular country are unique and the transfer of a planning structure from elsewhere is unlikely to be successful. To facilitate this procedure, Kadivar (1985) identified five working principles:

1. Water planning is a new profession which requires a synthesis of methodologies, based on disciplines such as: engineering and physical sciences, architecture, city and regional planning, economics, sociology, political science, law and the biological sciences.
2. Consultants in related fields such as engineering or oil extraction may not be qualified to develop water plans. Water planning requires water planners who are directly involved with high-level government officials and the political leadership in the planning process.
3. The water planning process involves three categories of personnel:
   (a) those who participate in the plan making process;
   (b) those who are users of the plans; and
   (c) those who observe the plan making process.
   These individuals and their roles need to be assessed in order to develop a framework for a national water planning process.
4. The result of the planning will lead to the execution of a series of projects. These projects may be structural measures such as the construction of facilities or non-structural measures such as management and legal techniques.
5. National water planning must begin with a preliminary screening of all existing or potential water resource development projects. On the basis of a careful assessment process, these must be identified, ranked, screened out or retained for detailed studies. The information from the retained projects is then assessed and the outcome of this screening will be a schematic plan of existing potential projects.

Regional projects

The planning procedure described is not scale specific and there are many benefits to be gained from planning development projects on a regional, rather than purely national scale:

> Technically speaking unilateral development of common water resources will never reach its full usefulness and totality without full knowledge of the shared main water parameters, i.e. geomorphology, hydrology, hydrogeology, hydrometeorology, agrometeorology, climate, soil and socio-

> economic conditions prevailing in the project area. Cooperation among countries will facilitate studies of such parameters required and the outcome will benefit all.
>
> (United Nations, 1986)

Despite success in such cooperation elsewhere, the number of regional water resource development projects in the Middle East is limited. However, the great potential for such projects is confirmed by the existence of major surface and groundwater basins such as the following:

1 the Palaeozic–mesozoic sandstone aquifer extending through Jordan, Saudi Arabia, Qatar, Kuwait, Bahrain, the United Arab Emirates, Oman and PDR Yemen;
2 the Um-Al-Rdhuma-Dammam limestone aquifer occurring in Saudi Arabia, Bahrain, Qatar, the United Arab Emirates, Oman and PDR Yemen;
3 the Euphrates River basin, shared by Turkey, Syria and Iraq;
4 the Yarmook River basin, shared by Syria and Jordan and, at present, Israel;
5 the Al-Asi-Al-Kabir River basin, shared by Syria and Lebanon; and
6 the watersheds of wadis Tuban, Bana and Beiham, shared by Yemen AR and PDR Yemen.

In addition there are non-conventional desalination problems, common to the Gulf Cooperation Council countries (United Nations, 1986)

The regional projects which have been developed have, for the most part, been executed and funded by various United Nations agencies and the Arab Centre for the Study of Arid Zones and Dry Lands (ACSAD). Among the more important projects are:

1 The Hammad Basin Project – this wide-ranging programme was initiated in 1975 by the four countries sharing the basin: Iraq, Jordan, Saudi Arabia and Syria. The aim is to produce an integrated study of available natural resources in the basin, with particular emphasis upon water.
2 Shared water resources in the Gulf States and the Arabian Peninsula (Dammam aquifer) – executed by the Food and Agriculture Organization of the United Nations (FAO), the project area covers 1.7 million km^2 and the objective is to develop a hydrogeological model for the Eastern Arabian basin.
3 International Drinking Water Supply and Sanitation Decade (1981–90) executed by ESCWA and funded by the United Nations Development Programme (UNDP) – the aims are to review the national plans in ESCWA countries for the Decade and to draw up a regional plan of action on the Decade's activities, to be undertaken by the Commission during that period.
4 Study and development of international water-works – executed and funded by ACSAD, the aim is to optimize utilisation of the rural water resources through development of the existing conventional water-works in the Arab countries.

5 Water resources data bank – this data bank for Arab countries is being executed and funded by ACSAD.
6 Regional development and application of components of a hydrological operational multi-purpose system – executed and funded by ACSAD, UNDP and Arab countries. The project is designed to improve hydrological data and to utilise rationally the available water resources.
7 Water resources mapping – this programme for the Arab countries is being executed and funded by ACSAD and the countries themselves.
8 Yemeni Joint Project for Natural Resources – the project is being executed by the United Nations Department of Technical Cooperation for Development (DTCD) and funded by UNDP and the Arab Fund for Economic and Social Development (AFESD). The programme covers the exploration and management of mineral and water resources in both Yemens and includes geological and hydrogeological surveys.

It is clear that the United Nations Agencies are playing a major role in developing the water sector in the Middle East, either by executing or funding water and sanitation projects. Water-related programmes have always been an integral part of the United Nations agencies over all development effort, but in particular of UNDP. Other key agencies active in funding, supervising or executing projects are: the World Health Organization (WHO), the World Meteorological Organization (WMO), FAO, the World Bank and the International Development Association (IDA). There are also several bilateral donors such as the Netherlands, the Federal Republic of Germany and the United States, together with several development banks and funds which have made contributions to the water sector, including AFESD, the Islamic Bank, the Kuwaiti Fund and the Abu Dhabi Fund.

National projects

Throughout the Middle East, improved standards of living have resulted in increasing demands for water. Therefore, water resources must be further developed, but, more importantly, management must be based on a comprehensive water resource master plan. This can be illustrated from the conclusions and recommendations presented in the ESCWA report (United Nations, 1986).

1 **Bahrain**
A National Water Plan would help in coping with the water situation and a strategy should be formulated with special emphasis on the rational utilisation of scarce water resources and the expensive non-conventional desalination of seawater; this would include conservation measures, strict regional water management, establishment of priorities in water use and closer coordination with agricultural, industrial and water development plans.
2 **PDR Yemen**
A National Water Plan be established and water programmes generated to

attract donor agencies to provide financial and technical assistance in the water sector. A water resource development programme be initiated to provide realistic knowledge of the country's water resources.

3 **Egypt**
The basic strategic principles for planning in the water sector are to develop appropriate administrative organisations to serve the new needs of the community. Accordingly, a need has arisen to reassess the institutional framework of water administration in Egypt.

4 **Jordan**
The existing National Water Plan should be enforced to ensure the provision of services and expanded to meet current and foreseeable requirements. Water resource development programmes should be established to attract donor agencies to provide technical and financial assistance.

5 **United Arab Emirates**
A National Water Plan should be established and enforced to ensure the provision of future required services and their expansion in the water sector.

The imbalance between water use and the available water resources necessitates the coordination of all activities in the water sector through the best use of water available.

Oman: an example of national water planning

The conclusions and recommendations in the ESCWA report (United Nations, 1986) for Oman were as follows:

1 Since Oman currently exploits its groundwater resources on a large scale, there are fears of seawater encroachment and rising salinity. Recharge through the construction of dams would minimise surface water losses to the sea or the desert and a new equilibrium would be maintained by recharging the fans' aquifers.
2 A National Water Plan would assist in ensuring the provision of water and sanitation and their expansion according to current and foreseeable requirements.
3 Owing to the rapidly growing imbalance between water use and available water resources, there is now a need to organise and coordinate all activities in the field of water by establishing control regulations and optimising water use.
4 A long-term programme should be established to develop skilled manpower in the water sector at all levels.

These points must be viewed in the context of the philosophy and strategy of development in Oman. The most conservative philosophy of development would be to limit the extraction of groundwater to the long-term rate of recharge in that particular hydrological area. However, already along parts of the Batinah coastal area and near Buraimi, rates of pumping greatly exceed those of recharge. As a result, salt water encroachment along the Batinah is becoming evident and in Al Ain the aquifer is being de-watered at an alarming rate (Dingman, 1984). It may be accepted that, for overriding political, economic or social reasons, water may be abstracted for at least a finite period in excess of possible recharge. However, the concomitance of such water mining must be realised:

> In developing a plan all possible contingencies must be evaluated and a series of plans developed. These plans must account for constant changes in patterns of water demand and development. The plans must be correlated with plans for economic and social growth.
>
> (Dingman, 1984)

These ideas must be viewed against overall national strategy as outlined by His Majesty Sultan Qaboos in 1985 for the current 5 year development plan. These objectives can be summarised as follows:

1 Further development in the telecommunications system as it plays a significant role in the overall development of the country.
2 Building new roads, for they are the pulse of life.
3 Further improving the quality and efficiency of defence forces through efficient coordination of human and financial resources.
4 Developing the agricultural sector to the stage of self-sufficiency, based on scientific research and studies.
5 Developing industries for local needs by encouraging public and private investment.
6 Concentrating on education and health; Omanis will learn in a university in their own environment.
7 Strengthening the role of the State Consultative Council which determines the actual needs and aspirations of the country and its people.
8 Strengthening the role of the media in telling people the real facts and providing positive criticism, based on facts.

Kadivar (1985) comments:

> In dealing with the first six objectives, water plays a critical role. Water in urban and rural areas is a determining pulse of life and it is the key factor in the level of people's health. The agricultural sector can only be developed if adequate water is found and utilized. Medium-scale industry requires water in varying quantities and of various qualities. Defence forces need camps supplied with water. Building roads and a better communications system stimulate further physical development which increases demand for water.

Thus, the six major goals set by His Majesty depended mainly on water and the national strategy, so defined, presented a great challenge to water planners.

In Oman, exploration indicates that water is very much a finite resource and therefore requires careful planning and management. However, the characteristic of water planning so far has been decentralised and uncoordinated exploration and development. As a result, certain areas have been considerably depleted, thereby greatly reducing future development possibilities. An uncoordinated water planning process militates against the effective and efficient use of the total resources of Oman. It is important therefore that there is not only a National Water Plan, but also a central controlling and management body.

Organization

Typically in most arid zone countries, many organisations are actively involved in the water resources sector. Water is crucial to so many facets of national life that genuine interests are widely dispersed. If there is also, as in the countries of the Arabian Peninsula, a rapidly growing population which demands increases in particularly the agricultural and industrial sectors, then there is bound to be competition between government agencies and other such bodies, not only to provide the necessary water, but to prevent damage or pollution to available supplies.

In Oman, the government structure of which is similar to other countries in the region, the different water-resources-related responsibilities can be illustrated by examining the roles of various relevant government departments and agencies, summarised by Johnson (1986). The primary responsibility for the municipal water supply, as interpreted from Royal Decrees 7/78 and 11/82, rests with the Ministry of Electricity and Water (MEW). MEW has been delegated the following responsibilities:

1 Plan, design, execute, operate and manage potable water supply systems for the Capital Area, towns and villages.
2 Plan, design, execute and assist local authorities in the operation of basic potable water supply systems in rural communities.
3 Provide technical support, as requested by agencies responsible for the potable water supply in Buraimi, Dhofar and Musandam.
4 Study and develop new sources of water supply, including groundwater extraction, surface supplies and desalination of seawater or brackish waters.

The Public Authority for Water Resources (PAWR) was set up by Royal Decree 63/79 as a relatively independent body, although it is chaired by the Minister of Electricity and Water and has its staff selected by him. The role of PAWR is essentially one of coordination through consultation with all Ministries with water resources responsibilities. For PAWR purposes, Oman is divided into five districts, with offices at Sohar, Seeb, Nizwa, Mudhairib and Salalah (figure 11.3). The main responsibilities of PAWR are as follows:

1 Propose policies, rules and regulations connected with water resources for government consideration.

2 Conduct research and undertake development projects to increase available water resources.
3 Carry out experimental projects, including hydrological exploration.
4 Train Omani experts in appropriate fields of science and technology.
5 Maintain a central data base of Oman's water resources, together with other programmes concerning the production and utilisation of water.
6 Establish its own financial and administrative regulations.
7 Assist the Water Resources Council in the coordination of all water resource activities.

In connection with responsibility (5), PAWR now operates the complete well permitting procedure. In the interests of conservation, 'no drill' areas have been identified and well digging is constrained within a certain distance of a falaj or another well. Each location needs to be carefully examined before a permit can be issued. Under responsibility (3), apart from the exploration drilling programme, the quality of water is monitored at the PAWR water analysis laboratory.

The importance of PAWR as the scientific, technical and coordinating agency is obvious and, as a result, its status has changed more than once. It was, for example, subsumed within a Council and, in 1989, was upgraded to a Ministry. Since agriculture is dependent upon irrigation, the Ministry of Agriculture and Fisheries (MAF) is heavily concerned within the water sector, although not bearing any primary responsibility for water resources. Nonetheless, some 70 per cent of available water is used in agriculture, as against the 30 per cent which is domestic and therefore comes directly under MEW. The responsibilities of MAF which are related to water are as follows:

1 Protect and repair the aflaj system.
2 Plan, design, execute, operate and manage an efficient supply system.
3 Study and develop new sources of supply for irrigation water, including aquifer recharge and groundwater extraction.
4 Collect water resource data.
5 Issue well drilling permits.

It can be seen that in the collection of data and the issuing of permits, there is an overlap between responsibilities of MAF and those of PAWR.

Two subdivisions of the Ministry of Communication (MOC), the Directorate General of Roads (DGR) and the Directorate General of Meteorology (DGM) have water resource interests. DGR is particularly concerned with data on surface flow for the design of roads, but more particularly hydrologic engineering structures such as bridges and culverts. DGM maintains a network of hydrometeorological observation stations and is also responsible for climatological and meteorological data collection and weather forecasting. Again, it is obvious that there needs to be close cooperation with PAWR if the data bank is to be developed and used effectively.

The Ministry of Regional Municipalities (MRM) is responsible for the waste water collection and disposal system for all cities and villages outside the Capital Area in which the responsibility rests with the Capital Area Municipality (CMA). MRM also maintains a health inspection programme in urban areas and this includes the inspection of the water supply and waste water. The quality of water for human consumption and also of waste water is part of the national environmental and occupational health programme which is the responsibility of the Ministry of Health (MOHE). Water samples are collected from storage tanks and wells on a weekly basis and subjected to chemical and bacteriological analysis in the MOHE water quality laboratory.

There is therefore a close coincidence of interest between MRM, MOHE and the Ministry of Commerce and Industry (MCI) which, under the Directorate General of Specifications and Measurements, developed water and waste standards for Oman and conducts chemical, physical and microbiological analyses of water, sewage and industrial wastes. The Directorate General of Industry, also within MCI, as part of the basic infrastructure of the country, establishes water supply works for industrial development sites.

The Ministry of Defence (MOD) has, through its Engineering Division (ED), responsibility for the installation of water supply and sewage treatment plants for military camps. Through its civilian section, MOD has its own programme for drilling wells and for constructing water distillation units.

The Ministry of Environment (MOE) exercises power widely in all aspects of the environment and includes among its responsibilities the enforcement of regulations and standards for the discharge of waste and waste water and the pollution of all water resources. The major concern of MOE is, however, with environmental impact statements, which must be approved before any project can proceed, together with the subsequent licensing and environmental monitoring.

Other Ministries and government bodies have a more restricted interest in water. The Ministry of Social Affairs and Labour (MSAL) provides potable water for new housing developments outside existing cities. In this, it overlaps with the Ministry of Housing (MOHO) which is involved in the supply of water to housing developments and also in the drilling of wells. The Ministry of Petroleum and Minerals is also, for obvious reasons, more constrained in its breadth of interest, but is important as a source of hydrogeological data and is also an important water user. Lastly, two Ministries are concerned with the legal aspects of water. The Ministry of Interior (MOI) acts as the first judge in settling disputes concerning water rights through the offices of the local Walis. The Ministry of Justice, AWQAF and Islamic Affairs (MOJ) maintains the register of awqaf properties such as religious or administrative trusts, for example of a falaj, including water rights. As indicated earlier, water rights and awqafs are of particular significance in water resource development, especially in Middle East countries.

Other bodies with a concern for water include the Office of the Minister of

State and Wali of Dhofar (WOD), which has a Directorate General of Water Supply and Transport, responsible for water provision for Salalah and the villages in the Jebel. The Diwan for Royal Court Affairs (DRCA) is in charge of the provision of potable and irrigation water to the Sultan's properties, throughout Oman.

With such a scarce resource as water and so many bodies, from one standpoint or another, concerned, competition and even conflict is always a possible outcome. At the lowest level of cooperation, the exchange of information and hydrological data, so important for effective national development, is likely to encounter difficulties. In an attempt to overcome these, a high-level council, the Water Resources Council (WRC), was established in 1975 by Royal Decree 45/75. In Omani terms, a council is a body established between Ministries to facilitate cooperation. His Majesty Sultan Qaboos is Chairman of the Council and other members are the Ministers of Agriculture and Fisheries, Electricity and Water, Communications, Environment, Housing, Interior, and Petroleum and Minerals, together with the Office of the Wali of Dhofar. The Council is the main authority for drafting the National Plan for Development of Water Resources and for the approval of all water-related plans and projects proposed by the various Ministries and agencies. The Council prepares an annual budget for water resources development, which is submitted to the Development Council, and it also establishes policy codes and licensing procedures to ensure orderly development. With so many wide-ranging water concerns throughout the government, the WRC is perhaps the only kind of body which could operate effectively. A Ministry having sole responsibility for water would be able to exercise more direct power, but would be involved in too many aspects of life to be efficient. Furthermore, many of the remaining Ministries would be effectively emasculated.

SUPPLY, DEMAND AND WATER BALANCE

The management of a national water system involves the coordination of all available sources on the supply side and the aggregation of the various water requirements. Demand can then be compared with potential supply and a possible water balance calculated for the particular time and budget. In the arid zone where demand is likely to outstrip supply, the determination of a water budget may necessitate prioritisation among users and allocation. Management at the national level subsumes the management of all the components such as recharge, irrigation and recycling. All the management procedures on both the supply and demand sides must be integrated into the national water management process. The number of variables and attendant problems is therefore extremely large and is compounded by the fact that the system in each country will be unique.

Supply

In analysing a water supply system, two key elements can be recognised: the sources and the infrastructure. The most obvious distinction among sources is between natural and artificial supplies, the latter having been produced through the activities of man. However, as discussed earlier, the hydrological cycle in the arid zone is essentially a control system and even this basic twofold division is not clear-cut. For example, groundwater may result from artificial recharge rather than natural infiltration. Any discussion on natural sources of supply highlights the contrast and sometimes the conflict between high-technology and low-technology approaches. In the truly arid zone, the major such source is groundwater and therefore aquifer management is the most vital component in the entire management structure. An initial handicap throughout much of the region is that the dimensions and parameters of the aquifers are, at best, imperfectly known. For example, unless there has been a reasonably detailed drilling programme allied to careful monitoring, variations in transmissivity throughout the aquifer can only be conjectured. Furthermore, such problems occur in the interpretation of shallow aquifers, while for deeper aquifers, even the recharge mechanism itself is speculative. Water is abstracted from aquifers by modern wells and pumps, but also by older wells and lifting devices and, in many areas, by the traditional aflaj or qanats. When increased abstraction is required, this raises the issue of whether modern methods should be employed or enhanced or the more traditional devices should be upgraded. In Oman, the accent is on falaj rejuvenation, but elsewhere in the Gulf, the focus tends to be on high-technology approaches. At a local level, the cooperation required to manage a falaj can be disrupted by the drilling of wells, particularly if they tap the same aquifer. The management of a falaj, involving hydrological, political and socio-economic concerns, presents a microcosm of the national system. Water distribution is controlled by social factors as much as agricultural or domestic requirements.

Surface water is of much less significance throughout the truly arid zone, but is important in semi-arid areas. Again there is a major contrast between modern hydraulics approaches and low-technology procedures such as rainfall harvesting. The channelling of runoff to natural depressions is an ancient skill still in evidence, particularly in the mountainous areas of the Middle East. Barrow (1987) provides a number of examples from other areas. The construction of permanent barrages and other such engineering works for storage or, more particularly, recharge has gained momentum in the arid areas. There are also more modern, low-technology approaches such as the use of gabions to form temporary structures.

Whether with groundwater or surface water, the more traditional and the low-technology techniques tend to be more environmentally sensitive and congruent with the natural systems. They are also comparatively flexible, inexpensive and easy to maintain by local labour. In contrast, high-technology approaches are usually costly to install and maintain and they may introduce a chain of unforeseen changes in the environment. At the local level, as indicated earlier, well drilling may result in the destruction of village cohesion. The emplacement of a major dam may induce a series of problems downstream with regard to irrigation, erosion, fishing and navigation. Such a catalogue has been evident in lower Egypt since the opening of the Aswan High Dam.

The origin of most surface water and groundwater is rainfall, but in certain areas such as Iran, this may be supplemented by snow melt. As a result, the flow and recharge regimes may be complex. Additionally, many arid coastal areas are subject to a heavy seasonal incidence of fog, mist and dew. It is increasingly realised that this is vital for the ecosystem (Anderson, 1988a) and in some areas the occult precipitation which results is thought to be sufficient to produce measurable aquifer recharge. The ecological system of the Dhofar Mountains in Oman appears to subsist to a very large extent upon such precipitation (Stanley Price *et al.* 1986).

Among the artificial sources, by far the most important is desalination of brackish or seawater. Depending on the process, extremely pure water may be produced and then blended as required with brackish water or the feed stock itself may be processed to the required purity. To obtain the various advantages, more hybrid desalination plants are now being constructed. From the management viewpoint, the main problems are the scale of production necessary and the capital and energy costs of desalinated water. As costs decline, probably through the use of solar energy and as small-scale units become viable, it will be possible to introduce greater flexibility into national water planning.

Recycling is the other increasingly important artificial water source. The bulk results from domestic or industrial use and, depending upon the degree of purification, can be used for irrigating municipal amenity areas or crops. However, in the arid zone, sewage treatment plants are restricted to only the larger urban areas in the more advanced countries. In the Middle East, Israel was the pioneer with the Dan treatment unit, Malta has one small treatment unit at Sant'Antnin but discharges most of its sewage to sea, while there is as yet no recycling in Oman. Industrial recycling is generally of significance only for the plant concerned and, in many cases, effluent is too polluted for recovery. The process of irrigation leads to what might be considered a specialised form of recycling. Unused water may provide surface runoff to other irrigated areas or may percolate to the aquifer. In either case, there will be an increase in dissolved solids, particularly salts. On a large scale, the effects of the various Turkish and Syrian irrigation schemes will be increasingly seen in the water quality of the Euphrates as it enters Iraq.

The scarcity of water in Israel has led to considerable interest being shown in the utilisation of urban waste water and brackish water (Buras and Darr, 1979). A non-linear mathematical model has been formulated and has shown that these sources have substantial values, ranging up to 65 per cent of the cost of imported freshwater. The model, when applied, indicated that the optimal allocation of water for the Central Negev in 1992 will call for 40 per cent of the forecast total water demand to be supplied by treated urban waste water and brackish water.

The supply infrastructure includes pumping stations, chlorination and treatment facilities, storage installations and pipelines. These all require careful planning with regard to capacity and cost. In the Middle East, in all but the longest established cities, the network was extremely restricted and the laying of pipelines to provide a modern service posed a wide range of problems. For example, the Capital Area of Oman needed some 1,000 km of pipe, varying in diameter from 100 mm to 1,200 mm, to lay the initial system. To prevent accidental damage, vandalism or sabotage, the pipelines are generally laid at anything from 1 metre to 6 metres deep, which poses problems, not only of land acquisition, but also of subterranean competition with other service pipelines. The effective management and monitoring of the completed network is vital, not only to satisfy consumers, but also to guard against significant losses through leakage.

Apart from the necessary provision, the main problems on the supply side involve conservation and pollution. With surface flow and surface storage, the major losses occur through evaporation and seepage. Significant savings can be achieved through covering storage tanks and through lining channels, particularly those which are only used occasionally. In the case of ephemeral flows and flash floods, a greater problem may be losses to the sea. The development of engineering structures to impede or spread the flow to induce recharge can therefore be seen as an integral part of conservation. The main constraint on subterranean losses is likely to be through the efficient lining of well shafts and aflaj tunnels. Within the distribution infrastructure, the larger leakages can be detected and traced through a careful monitoring of the system. Pollutants can enter the national water system in many ways and can cause a range of damage from corrosion to ill health in the population. If such substances were to enter the intake pipes of a desalination plant, the effect on the national water supply could be disastrous. The realisation of this in the Persian Arabian Gulf was shown by the arrays of booms and similar devices which were constructed under threat from the 1981 oil slick.

Toxic substances and other pollutants may enter the water supply as a result of the normal chemical weathering of rocks, or through multiple use during irrigation. In settled and particularly urban areas, untreated sewage and industrial waste can lead to pollution. All such problems are likely to be exacerbated through flood damage. However, the most widespread problem occurs in coastal areas where excess abstraction has reduced the seaward pressure gradient and saline incursions occur.

Demand

Whereas the available water supply can be comparatively easily calculated, demand has to be anticipated. The main elements of demand are municipal and domestic, industrial and agricultural, but growth in each of these depends not only upon national strategy, but upon other variables which are less easy to forecast such as population growth, urban migration and per capita water demand. Nationally, the population may increase at perhaps 3 per cent per annum but locally there may be abnormal growth as a result of government activity, the establishment of industry or other enhancing factors. Furthermore, even in these cases of excessive growth, the rate will vary over time. Per capita demand will vary according to such mundane factors as the proportion of the population with homes attached to the infrastructure, the proportion using standpipe supply and the proportion served by bowsers. It also rises with improvements in the general standard of living, whereas in nomadic society, usage may be in the order of 10–30 litres per day, in a village this is more likely to be 60–80 litres and in a modern housing development, anything from 400–800 litres per day. A further complicating factor is that various peripheral uses in settled areas such as for the irrigation of ornamental gardens, for the maintenance of livestock or for the use of industrial and commercial institutions are frequently aggregated with the domestic to produce the per capita consumption, both of which are uncertain and need to be estimated. Furthermore, these factors are likely to vary considerably from region to region within a country. In the case of the arid realm, with burgeoning populations and rapidly rising consumption rates, the statistics are particularly volatile and unreliable.

The installed capacities of the different elements within the water system must be planned in accordance with anticipated demand, but must also take into

Table 11.9 Planned expenditure on drinking water and sanitation ($ millions)

<table>
<tr><th></th><th colspan="2">Water supply</th><th colspan="2">Sanitation</th><th></th></tr>
<tr><th></th><th>Urban</th><th>Rural</th><th>Urban</th><th>Rural</th><th>Total</th></tr>
<tr><td>Egypt</td><td></td><td></td><td></td><td></td><td></td></tr>
<tr><td>1981–5</td><td colspan="2">600</td><td colspan="2">1,400</td><td>2,000</td></tr>
<tr><td>1986–90</td><td colspan="2">700</td><td colspan="2">1,000</td><td>1,700</td></tr>
<tr><td>1981–90</td><td colspan="2">1,300</td><td colspan="2">2,400</td><td>3,700</td></tr>
<tr><td>Jordan</td><td></td><td></td><td></td><td></td><td></td></tr>
<tr><td>1981–5</td><td colspan="2">470</td><td colspan="2">270</td><td>740</td></tr>
<tr><td>Iraq</td><td></td><td></td><td></td><td></td><td></td></tr>
<tr><td>1981–90</td><td>544</td><td>967</td><td>1,990</td><td>580</td><td>4,081</td></tr>
<tr><td>Libya</td><td></td><td></td><td></td><td></td><td></td></tr>
<tr><td>1981–5</td><td colspan="2">770</td><td colspan="2">1,300</td><td>2,070</td></tr>
<tr><td>Saudi Arabia</td><td></td><td></td><td></td><td></td><td></td></tr>
<tr><td>1981–5</td><td colspan="2">1,887</td><td colspan="2">2,849</td><td>4,736</td></tr>
<tr><td>1986–90</td><td colspan="2">1,887</td><td colspan="2">2,849</td><td>4,736</td></tr>
<tr><td>1981–90</td><td colspan="2">3,774</td><td colspan="2">5,698</td><td>9,472</td></tr>
</table>

(From International Drinking Water Supply and Sanitation Decade Directory, 1984)

Table 11.10 Urban population water supplies (population supplied, millions)

	Urban supply	*Private supply*	*Public standposts*	*Sewer connection*	*Septic tanks & pit privies*
Algeria					
1980	10.0	9.0	1.0	–	–
1985	13.5	12.8	0.7	12.8	0.7
1990	17.5	17.2	0.4	17.2	0.4
Iraq					
1980	8.3	–	–	–	–
1990	11.3	–	–	–	–
Jordan					
1980	1.6	1.2	0.3	0.3	1.2
1985	2.0	1.7	0.2	1.1	0.7
1990	2.2	2.1	0.1	1.6	0.5
Libya					
1980	2.6	2.5	0.1	1.2	1.5
1985	3.6	3.5	0.1	2.0	1.5
1990	3.8	–	–	–	–
Saudi Arabia					
1980	6.4	2.2	3.6	1.2	3.9
1985	7.9	5.0	2.5	3.4	3.1
1990	9.4	7.8	1.6	6.5	2.3
Syria					
1980	3.4	3.4	0.0	2.6	0.0
1985	4.1	4.1	0.0	3.3	0.0
1990	4.9	4.0	0.0	4.4	0.0
Tunisia					
1980	3.5	2.5	–	1.6	1.9
1985	3.9	–	–	2.4	–
1990	4.3	4.3	0.0	3.5	–

(From International Drinking Water Supply and Sanitation Decade Directory, 1984)

account the peaking factor. For each, the peaking time factor is likely to vary. Installed capacity and storage are designed to take into account monthly peak demand, the distribution is constructed for hourly peak demand, while pumping capacity and transmission pipelines are geared to maximum daily demand (Dawood and Subramanian, 1985). Given the general fragility of the system, and the low reliability of certain components, it is also vital to build in an emergency surplus capacity. Finally, any such system needs to be designed with potential future demand in mind and forecast of this must be highly conjectural.

The best available for the Middle East shows that by the year 2000 conditions will be severe. With statistics on total available water, from both surface and ground sources, together with current rates of population increase, water availability per capacity in the year 2000 can be calculated (Barney, 1980) and there are some very large shortfalls. While Algeria will have just sufficient, Morocco and Tunisia will experience shortages. However, the extreme cases occur with the Arabian Peninsula states and Egypt. The most satisfactorily placed country is Iraq, but upstream developments in Turkey and Syria may alter the picture (Anderson, 1988c).

Both the accuracy of demand estimates and demand priorities vary considerably from country to country throughout the arid zone. For example, in Saudi Arabia the rural municipalities suffer from poor water conditions as a result of environmental factors, the explosive population growth, the lack of treatment facilities and the absence of proper maintenance and operating procedures (Manwaring *et al.*, 1980). In Oman, direct measurement of agricultural water usage is a major problem and the pattern and calculation of demand is inferred from satellite imagery. The very large planned expenditure in the decade 1981–90 by a sample of Middle Eastern countries on potable water supplies and sanitation services illustrates the priority accorded these basics (table 11.9). Even with these large outlays, services will still not be complete by 1990 (table 11.10) (World Water, 1984).

Each state in the arid zone of the United States has its own detailed water plan. A state such as Arizona mounts an annual watershed symposium to consider all aspects of state water management. Agricultural use is accurately monitored and industry in general takes a larger share than in most other parts of the arid world. Thus, problems may range from a consideration of further sources of cooling water for power plants (Jurym *et al.*, 1979) to the impact of second-home development on availability (Bond and Dunikoski, 1977).

Priorities

Given competing national and sectional interests, prioritisation presents a key problem on the demand side. Overall national strategy sets the general aims and implied priorities, but, in attempting to realise these, there may well be conflict at lower levels of policy and planning. The economic return from the use of the water is clearly a major concern, but is not overriding. As discussed earlier, the distribution of water in an arid country exercises a major influence on the political, economic and social systems. Far more revenue may accrue from a piece of land if it were used for a leisure complex than if it remained under agriculture, but the decision to support local farmers may well be a national priority. Thus, although the return may be far less, water may be used to irrigate subsistence crops, rather than ornamental gardens in a tourist complex. Whatever the priorities, however, there is bound to be competition for a scarce resource and inequities will lead to conflict.

In the arid realm, competition is most often acute in the agricultural sector. As wells are dug deeper, the water table is lowered and more shallow wells are left dry. Eventually, the aquifer is totally exhausted or, depending on the situation, seawater incursions may occur. In California, as a result of such competition, farmers cooperated and petitioned the State legislature to create a local government agency, the Kerm County Water Agency, with professional secretariat and elected officers. The maximum permissible rates of pumping were calculated and meters installed on every well in the county. The pumping tax charged provided an incentive for farmers to use water-saving technology and the income raised was used for artificial recharge projects. This initiative illustrates the point that the problem is more political than hydrological. To calculate safe extraction rates is difficult, but to persuade farmers to give up rights and accept regulation is likely to pose even greater problems.

As shown by the use of well and falaj water in the same village, competition between commercial and traditional agriculture is also potentially damaging. Additionally, commercial agriculture being, by definition, on a larger scale and producing more obvious economic benefits might be favoured from the production viewpoint. However, from the social and even cultural aspect, the survival of traditional agriculture may be considered more important. Indeed, if it is not supported, throughout much of the Middle East the rural areas will be depopulated and will become effectively socially and economically obsolete.

Conservation

The other major problem, more acute than on the supply side, is conservation. Recycling, an increasingly important element in any conservation programme, can by its nature be classified under both supply and demand. The policy on waste water should include the identification of essential sources, the determination of uses and possible users, the establishment of criteria for treatment for the various uses, safety levels and measures to maintain them, systems for the control of water quality and its application, the education of the general public on the uses and potential hazards of recycled water, and the establishment of legislation on water reuse (United Nations, 1985).

Within the domestic and municipal sector, conservation can be achieved through public awareness campaigns, metering and the levying of appropriate charges, the installation of water-saving devices and low-technology techniques for obtaining more benefits from the same water supply. Leakage detection throughout the system is also crucial, since this sector accounts for the major part of the distribution network. Additionally, water use targets should be set so that the public and other users become water conscious.

In the agricultural sector, the key concept is of irrigation efficiency. Minimum water requirements need to be identified for the major crops and converted into minimum required water allotment per hectare. Water application techniques need to be examined so that, where feasible, inundation can be replaced by drop or sprinkler irrigation. Furthermore, storage water surfaces should be covered, channels should be lined, land accurately levelled and water reused. Given reasonable conservation methods, a water use target could then be set.

As appropriate for each specific industry, water conservation technology consistent with a reasonable economic return and with public interests should be introduced. Again, with these in place, reasonable water use targets could be set. However, wastage in most water systems results from leakages and the detection and treatment of these is a key conservation measure. Unfortunately, location and sealing of leaks require reasonably sophisticated technology and are comparatively expensive. As a result, leakages and losses in the Capital Area of Oman amount to some 30 per cent of the total water throughput.

The water balance

A water balance is a quantitative summary of all the inputs to and outputs from a water system. Inputs, principally in the form of rainfall and recharge are compared with the outputs resulting from abstraction and other losses. This will indicate for an aquifer the safe yield, whereby inputs are balanced by extraction and outflow and beyond which there will be a deficit, which, if prolonged, will lead to the mining and degradation of the aquifer. As an example, the water balance averaged for the 5 year period to 1976 for Qatar is presented in table 11.11. It shows a significant deficit for the northern provinces and a minor increment to storage for the southern. The progressive decline in the level of the water table in the northern wellfields is likely to result in saline incursions.

As a result of this adverse balance, the Government of Qatar decided to expand its desalination capacity, to close the northern wellfields, to use quantities of brackish water for blending, to investigate the construction of recharge wells and to examine the possibility of recycling sewage water from Doha. Increasingly such water balance planning is facilitated by models. For example, a water supply planning model has been formulated for use in the arid underdeveloped Sahel–Sudan region of West Africa (Kirshen, 1979). Its purpose is to determine either the least-cost set of sources to meet fixed demands or the sources and water uses that maximise the net benefits of water use. It can be adapted for either large- or small-scale development and the function can be expressed as either the minimisation of supply costs or as the maximisation of net benefits of water use.

However the balance is modelled, with growing populations and increasing per capita demands, in the arid zone there are always likely to be shortfalls. These

Table 11.11 Qatar annual water balance (MCM)

Groundwater zone	*Recharge*	*Extraction*		*Outflow*	*Storage Change*
		Agric.	*Municipal*		
Northern Province	17.6	34.4	4.5	6.3	−27.6
Southern Province	14.2	9.0	0.1	3.8	+1.3
Total	31.8	43.4	4.6	10.1	−26.3

(From UNDP, 1977)

may be overcome by the high-cost alternative, desalination, or by desalination and the blending of groundwater in ratios between 4:1 and 2:1. Recycling is also comparatively expensive, but is more attractive as the need for sewage treatment installations becomes more apparent.

Recharge from storm water runoff presents a range of technical problems and, with the lack of detailed geological knowledge in much of the arid zone, presents risks. There have been very few controlled experiments to judge the effectiveness of such enhanced recharge. Recharge by distillate from desalination is an expensive option with the same risks attached.

Water can also be saved by all the conservation procedures discussed in both the context of the supply and demand side. More far-reaching possibilities would range from the improvement of irrigation efficiency to the conversion of agriculture to a capital-intensive operation. Further drastic possibilities include the rationing of water and the raising of water prices. A stepped tariff might be introduced to protect small users and to encourage commercial operations to introduce conservation techniques. The revenue generated would of course allow the government to fund extra development programmes. In the short term, deficits can be overcome by improving storage facilities, the procedure traditionally practised in Malta. Since in arid climates stored water must be protected from evaporation losses, the increase in storage facilities is an expensive alternative.

The only other possibility offering substantial increases in supply is through the development of new aquifers, either internally or externally. Internally, this is likely to entail at least significant transport costs since most accessible aquifers near populated areas have already been exploited. Deep aquifers may provide the answer in some cases, but the bulk of current opinion is that these are not being replenished and therefore such withdrawal is water mining. The import of water must also be considered, but the costs of pipeline construction are likely to be such that only the best-quality water would be imported and to rely on an external source for potable supplies results in strategic vulnerability. Tankers are strictly limited in their capacity and, for most countries, can only be considered for emergency supplies.

Finally, various new technologies may help to alleviate the situation in the future. The development of halophytic crops which can be watered directly by seawater will lead to immense savings in the agricultural sector. Cheaper desalination techniques, using solar power, may allow distillate to be used more flexibly. The exploitation of secondary aquifers, which exist in sedimentary, metamorphic and igneous rocks, has been successfully pioneered and may be an important factor in future groundwater development (Hoag and Ingari, 1985).

Flood hazards

In arid areas, the irregularity, frequently high intensity and short duration of rainfall produces a distinctive flood hydrograph, which is characterised by extremely steep rising and recession limbs. Since the entire event occurs over a very short time period, commonly less than 24 hours, changes in discharge can be extremely dramatic. For Wadi Dayqah in Oman, for example, the highest peak flow which was recorded in March 1982 is over 100,000 times the lowest recorded in July 1985 (Anderson and Curry, 1987). The destructive force of such floods can of course result in widespread damage, including the disruption of the infrastructure which embraces the sewage and water systems. However, the destruction caused may be considered less important than the health hazards occasioned by the resulting pollution.

Not only are lives, homes, commercial and industrial premises at risk, but also agricultural land and irrigation systems. Thus, legislation on controlling development in areas liable to flooding is required and, apart from supply and demand, is the other major responsibility of a national water authority. The legislation needs to be supported by evidence from flood hazard mapping. For urban areas, it is vital to know the return periods for floods on land near wadi channels. Although the isopleths for 10 year, 20 year or 50 year floods only indicate probabilities, they do provide valuable guidelines (Cooke *et al.*, 1982). Between 1988 and 1990, PAWR mounted just such a programme in Oman to relate channel form to flood return period and to map degrees of flood hazard. The chief concern is the potential flooding in the Capital Area, Salalah and Nizwa (Anderson, 1988a).

REGULATION

The control of water sources, the implementation of conservation measures, the allocation of supplies to consumers and, as a result, the maintenance of a national water balance, together with the provision of flood control measures, all imply control. The operation of a national water body therefore needs to be set clearly in the context of national law.

In Oman, the Water Resources Council has been given the following responsibilities according to Royal Decree 76/77 and the Water Resources Development Law, which specifies in Article 2:

> To set objectives and prepare a Policy for Water Resources development, to submit needed proposals for a long-term Water Plan in harmony with the economic development plans of the country. This Plan will be submitted to the Development Council for approval.
> To coordinate the activities of Ministries and governmental departments in so far as the implementation of the Water Plan is concerned.
> To issue rules and regulations concerning water resources development and preservation.

A complete set of such policies, rules and regulations would constitute a Water Code, and in Oman the authority to promulgate such a code derives specifically from the following articles of the same law, already quoted:

Article 10

(a) The Water Resources Council shall be responsible for the issuing of licences and permits according to conditions it will set later, for the purpose of controlling the drilling of wells or the exploration for the use of water from any source.
The use of water, its storage and its recharge will be subject to the conditions specified in these licences, and permits as issued by the Council from time to time.

(b) Whoever violates the procedures, regulations and conditions of licences, as issued by the Council and whoever makes use of any water resources for any reason without having a licence as mentioned in (a) above and whoever misuses the Water Resources, wastes or pollutes them will be subject to a fine.

In accordance with Sharia Law, an exception from (a) and (b) above is water to be used for drinking by humans or animals or for reasonable domestic uses.

Article 11

The Water Resources Council Secretariat will keep a record of the water rights, according to different permits, in a Water Register.

Article 12

The Water Resources Council will seek and establish an Executive Department to administer the Water Resources Development in the Sultanate, as well as to implement the different permits.

Article 13

The Water Resources Council shall prepare a Water Code and issue regulations for implementation of this law to achieve its aims.

A Water Code would include articles on policy, licensing, the issuing of permits, water rights and regulations. In the case of Oman, there are additional sections on administration and management plans.

Water policy relates directly to national strategy and therefore reflects, in the arid zone, the basic national water problems. Fundamental to this would be the establishment of priorities for allocation. In the case of Oman these are as follows:

1 Water for human consumption.
2 Water for animal consumption.
3 Water for domestic use.
4 Water for commercial uses in agriculture or industry.
5 Water for landscape irrigation.
6 Water for recreational purposes.

Among the various competing commercial uses, precedence might be given according to economic return or to other factors such as those of cultural importance which might be considered long-term benefits for the country. For example, in Oman there is a definite policy with regard to aflaj:

> The government shall select several Aflaj to be maintained as examples of Oman's national heritage; the Aflaj thus designated shall be exempt from the stipulations of the Water Code and they shall be considered Protected National Sites in accordance with Royal Decree 26/79.

To ensure correct practice, licences would be required for both drilling wells and monitoring water abstraction. Details of the procedure to be adopted and the reports to be submitted would be included in the Water Code. Permits would be needed for exploration, water use, water recharge and well ownership. In each case, full reports of the operation would be requested.

Before water use permits are issued, it is important that, particularly in rural areas, water rights are established. The basis on which these were established may well be lost in time, but a legal determination must be made and, at a given date, the new system brought in and any pre-existing rights must be judged null and void. Clearly, there also need to be controls on the transfer of water rights.

Such water regulations all require enforcement and details of this would be set out in a Water Code. Details of inspections, a classification of possible violations and a tariff of penalties would need to be included. Water law, as set out in a Water Code, might also consider data gathering, rights of non-users of water, protection against any harmful effects of water, judicial aspects, the classification of water, and approval for any new use techniques (Cano, 1978).

All such facets of the water system need to be enshrined in law in that as water becomes an even more scarce resource, conflict is bound to occur. In the arid Western United States, one doctrine which has been developed as a result of scarcity is implied reservation. When Congress sets aside public land for a particular use, it implicitly reserves water rights for the benefit of that land. These rights allow the abstraction of sufficient water to accomplish the intended purpose. Economic, social and political issues are intertwined, but two basic problems are likely to arise:

1. how the existence or non-existence of a federally reserved water right will be determined in particular situations; and
2. once the existence of such a right has been established, how the reservation will be quantified (Murray, 1977).

12

CONCLUSIONS

The arid realm is an enormous region covering a third of the Earth's land surface and supporting a population of 400 million. Arid lands have been presented as having only one common feature, that is, for at least some of the time, a shortage of water. All other attempts to characterise biological conditions, geomorphological features or even human resources have encountered great difficulty due to the diversity of environments. Nevertheless, in addition to combating aridity, the inhabitants of arid lands also have to face great spatial and temporal diversity of water resources, presenting acute problems for the development and maintenance of water supplies. Rainfalls are often highly localised and intensive, producing surface runoff characterised by a high peak discharge and high sediment loads. The recharge of groundwater found close to the surface is subject to the vagaries of rainfall while the extent and water balance of deep aquifers are frequently unknown.

It is appropriate then to consider the shortage of water resources of the arid realm as the one serious environmental problem faced by all arid lands. While accepting that there are no universal causes or solutions to water supply problems, this theme has been used to study the different technologies and management strategies that have been employed to combat water shortages.

Difficulties created by the variable nature of the hydrological cycle are accentuated, however, by confusion over the occurrence and causes of water shortages. Terms such as 'aridity', 'drought' and 'climatic change' are used with little rigour and yet are often cited as the cause of hardship and famine in the region. It is, however, difficult to establish 'norms' when there are few observations through time and where established 'stations' are separated by great distances. Coupled to this are practical problems of data collection in a harsh environment where heat, dust, remoteness and the occasional very wet period lead to mechanical breakdown of instruments and data losses. There are new developments in the field of remote sensing, use of automatic devices and more robust instrumentation but the historic data record remains patchy and incomplete. It follows that 'averages' are difficult to quantify and uncertainty surrounds the identification of trends and the occurrence of abnormal conditions.

Uncertainty over water supplies is exacerbated by the growing imbalance between supply and demand due to the growing urban populations, rising standards of living, industrialisation, growth of irrigation and the provision of sanitation and potable water supplies. The per capita water demands of industrialised cities may soon be approaching 1,000 litres while the rural inhabitants of many arid lands presently survive on only 10 to 20 litres each day. The drive for better health and greater agricultural and industrial production has a pronounced impact on water demands and potable water is now seen by the United Nations as a basic human right. Thus the water supply problem has been called a 'people' or more rightly a 'development' problem. The challenge facing arid nations in particular is to provide, when demand is accelerating, sufficient water resources during periods of scarcity without environmental degradation.

Various techniques have been developed for the regulation and increase of water supplies ranging from cloud seeding through to conservation. Although most have an ancient history, some such as cloud seeding and desalination are essentially 20th century technologies. Cloud seeding continues to receive attention and continues to be criticised for the lack of scientific proof. Given the complexities of the method and the vagaries of cloud systems, it is likely that there will always be those who question the outcome of this method. It does, however, require specialised knowledge and equipment and is unlikely to be employed routinely in many arid lands but will be resorted to in times of crisis, i.e. drought. Desalination in contrast is becoming ever more attractive as a supplier of potable water. Unfortunately it still requires significant capital investment, a cheap energy supply and produces large amounts of salt waste. At present desalination is mostly found in oil-rich or economically developed nations but there are prospects of less expensive, scale-neutral plants being developed as reverse osmosis challenges distillation for supremacy. It is then possible that desalination will become more widespread but only for potable water supplies given the higher costs of this technology compared to more traditional water supplies.

Irrigation, dams and interbasin water transfers can be considered together as they have an ancient ancestry but have proliferated during the 20th century and are currently being re-evaluated because of their high costs and associated environmental damage. Irrigation has become an exploitive technology with insufficient regard for sustainability. Emphasis is upon production and financial returns with soil salinisation, rising groundwater tables and increases in disease often the result. Environmental problems are exacerbated in large-scale projects which are unable to be modified in the light of experience, where bureaucracy is inefficient and training is inadequate. Arid lands accentuate these difficulties due to naturally occurring salts, the variability of rainfall and inadequate data bases upon which to design large and complex systems. In many parts of drylands population densities are low; if irrigation were to be introduced this would require the migration of farmers or sedentarisation of pastoralists, possibly with little appropriate experience and ensuing social disruption.

Small-scale production can tackle some of these problems through less social disruption, appropriate levels of technology and production. However, against this must be weighed the difficulties of smaller enterprises attracting finance, regulating water supplies, providing seeds and chemicals and making a significant contribution to the economy. Irrigation will continue to operate at a variety of scales using a range of technologies but the greatest asset of any scheme will be flexibility to respond to changing human and environmental circumstances.

Dams have many purposes including water storage, groundwater recharge, hydroelectric power production, flood control, maintenance of low flows and improving navigation. Their adverse environmental effects are seismic disturbances, alteration of the hydrological regime downstream, evaporation losses, the threat of dam failure and an increase in water-related diseases. As most suitable sites are rapidly used up so the attractions of large-scale, multi-purpose surface dams is being questioned. Whether they can truly alleviate floods is uncertain and their longevity is frequently overestimated as soil erosion in the upper catchments results in much more rapid rates of sedimentation than expected. Interbasin transfers of water also suffer from enormous financial costs and threaten significant regional environmental changes. Alternatives are therefore sought and small-scale dam construction and groundwater storage are looking increasingly attractive.

Small dams can be built with local skills and materials, they are less of a threat to the environment and even HEP is becoming possible through technological advances. They are, however, only suitable for rural supplies. Large urban areas, intensive agriculture and industrial estates are unlikely to be satisfied. Groundwater can make a more suitable medium for the storage of water though reduced evaporation losses, possible improvements in water quality and reduced costs of construction, although financial savings may not be great for artificial recharge when the costs of recharge dams, surface lagoons and deep recharge wells have to be included.

The exploitation of groundwater resources is hampered by vague figures on aquifer water balances and recharge rates although this source is increasingly attractive with larger pumping capacity. In many areas, however, overpumping has resulted in declining water quality, land subsidence and seawater intrusions. Groundwater resources therefore need to be accurately assessed, monitored and extraction carefully controlled. Water conservation is important and the recycling of waste water has been adopted by several arid land countries using artificial and natural processes. The costs of purification can be significant, however, depending upon the amount of contamination and there are cultural obstacles preventing the reuse of such water other than for municipal gardens and industrial processes. In addition, the purification process must be effective, particularly when treated water is fed back into surface or ground waters, to prevent health risks.

There are then a range of techniques, operating at a variety of scales to enhance water supplies in arid lands and providing a variety of choices to those planning water resources. Case studies of the Sultanate of Oman, Malta and

Israel have shown that there are then a number of permutations depending upon financial, social and environmental circumstances. Water management cannot be divorced from national planning and in arid lands the need to consider water as a strategic resource is only too apparent. The desire to develop nationally 'secure' water supplies is, however, constrained for many countries dependent upon exogenous surface flows (Nile, Niger, Tigris, etc.) or where major aquifers are recharged in regions beyond national boundaries. The geopolitical nature of water supplies is then an acute problem for many arid lands. Indeed in the Middle East the debate on water geopolitics has already achieved a high profile.

Attempts to balance supply and demand have by and large focused upon increasing supplies through exploiting new resources or regularising surface flows. Comparatively little consideration is given to controlling demand through population control, water pricing and metering, water extraction permits, limiting irrigation, restricting industrialisation or urban expansion. In part this is due to the greater priority given to economic development and the improvement of health and sanitation. Hence an increasing demand for water is inevitable, but as water conservation becomes more important so attention will inevitably focus on reducing demand through education, financial measures and legislation.

The river basin has been suggested as the most suitable geographical framework for water planning and while in some arid lands this is appropriate, for others it is meaningless. When surface flows are ephemeral and unpredictable, groundwater regions can be a more appropriate unit for water balance computations. There is then no universal framework or solution for water management in arid lands, and the thrust of this text has been to review case studies without being too prescriptive. We have largely focused upon technical issues of water supply but chapters 10 and 11 have illustrated the range of management permutations. In an environment where prediction faces particularly acute difficulties it is necessary to be able to respond to the unforeseen, both threats and opportunities. Flexibility is then the key, both in scale of approach and decision making. Some may see this as a recipe for 'mayhem', a disintegration of authority where decisions are never made or carried through. Flexibility is the ability to reassess plans in the light of new information. An essential component of any water strategy should then be monitoring and mechanisms to review such data.

A second equally vital element is the realisation that water is not infinite. It is a resource that needs conserving and if necessary rationing. Preventing environmental degradation is perhaps the highest priority for any water development scheme for the impoverishment of existing water resources through overexploitation threatens the future for many arid lands. While cloud seeding and desalination offer 'high-tech' solutions they cannot replace the natural replenishment of water resources. There is then an urgent need to complete water inventories and to prevent further loss through overpumping and contamination.

REFERENCES

Adams, A. 1978 The Senegal river Valley, what change?, Review of African Political Economy 10: 33–59.

Adams, W.M. & Carter, R.C. 1987 Small scale irrigation in sub-saharan Africa, Progress in Physical Geography 11(1): 1–27.

Adams, W.M. & Grove, A.T. (eds) 1984 Irrigation in Tropical Africa, *Cambridge African Monographs 3*, CUP, 148 pp.

ADAS 1976 Sandland irrigation 1975, Gleadthorpe experimental husbandry farm, MAFF, HMSO.

Agnew, C.T. 1980 The water balance approach to the development of rainfed agriculture in S.W. Niger, PhD Thesis, University of East Anglia.

Agnew, C.T., 1982 Water availability and the development of rainfed agriculture in S.W. Niger, Transactions of the Institute of British Geographers 7: 419–57.

Agnew, C.T. 1983 *Pastoralism in the Sahel*, U204, Third World Studies Course, Open University, Milton Keynes.

Agnew, C.T. 1984a Report on the results of an automatic weather station at Lake Ichkeul, in Stevenson, A.C. (ed.) *Studies at the Ichkeul National Park, Tunisia*, Ecology and Conservation Unit, University College London.

Agnew, C.T. 1984b Checkland's soft systems approach – a methodology for geographers? Area 16(2): 167–74.

Agnew, C.T. 1985 Evaporation and evapotranspiration, in Goudie, A. (ed.) *The encyclopaedic dictionary of physical geography*, Basil Blackwell, Oxford, 528 pp.

Agnew, C.T. 1986 Moisture in the Wahiba sands, in Dutton, R. (ed.) *Oman Wahiba Sands Project*, Royal Geographical Society, London.

Agnew, C.T. 1988a Soil hydrology in the Wahiba sands, Journal of Oman Studies 3: 191–200.

Agnew, C.T. 1988b Dewfall and atmospheric conditions. Journal of Oman Studies 3: 213–216.

Agnew, C.T. 1989 Sahel drought, meteorological or agricultural? International Journal of Climatology 9: 371–82.

Agnew, C.T. 1990a Spatial aspects of drought in the Sahel, Journal of Arid Environments 18: 279–93.

Agnew, C.T. 1990b Green belt around the Sahara, Geographical Magazine LXII(4): 26–30.

Agnew, C.T. 1991 Evaluation of a soil water balance model for the analysis of agricultural drought in the Sahel, in Sivakumar, M.V.K. *et al.* (eds) *Proceedings of Workshop on Soil Water Balance in the Sudano Sahelian zone, February 18–23*, IAHS Publication 199, Wallingford, pp. 583–92.

Agnew, C.T. & Anderson, E.W. 1988 Dewfall and atmospheric conditions, Journal of Oman Studies Special Report 3: 213–16.

Agnew, C.T. & O'Connor, A.M. 1990 The Meteorological scapegoat, Geographical Analysis Sept., 1–4.

Ahlcrona, E. 1986 Monitoring the impact of climate and man on land transformation, Report 66, Lunds Universitets Naturgeografiska Institution, Sweden.

Alam, M. 1990 Water resources of the Middle East and North Africa with particular reference to deep artesian groundwater resources of the area, Water International 14: 122–7.

Allan, J.A. 1985 Irrigated Agriculture in the Middle East: the future, in Beaumont, P. & McLachlan, K. (eds) *Agricultural development in the Middle East*, Wiley, Chichester.

Allen, C.H. 1987 *Oman*, Westview Press, Boulder.

Amin, M.A. 1978 Problems of schistosomiasis, in Worthington, E.B. (ed.) *Arid land irrigation in developing countries*, Pergamon Press, Oxford, pp. 407–11.

Anderson, D.E., Verma, S.B. & Rosenberg, N.J. 1984 Eddy correlation measurements of CO_2, latent heat, and sensible heat fluxes over a crop surface, Boundary Layer Meteorology 29: 263–72.

Anderson, E.W. 1986 Water Problems in the Sultanate of Oman, Geojournal 13: 3.

Anderson, E.W. 1988a Flood studies: phase one, Council for Conservation of the Environment and Water Resources, Sultanate of Oman.

Anderson E.W. 1988b Water: the next strategic resource, in Starr, J.R. & Stoll, D.C. (eds) *The Politics of scarcity*, Westview Press, Boulder.

Anderson, E.W. 1988c Water resources and boundaries in the Middle East, in Blake, G.H. & Schofield, R.N. (eds) *Boundaries and state territory in the Middle East and North Africa*. MENAS, Wisbech.

Anderson, E.W. 1988d Preliminary dew measurements in the Eastern Prosopis belt of the Wahiba Sands, Journal of Oman Studies Special Report 3: 201–12.

Anderson, E.W. 1990 Afghan refugees: the geopolitical context, in Anderson, E.W. & Dupree, N.H. (eds.) *The cultural basis of Afghan Nationalism*, Pinter, London.

Anderson, E.W. 1991a White oil, Geographical Magazine February LXIII, II: 10–14.

Anderson, E.W. 1991b Making waves on the Nile. Geographical Magazine LXIII, IV: 10–13.

Anderson, E.W. 1991c The violence of thirst. Geographical Magazine LXIII, V: 31–4.

Anderson, E.W. & Cox, N.J. 1984 The relationship between soil creep rates and certain controlling variables in a catchment in Upper Weardale, Northern England, in Walling, D.E. & Burt, T.P. (eds) *Catchment experiments in fluvial geomorphology*, Geo Books, Norwich.

Anderson, E.W. & Curry, W. 1987 A view of the Wadi Daygah Gorge, Sultanate of Oman, Council for Conservation of Environment and Water Resources.

Anderson, E.W. & Schembri, P.J. 1989 Coastal zone survey of the Maltese Islands, Beltissebh, Government of Malta.

Anderson, J.R., Hardy, E.C., Roach, J.T. & Witmer, R.E. 1976 Landuse and land cover classification system for use with remote sensor data, USGS Professional Paper 964.

Anderson, T.D. 1984 *Geopolitics of the Caribbean*, Praeger, New York.

Anochili, B.C. 1978 *Food crop production*, Macmillan, New York.

Anon 1975 Ground water and wells, Johnson Division UOP Inc., Minnesota, 440 pp.

Anon 1985 Water resources: technology and application, Arab–British Chamber of Commerce Report 6.

Arab–British Chamber of Commerce 1985 Water resources: Technology and application, Chamber Economic Reports 6.

Arkley, R.J. 1963 Relationships between plant growth and transpiration, Hilgardia 34(13): 559–624.

Arora, V.K., Prihar, S.S. & Gajri, P.R. 1987 Synthesis of a simplified water use simulation model for predicting wheat yields, Water Resources Research 23(5): 903–10.

Arnon, I. 1972 *Crop production in dry regions*, Leonard Hill, London.

El-Ashry, M.T. 1985 Salinity pollution from irrigated agriculture, Journal of Soil and Water Conservation (Jan.–Feb.): 48–52.

Australian Water Resources Council 1975 Groundwater resources of Australia, Australian Government Pub. Services, Canberra.

Bach, W. 1984a Carbon dioxide and climatic change, Progress in Physical Geography 8: 83–93.

Bach, W. 1984b CO_2 sensitivity experiments using general circulation models, Progress in Physical Geography 8: 583–609.

Bach, W. 1988 Modelling the climatic response to greenhouse gases, in Gregory, S. (ed.) *Recent climatic change*, Belhaven Press, London, pp. 7–19.

Bach, W. 1988 Modelling the climatic response to greenhouse gases, in gregogy, S. (ed.) *Recent climatic change*, Belhaven Press, London, pp. 7–19.

Baillon, R.P. 1987 Anthropology and irrigation, MSc Thesis, Dept. of Anthropology, University College London.

Baker, R. 1974 Famine: the cost of development? Ecologist 4: 170–5.

Baldy, C. 1976 *Etude agrometeorologique Haute Volta*, WMO, Geneva.

Balek, J. 1977 & 1983 *Hydrology and Water resources in Tropical Africa*, Elsevier, Oxford, 208pp.

Al-Baluchi, S.S. 1985 Water usage in oil and gas sectors, in Yassin, A.A. (ed.) *Development and rationalization of water resources in Oman*, Muscat, Oman Water Policy Planning Conference October 27.

Bandyopadhyaya, J. 1983 *Climate and world order*, Humanities Press, Atlantic Highlands, New Jersey.

Barker, R.D. & Griffiths, D.H. 1981 Surface geophysical methods in hydrogeology, in Lloyd, J.W. (ed.) *Case studies in groundwater resources evaluation*, Clarendon Press, Oxford, pp. 29–44.

Barnett, T. 1977 *The Gezira scheme*, Frank Cass, London, 192 pp.

Barney, G.O. 1980 The Global 2000 Report to the President. The technical report No. 2, The Council on Environmental Quality, Department of State, pp. 137–59.

Barrett, E.C. & Martin, D.W. 1981 *Use of satellite data in rainfall monitoring*, Academic Press, London.

Barrow, C. 1987 *Water resources and agricultural development in the tropics*, Longman Scientific and Technical, Harlow, 356 pp.

Barry, R.G. & Chorley, R.J. 1978 *Atmosphere, weather and climate*, Methuen, London.

Basu, D.N. & Ljung, P. 1988 Irrigation management and scheduling: study of an irrigation system in India, in O'Mara, G.T. (ed.) *Efficiency in irrigation: the conjunctive use of surface and groundwater resources*, A World Bank symposium, Washington, DC, pp. 178–96.

Batisse, M. 1988 Progress and perspectives: a look back at UNESCO arid zone activities, in Whitehead, E.E. *et. al.* (eds.) *Arid lands today and tomorrow*, Westview Press, Boulder, pp. 21–30.

Battarbee, R.W., Titcombe, C., Donnelly, K. & Anderson, J. 1983 An automated technique for the accurate positioning of sediment core sites and the bathymetric mapping of lake basins, Hydrobiologia 103(71): 71–4.

El-Baz, F. (ed.) 1984 *Deserts and arid lands*, Martinus Nijhoff, Lancaster, 222 pp.

Bear, J. 1979 *Hydraulics of groundwater*, McGraw-Hill, Israel, 567 pp.

Beaumont, P. 1981 Water resources and their management in the Middle East, in Clarke, J.I. & Bowen-Jones, H. (eds) *Change and development in the Middle East*, Methuen, London.

Beaumont, P. 1988 Water scarcity as a limiting factor to development in the Middle East. Paper presented at Insitute of Applied Economics and Geography, High Council of Scientific Research, Salamanca, Spain, December.

Beaumont, P. 1989 *Environmental management and development in drylands*, Routledge, London.

Beaumont, P., Blake, G.H. & Wagstaff, J.M. 1988 *The Middle East* (2nd edn), David Fulton, London.

Begg, J. & Turner, N. 1976 Crop water deficits, Advances in Agronomy 28: 161–217.

Bell, J.P. 1976 Neutron probe practice, Institute of Hydrology Report 19, Wallingford.

Bemont, F. 1961 L'Irrigation en Iran, Annales Geographie 70: 587–620.

Benson, M.A. 1964 Factors affecting the occurrence of floods in the South-west, USGS Water supply paper 1580-D.

Beran, M.A. & Rodier, J.A. 1985 Hydrological aspects of drought, Studies and Reports in Hydrology 39, UNESCO–WHO.

Berber, F.J. 1959 *Rivers in international law*, Stevens, London.

Bidinger, F.R., Mahalakshmi, V., Talukdar, B.S. & Alagarswamy, G. 1982 Improvements of drought resistance in pearl millet, in *Drought resistance in crops with emphasis upon rice*, Philippines, International Rice Research Institute, 414 pp.

Bielorai, H. 1973 Predicting irrigation needs, in Yaron, B., Danfors, E. and Vaadia, Y. (eds) *Arid zone irrigation*, Chapman and Hall, London, pp. 359–68.

Biswas, A.K. (ed.) 1978 *U.N. water conference*, Pergamon Press, Oxford.

Biswas, A.K. 1979 North American water transfers: an overview, in Golubev, G. & Biswas, A.K. (eds) *Interregional water transfers*, Pergamon Press, Oxford, pp. 79–90.

Biswas, A.K. 1980 Water: A perspective on global issues and politics, in Biswas, A.K. *et al.* (eds) *Water management for arid lands in developing countries – training workshop*, Pergamon Press, Oxford, pp. 9–28.

Biswas, A.K. 1981 Water for the third world, Foreign Affairs 60(1): 148–66.

Biswas, A.K. 1988 Role of wastewater reuse in water planning and management, in Biswas, A.K. & Arar, A. (eds.) *Treatment and reuse of wastewater*, Butterworths, London, pp. 3–17.

Black, T.A. & McNaughton, K.G. 1971 Psychrometric apparatus for Bowen ratio determination over forests, Boundary Layer Meteorology 2: 246–54.

Blake, G., Dewdney, Y. & Mitchell, J. 1987 *The Cambridge atlas on the Middle East and North Africa*, CUP.

Blaney, H. 1954 Consumptive use requirements for water, Agricultural Engineering 35: 870–80.

Blaney, H. & Criddle, W. 1950 Determining water requirements in irrigated areas from climatological and irrigation data, USDA (SCS) TP-96.

Blaney, H. & Criddle, W. 1962 Determining use and irrigation water requirements, USDA (ARS) Technical Bulletin 1275.

Boden, T., Kanciruk, P. & Farrell, M.P. 1990 *Trends '90. A compendium of data on global change*, Information Analysis Centre, Oak Ridge, Tennesse.

Bodhaine, G.L. 1968 Measurement of peak discharge at culverts by indirect methods, USGS.

Boggie, R. & Knight, A.H. 1962 An improved method for the placement of radioactive isotopes in the study of root systems of plants growing in deep peat, Journal of Ecology 50: 461–2.

Bolin, B. *et al.* 1986 *The greenhouse effect, climate change and ecosystems*, J. Wiley, Chicester.

Bond, M.E. & Dunikoski, H. 1977 The impact of second home development on water availability on North Central Arizona, Eisenhower Consortium Institutional Series Report 1.

Boonyatharokul, W. & Walker, W.R. 1979 Evapotranspiration under depleting soil moisture, *Proceedings of the American Society of Civil Engineers, Irrigation and Drainage Division*, 105: 391–402.

Bos, M.G. 1977 Some influences of project management on irrigation efficiencies, in Worthington, E.B. (ed.) *Arid land irrigation in developing countries* Pergamon Press, Oxford, pp. 351–60.

Bosart, L.F. 1985 Weather forecasting, in Houghton, D.D. (ed.) *Handbook of applied meteorology*, J. Wiley, New York, pp. 205–79.

Bouycous, G.J. & Mick, A.H. 1941 Comparison of absorbant materials employed in the electrical resistance method of making a continuous measurement of soil moisture under field conditions, *Proc. Soil Science Society America* 5, 77–9.

Boyko, H. 1968 Principles of saline and marine agriculture, in Boyko, H. (ed.) *Saline irrigation for agriculture and forestry*, W. Junk, The Hague, pp. 3–10.

Brady, N.C. 1974 *The nature and properties of soils*, Collier Macmillan, London.

Braham, Jr, R.R. (ed.) 1986 Precipitation enhancement – A scientific challenge, Meteorological Monographs of American Meteorological Society 21: 43.

Braham, Jr, R.R., Battan, L.J. & Byers, H.R. 1957 Artificial nucleation of cumulus clouds, Cloud and Weather Modification Meteorology Monograph 11: 47–85.

Brahmananda, V., Satyamurty, R. and deBrito, J. 1986 On the 1983 drought in N.E. Brazil, Journal of Climatology 6: 43–51.

Brazel, A.J. & Brazel, S.W. 1988 Desertification, desert dust and climate. Environmental problems in Arizona, in Whitehead, E.E. *et al.* (eds) *Arid lands today and tomorrow*, Westview Press, Boulder, pp. 35–42.

Breuer, G. 1980 *Weather modification*, CUP.

Brown, K. 1988 Ecophysiology of *Prosopis cineraria* in Wahiba Sands with reference to its reafforestation potential. Journal of Oman Studies Special Report 3: 257–70.

Brunsden, D. & Cooke, R.U. 1986 *The Wahiba sands project 1985/86*, RGS, London.

Brutsaert, W. & Strickler, H. 1979 An advection aridity approach to estimate actual regional evapotranspiration, Water Resources Research 15(2): 443–50.

Bryson, R.A. 1973 Drought in Sahelia: who or what is to blame? Ecologist 3(10): 336–70.

Bryson, R.A. 1974 A perspective on climatic change, Science 184: 753–60.

Bryson, R.A. 1989 Will there be a global greenhouse warming? Environmental Conservation 16(2): 97–9.

Bucks, D.A., Nakayama, F.S. & Warrick, A.W. 1982 Principles, practices, and potentialities of trickle (drip) irrigation, in Hillel, D. (ed.) *Advances in irrigation*, Academic Press, London, 220–98.

Budyko, M.I. 1978 The heat balance of the earth, in Gribben, J. (ed.) *Climatic change*, Cambridge University Press, London, pp. 85–113.

Bull, G.A. 1960 Comparison of raingauges, Nature 185: 437–8.

Bunting, A.H., Dennett, M.D., Elston, J. & Milford, J.R. 1976 Rainfall trends in the West African Sahel, Quarterly Journal of the Royal Meteorological Society 102: 59–64.

Bunyard, P. 1985 World climate and tropical forest destruction, The Ecologist 15(3): 125–36.

Buras, N. & Darr, P. 1979 An evaluation of marginal waters as a natural resource in Israel, Water Resources Research 15(6): 1349–53.

Burden, D.S. & Selim, H.M. 1989 Correlation of spatially variable soil water retention for a surface soil. Soil Science 148(6): 436–47.

Burman, R.D., Cuenca, R.H. & Weiss, A. 1983 Techniques for estimating irrigation water requirements in Hillel, D. (ed.) *Advances in irrigation* Academic Press, London.

Burr, G.O. *et al.* 1957 The sugar cane plant, Annual Review of Plant Physiology 8: 275–308.

Burt, J.E., Hayes, J.T., O'Rourke, P.A., Terjung, W.H. & Todhunter, P.F. 1980 Water: a model of water requirements for irrigated and rainfed agriculture, Publications in Climatology XXXIII(2).

Byers, H.R. 1974 History of weather modification, in Hess, W.N. (ed.) *Weather and climate modification* J. Wiley, London, pp. 3–44.

Byres, T., Crow, B. & Mae-Wan Ho 1984 *The green revolution in India*, OU 273, Third World Studies Course, Open University, Milton Keynes.

Cairncross, S. 1987 The benefits of water supply, in *Conference on Water and Engineering for Developing Countries, Developing World Water*, Grosvenor Press International, Hong Kong, pp. 30–4.

Campbell, G.S. 1982 Irrigation scheduling using soil moisture measurements, in Hillel, D. (ed). *Advances in irrigation*, Academic Press, London, pp. 25.

Campbell, G.S. & Unsworth, M.H. 1979 An inexpensive sonic anemometer for eddy correlation, Journal of Applied Meteorology 18: 1072–7.

Cannell, C.H. & Dregne, H.E. 1983 Regional setting, in Dregne, H.E. & Willis, W.O. (eds.) *Dryland agriculture*, American Society of Agronomy, Number 23, Madison, WI, pp. 6–16.

Cano, G.J. 1978 Water law and legislation: how to use them to obtain optimum results from water resources, in *Proceedings of the United Nations Conference Mar del Plata on Water Management and Development, Argentina, March 1977*, Pergamon, New York.

Canter, L.W. 1985 *Environmental impact of water resources projects*, Lewis Publ., Chelsea, Michigan, 352 pp.

Cantor, L.M. 1967 *A world geography of irrigation*, Oliver and Boyd, London, 252 pp.

Carruthers, I. 1978 Contentious issues in planning irrigation schemes, in Worthington, E.B. (ed.) *Arid land irrigation in developing countries*, Pergamon Press, Oxford, pp. 301–8.

Carruthers, I. (ed.) 1983 *Aid for the development of irrigation*, OECD, Paris.

Carruthers, I. & Clark, C. (eds) 1983 *The economics of irrigation*, Liverpool University Press.

Caswell, M.F. 1989 The adoption of low value irrigation technologies as a water conservation tool, Water International 14: 19–26.

Caviedes, C.N. 1988 The effects of ENSO events in some key regions of the South American continent, in Gregogy, S. (ed.) *Recent climatic change*, Belhaven Press, London, pp. 252–66.

Centre of Near & Middle Eastern Studies 1987 *Qanats: History, development and utilisation*, Conference at SOAS, University of London.

Chaliand, G. & Ragaeu, J. 1985 *Strategic atlas*, Penguin, Harmondsworth.

Chambers, R. 1980 Basic concepts in the organisation of irrigation, in Coward, E.W. (ed.) *Irrigation and agricultural development in Asia*, Cornell University Press, London, pp. 28–50.

Channabasappa, K.C. 1982 an overview of desalination systems for brackish water and their applicability to desert conditions, *Proceedings of Conference on Alternative strategies for desert development and management. Sacramento*, Pergamon Press, New York.

Chapin, G. & Wasserstrom, R. 1983 Pesticide use and malaria resurgence in Central America and India, The Ecologist 13(4): 115–26.

Charney, J. & Stone, P. 1974 Drought in the Sahel: a biogeophysical feedback mechanism, Science 187: 434–5.

Chesworth, P. 1990 History of water use in Sudan and Egypt, in Howell, P.P. & Allan, J.A. (eds) *The Nile, Conference at RGS 2–3 May*, SOAS, London, pp. 40–58.

Chidley, T.R.E. 1981 Assessment of groundwater recharge, in Lloyd, J.W. (ed.) *Case studies of groundwater resources evaluation*, Clarendon Press, Oxford, pp. 133–49.

Child, J. 1985 *Geopolitics and conflict in South America*, Praeger, New York.

Childs, E.C. 1969 *The physical basis of soil water phenomena*, J. Wiley, London, 493 pp.

Chorley, R.J. & Kennedy, B.A. 1971 *Physical geography*, Prentice Hall, London.

Chow, Ven te 1964 *Handbook of applied hydrology*, McGraw-Hill, New York.

Clark, W. & Munn, R. 1986 (eds) *Sustainable development of the biosphere*, CUP.

Clarke, C.L. & Anderson, J. 1987 The feasibility report, in Rydzewski, J.R. (ed.) *Irrigation development planning*, J. Wiley, London pp. 239–60.

Clarke, T. 1978 *The last caravan*, G.P. Putnam's & Sons, New York.

Clements, F.A. 1980 *Oman, the reborn land*, Longman, New York.

Cocheme, J. & Franquin, P. 1967 An agroclimatology survey of a semi arid area in Africa south of the Sahara, WMO Technical Note 86, Geneva.

Coe, J.J. 1988 Resoponses to some adverse external effects of groundwater withdrawals in California, in O'Mara, G.T. (ed.) *Efficiency in irrigation: the conjunctive use of surface and groundwater resources*. A World Bank Symposium, Washington, DC. pp. 51–7.

Cohen, S. 1973 *Geography and politics in a world divided*, OUP, New York.

Collinge, V.K. & Kirby, C. (eds.) 1987 *Weather radar and flood forecasting*, J. Wiley, Chichester.

Collins, R.O. 1988 The Jonglei canal; illusion or reality? Water International 13: 144–53.

Collins, R.O. 1990 *The waters of the Nile: hydropolitics of the Jonglei canal*, OUP.

Colman, E.A. & Hendrix, T.M. 1949 The fibreglass electrical soil moisture instrument, Soil Science 67: 425–38.

Committee on Atmospheric Sciences (CAS) 1966 *Weather and climate modification*, Publication 1350, National Academy of Sciences, Washington, DC.

Committee on Atmospheric Sciences (CAS) 1980 *Weather and climate modification: problems and progress*, Grand River Books, Detroit.

Committee on Safety Criteria for Dams 1985 *Safety of dams*, National Research Council, National Academy Press, Washington, DC, 276 pp.

Conant, M.A. 1982 *The oil factor in U.S. foreign policy 1980–1990*, Lexington Books, Lexington, CT.

Conway, G.R. & Barbier, E.B. 1990 *After the green revolution*, Earthscan, London.

Cooke, R.U. 1981 Salt weathering in deserts, Proceedings of the Geological Association 92: 1–16.

Cooke, R.U., Brunsden, D., Doornkamp, J.C. & Jones, D.K.C. 1982 *Urban geomorphology in dry lands*, OUP, Oxford.

Cooke, R.U. & Warren, A. 1973 *Geomorphology in deserts*, Batsford, London.

Cooper, W.A. & Lawson, R.P. 1984 Physical interpretation of results from the HIPLEX-1 experiment, Journal of Climate Applied Meteorology 23: 523–40.

Copans, J. 1983 The Sahelian drought: social sciences and the political economy for underdevelopment, in Hewitt, K. (ed.) *Interpretations of calamity*, Allen & Unwin, London, pp. 83–97.

Cotton, W.R. 1986 Testing, implementation and evolution of seeding concepts; A review, in Braham, Jr, R.R. (ed.) *Precipitation enhancement – A scientific challenge*, Meteorological Monographs of American Meteorological Society 21(43): 139–49.

Courel, M.F., Kandel, R.S. & Rasool, S.I. 1984 Surface albedo and the Sahel drought, Nature 307: 528–31.

Coward, E.W. (ed.) 1980 *Irrigation and agricultural development in Asia*, Cornell University Press, London, 379 pp.

Creighton, J. 1982 Values and public involvement, in Creighton, J. & Priscoli, J.D. (eds.) *Public involvement techniques: a reader of ten years experience at the Institute of Water Resources, Fort Belvoir, Virginia*, US Army Corps of Engineers.

Crundwell, M.E. 1986 A review of hydrophyte evapotranspiration. Rev. Hydrobiol. Trop. (3–4): 215–32.

Cummings, B.J. 1990 *Dam the rivers, damn the people*, Earthscan, London.

Cunnington, W.M. & Rowntree, P.R. 1986 Simulation of the Saharan atmosphere – dependence upon moisture and albedo, Quarterly Journal of Royal Meteorological Society 112: 971–99.

Curtis, D., Hubbard, M. & Shepard, A. 1988 *Preventing famine*, Routledge, London.

Dagg, M. 1968 Evaporation pans in East Africa, *Proc. of 4th Specialist Meeting on Applied Meteorology in East Africa, Nairobi.*

Dahl, G. & Hjort, A. 1976 *Having herds*, Stockholm Studies in Anthropology, University of Stockholm, 335 pp.

Dahlberg, K.A. 1979 *Beyond the green revolution*, Plenum Press, London.

Dalrymple, T.D. & Benson, M.A. 1967 Measurement of peak discharge by slope area method, USGS.

DaMota, F. 1978 A dependable agroclimatological water balance, Agricultural Meteorology 19, 209–13.

Dancette, C. 1980 Water requirements and adaptations to the rainy season of millet in Senegal. *Proceedings of Int. Workshop on Agroclimatological Resarch. Needs of Semi-Arid Tropics, November 1978. ICRISAT, Hyderabad*, pp. 106–20.

Dastane, N.G. 1974 Effective rainfall in irrigated agriculture, FAO Irrigation and Drainage Paper 25, Rome.

Davies, J. 1987 Desalination distillation or reverse osmosis? *Conference on Water and Engineering for Developing Countries, Developing World Water*, Grosvenor Press International, Hong Kong, pp. 222–4.

Davis, L.G. 1978 Operational weather modification prospects, in Davis, R.J. & Grant, L.O. (eds.) *Weather modification and the law*. American Association for the Advancement of Science 20, Westview Press, Bolder, pp. 11–21.

Davis, R.J. 1974 Weather modification litigation and statutes, in Hess, W.N. (ed.) *Weather and climate modification*, J. Wiley, London, pp. 767–86.

Davis, R.J. & Grant, L.O. (eds) 1978 *Weather modification and the law*, American Association for the Advancement of Science 20, Westview Press, Boulder.

Davis, S.N. 1974 The hydrogeology of arid regions, in Brown, G.W. (ed.) *Desert Biology*, Vol. 2, Academic Press, London, pp. 1–30.

Davy, E.G., Mattei, F. & Solomon, J.I. 1976 An evaluation of climate and water resources for development of agriculture in the Sudano-Sahelian zone of West Africa, WMO Special Environmental Report 9, Geneva.

Dawood, A.A. & Subramanian, N.P. 1985 Problems in the planning of water supply projects, *Water Policy Planning Conference, Muscat, Oman, October 27.*

Day, A.D. & Interlap, S. 1970 Some effects of soil moisture stress on the growth of wheat, Agronomy Journal 62: 27–9.

Decker, W.L. 1978 Operational weather modification prospects, in Davis, R.J. & Grant, L.O. (eds) *Weather modification and the law*, American Association for the Advancement of Science 20, Westview Press, Boulder, pp. 23–7.

Dedrick, A.R., Erie, L.J. & Clemmen, A.J. 1982 Level basin irrigation, in Hillel, D. (ed.) *Advances in irrigation*, Academic Press, London, pp. 105–45.

Dewaulle, J.C. 1973 Resultats de six ans d'observations sur l'erosion au Niger, Revue Bois et Forets des Tropiques 150 (Juillet–Aout).

Delyannis, A. & Delyannis, E. 1980 *Seawater and desalting*, Springer-Verlag, Heidelberg, 189 pp.

Denmead, O.T. & Shaw, R.H. 1962 Availability of soil water to plants as affected by soil moisture content and meteorological conditions. Agronomy Journal 54: 385–90.

Dennis, A.S. 1980 *Weather modification by cloud seeding*, Academic Press, London.

De Wilde, J. 1967 *Experience with agricultural development in tropical Africa*, Johns Hopkins University Press.

Dhar, O.N., Rakhecha, P.R. & Kulkarni, A.K. 1979 Rainfall study of severe drought years of India, *Symp. Hydrological Aspects of Droughts, New Delhi, Indian Natl Comm. for Int. Hydrol. Prog.*

Dickey, G.B. 1961 *Filtration*, Van-Nostrand Rheinhold, New York.

Dickinson, R.E. 1986 Impact of human activities on climate, in Clark, W. & Munn, R. (eds) *Sustainable development of the biosphere*, CUP, pp. 252–89.

Dingman, R.J. 1984 Development of a water resources plan for the Sultanate of Oman, Public Authority for Water Resources, Oman.

Dogra, B. 1986 The Indian experience with large dams, in Goldsmith, e. & Hildyard, N. (eds) *The social and environmental effects of large dams; volume 2 case studies*, Wadebridge Ecological Centre, Cornwall, pp. 201–8.

Donaldson, D. 1987 Community participation in urban water supply and sanitation systems, *Conference on Developing World Water*, Grosvenor International, London, pp. 86–9.

Doorenpos, J. & Pruitt, W. 1975 & 1977 Crop water requirements, Irrigation and Drainage Paper 24, FAO, Rome.

Downing, J., Berry, L., Downing, L., Downing, T. & Ford, R. 1987 Drought and famine in Africa 1981: 1986; The US response, Clark University.

Dracup, J., Lee, K. & Paulson, E. 1980 On the definition of droughts, Water Resources Research 16(2): 297–302.

Drake, C. 1988 Oman: tradition and modern adaptations to the environment, Focus (summer) 15–20.

Dregne, H.E. 1976 *Soils of arid regions*, Elsevier, Oxford.

Dregne, H.E. 1983 *Desertification of arid lands*, Harwood Academic Publishers, London.

Dresner, L. & Johnson, J. 1980 Hyperfiltration, in Spiegler, K.S. & Laird, A.D.K. (eds) *Principles of desalination, part B*, Academic Press, London, 464 pp.

Druyan, L.M. 1989 Advances in the study of Sub-Saharan drought, International Journal of Climatology 9: 77–90.

Dugdale, G., McDougall, V.D. & Milford, J.R. 1991 Rainfall estimates in the Sahel from cold cloud statistics: accuracy and limitations of operational systems, in Sivakumar *et al.* (eds) *Proceedings of Workshop on Soil Water Balance in the Sudano-Sahelian Zone*, February 18–23, IAHS Publication 199, Wallingford, pp. 65–74.

Dunin, F.X. 1976 Infiltration; its simulation for field conditions, in Rodda, J. (ed.) *Facets of hydrology* J. Wiley, New York.

Durrant, M.J., Love, B.J.G., Messem, A.B. & Draycott, A.P. 1973 Growth of crop roots in relation to soil moisture extraction, Annals of Applied Biology 74: 387–94.

Eagelson, P. 1978 Climate, soil and vegetation: introduction to water balance dynamics, Water Resources Research 14(5): 705–12.

Eeles, C.W.O. 1969 Installation of access tubes and calibration of neutron moisture probes, Institute of Hydrology Report 7, Wallingford.

Ehlers, E. 1978 Social and economic consequences of large scale irrigation developments, the Dez irrigation project, Khuzestan, Iran, in Worthington, E.B. (ed.) *Arid land irrigation in developing countries*, Pergamon Press, Oxford, pp. 85–97.

Ellis, F.B. & Barnes, B.T. 1973 Estimation of the distribution of living roots of plants under field conditions, Plant and Soil 39: 81–91.

Evans, J.V. 1979 The potential for desalinating seawater, in Hallsworth, E.G. & Woodcock, J.T. (eds.) *Land and water resources in Australia*, Australian Academy of Technical Sciences.

Evans, T. 1990 History of Nile flows, in Howell, P.P. & Allan, J.A. (eds) *The Nile, Conference at RGS, 2–3 May*, SOAS, London, pp. 5–39.

Evenari, M., Noy-Meir, I. & Goodall, D.W. (eds) 1986 *Ecosystems of the world: hot deserts and arid shrublands*, Elsevier, Oxford.

Everett, J.R., Russell, O.R. & Nichols, D.A. 1984 Landsat surveys of S.E. Arabia, in El-Baz, F. (ed.) *Deserts and arid lands*, Martinus Nijhoff, Lancaster, pp. 157–70.

Eyre, L.A. 1988 Population pressure on arid lands: is it manageable? in Whitehead, E. *et al.* (eds) *Arid lands today and tomorrow*, Westview Press, Boulder, pp. 989–96.

Fair, J.R. 1987 *Distillation, whither not whether*, Institute of Chemical Engineers Symposium Series No 104, Distillation and Absorption, p. A613–24.

Al-Faisal, M. 1982 New water resources for desert development from icebergs, *Conference on Alternative Strategies for Desert Development and Management Vol. III: Water*, Pergamon Press, Oxford, pp. 704–11.

FAO/UNESCO 1973 *Irrigation, drainage and salinity*, Hutchinson, London, 510 pp.

FAO 1975 Water: available but mismanaged, CERES 8(4).

FAO 1979 Agriculture towards 2000, Rome.

FAO 1986 Early agrometeorological crop yield assessment, Plant Protection Paper 73, Rome. 150 pp.

Farid, M.A. 1978 Irrigation and malaria in arid lands, in Worthington, E.B. (ed.) *Arid land irrigation in developing countries*, Pergamon Press, Oxford, pp. 413–19.

Farmer, G. 1989 Rainfall, IUCN Sahel Studies, Nairobi: 1–25.

Farmer, G. & Wigley, T.M.L. 1985 Climatic trends for tropical Africa, Climatic Research Unit, University of East Anglia, 136 pp.

Faure, H. & Gac, J. 1981 Will the Sahelian drought end in 1985? Nature 291: 475–8.

Feddes, R.A., Kabat, P., Van Bakel, P.J.T., Bronswijk, J.J.B. & Halbertsma, J. 1988 Modelling soil water dynamics in the unsaturated zone; state of the art, Journal of Hydrology 100: 69–111.

Fedorov, Ye. K. 1974 Modification of meteorological processes, in Hess, W.N. (ed.) *Weather and climate modification*, J. Wiley, London pp. 387–409.

Flohn, H. 1987 Rainfall teleconnections in northern and eastern Africa, Theoretical and Applied Climatology 38: 191–7.

Folland, C. 1987 Sea temperatures predict African drought, New Scientist 1 Oct., p. 25.

Franke, R. & Chasin, B. 1980 *Seeds of famine*, Allanheld, Osman & Co., Montclair, New Jersey.

Franke, R. & Chasin, B. 1981 Peasants, Peanuts, Profits and Pastoralists, The Ecologist 11(4): 156–67.

Frankenberg, P. 1986 The problem of the aridity line, Applied geography and development, Institut für Wissenschaftliche Zusammenarbeit, Tubingen: 37–55.

Frasier, G.W. 1988 Technical, economic and social considerations of water harvesting and runoff farming, in Whitehead, E. *et al.* (eds) *Arid lands today and tomorrow*, Westview Press, Boulder, pp. 905–18.

Free, G., Browning, G. & Musgrave, G. 1940 Relative infiltration and related physical characteristics of certain soils, US Dept. of Agriculture Technical Bulletin 729, Washington, DC.

Frost, D.H. 1987 Developing water resources in low rainfall areas, *Conference on Developing World Water*, Grosvenor International, London, pp. 75–7.

Fuchs, M. 1973 The estimation of evapotranspiration, in Yaron, B., Danfors, E. and Vaadia, Y. (eds) *Arid zone irrigation*, Chapman and Hall, London, pp. 241–7.

Fuehring, H., Mazaheri, A., Bybordi, M. & Khan, A. 1966 Effects of soil moisture depletion on crop yield and stomatal infiltration, Agronomy Journal 58: 195–8.

Fuller, W.H. 1974 Desert soils, in Brown, G.W. (ed.) *Desert biology*, Vol. 2, Academic Press, London, pp 32–101.

El-Gabaly, M. 1976 Problems and effects of irrigation in the Near East region, in *Arid lands irrigation in developing countries, COWAR symp.*, Academic Sci. Res. Tech. Cairo.

Gabriel, K.R. & Neumann, J. 1962 A Markov chain model for daily rainfall occurrence at Tel Aviv, Quarterly, Journal of the Royal Meteorological Society 88: 90–5.

Gagin, A. & Neumann, J. 1974 Rain stimulation and cloud physics in Israel, in Hess, W.N. (ed.) *Weather and climate modification*, J. Wiley, London, pp. 454–94.

Gairon, S. & Hadas, A. 1973 Measurement of the water status of soil and soil characteris-

tics relevant to irrigation, in Yaron, B., Danfors, E. & Vaadia, Y. (eds) *Arid zone irrigation*, Chapman and Hall, London, pp. 215–40.
Garcia, R.V. 1981 *Drought and man*, Pergamon Press, Oxford.
Gardner, W.R. 1960 Dynamic aspects of water availability to plants, Soil Science 80(2): 63–73.
Garnier, B.J. 1967 *Weather conditions in Nigeria*, Climatological Research Series 2, Department of Geography, McGill University.
Gay, L.W. & Stewart, J.B. 1974 Energy balance studies in coniferous forests, Institute of Hydrology Report 23, Wallingford.
Ghandour, F. 1985 Water treatment and management, Arab–British Chamber of Commerce Report 6, pp. 26–31.
Gibb, Sir Alexander & Partners 1976 Water resources survey of Northern Oman, Final Report, The Institute of Hydrology, Wallingford.
Gibbon, D. & Harvey, J. 1977 Subsistence farming in the dry savanna of Western Sudan, School of Development Studies Discussion Paper 18, University of East Anglia.
Gifford, G.F. 1976 Applicability of some infiltration formulae to rangeland infiltrometer data, Journal of Hydrology 28: 1–12.
Glaeser, B. 1987 *The green revolution revisited*, Allen and Unwin, Hemel Hempstead.
Glantz, M. 1976 *The politics of natural disaster*, Praeger, New York.
Glantz, M.H. 1987 Drought in Africa, Scientific American 256(6): 34–40.
Glantz, M. 1989 Drought, famine, and the seasons in sub-Saharan Africa in Huss-Ashmore, R. & Katz, S. (eds) *Anthropological perspectives on the African famine*, Gordon and Breach, New York.
Glantz, M.H. & Katz, R. 1987 African drought and its impacts, Desertification Control Bulletin 14, UNEP, 23–30.
Gleick, P.H. 1988 Regional hydrologic impacts of global climatic changes, in Whitehead, E.E. *et al.* (eds) *Arid lands today and tomorrow*, Westview Press, Boulder, pp. 43–59.
Goldreich, Y. 1988 Temporal changes in the spatial distribution of rainfall in the central coastal plain of Israel, in Gregory, S. (ed.) *Recent climatic change*, Belhaven Press, London, pp. 116–23.
Goldsmith, E. & Hildyard, N. 1984 *The social and environmental effects of large dams*, Vol. 1, Wadebridge Ecological Centre, Cornwall, 346 pp.
Golladay, F. 1983 Meeting the needs of the poor for water supply and waste disposal, World Bank paper, Appropriate Technology for Water Supply and Sanitation Vol. 13, Washington, DC, 52 pp.
Golubey, G. & Biswas, A.K. (ed.) 1979 Interregional water transfers, in Biswas, A.K. (ed.) *Water development, supply and management*, Vol. 6, Pergamon Press, Oxford, pp. 59–217.
Golubev, G. & Vasiliev, O. 1979 Interregional water transfers as an interdisciplinary problem, in Golubev, G. & Biswas, A.K. (eds) *Interregional water transfers*, Pergamon Press, Oxford, pp. 67–77.
Good, R. 1970 *The geography of flowering plants*, J. Wiley, New York.
Gordon, E.D. 1987 Social aspects of irrigation development, in Rydzewski, J.R. (ed.) *Irrigation development planning*, J. Wiley, London, pp. 161–76.
Goudie, A. (ed.) 1985 *Encyclopaedic dictionary of physical geography*, Blackwell, Oxford, 528 pp.
Goudie, A. 1986 *The human impact*, 2nd edn, Blackwell, Oxford.
Goudie, A. & Wilkinson, J. 1977 *The warm desert environment*, CUP.
Graf, W.L. 1988 *Fluvial processes in dryland rivers*, Springer-Verlag, London.
Grainger, A. 1990 *The threatening desert*, Earthscan, London.
Greb, B.W. 1983 Water conservation: Central Great Plains, in Dregne, H.E. & Willis, W.O. (eds) *Dryland agriculture*, American Society of Agronomy, Number 23,

Madison, WI, pp. 57–72.

Gregory, K.J. & Walling, D.E. 1973 *Drainage basin form and process*, Edward Arnold, London.

Gregory, P.J. 1965 Rainfall over Sierra Leone, Research Paper 2, Department of Geography, University of Liverpool.

Gregory, P.J. 1991 Soil and plant factors affecting the estimation of water extraction by crops, in Sivakumar, M.V.K. *et al.* (eds) *Proceedings of Workshop on Soil Water Balance in the Sudano-Sahelian Zone, February 18–23*, IAHS Publication 199, Wallingford, pp. 261–74.

Gregory, P.J. & Squire, G.R. 1979 Irrigation effects upon roots and shoots of pearl millet, Experimental Agriculture 15: 161–8.

Gregory, S. 1972 Climate, in Wreford Watson, J. & Sissons, J.B. (eds) *The British Isles: A systematic geography*, Nelson, London, pp. 53–73.

Gregory, S. 1986 The climatology of drought, Geography 71: 97–104.

Gregory, S. 1988 El-Nino years and the spatial pattern of drought over India 1901–70, in Gregogy, S. (ed.) *Recent climatic change*, Belhaven Press, London, pp. 226–36.

Griffith, D.A. 1984 Assessing results of weather modification programs, in Kriege, D.F. (ed.) *Operational cloud seeding projects in the Western US* American Society of Civil Engineers, New York, pp. 18–27.

Griffths, J.F. 1972 *Climates of the World: Volume 10 – Africa*, Elsevier Scientific, Amsterdam.

Griffiths, W.J. 1984 Irrigation in Nigeria, madness of fowl, marriage with cat? in Adams, W.M. & Grove, A.T. (eds) *Irrigation in Tropical Africa*, Cambridge African Monographs 3, CUP, pp. 4–13.

Grindley, J. 1969 The calculation of actual evaporation and soil moisture deficit over specified catchment areas, Hydrological memoir 38, Meteorological Office, Bracknell.

Grove, A.T. 1977 The geography of semi-arid lands, Philosophical Transactions of the Royal Society of London B 278: 457–75.

Grove, A.T. 1986 The state of Africa in the 1980s, Geographical Journal 152(2): 193–203.

Gupta, S.C. & Larson, W.E. 1979 Estimating soil water retention characteristics from particle size distribution, organic matter percent and bulk density, Water Resources Research 15: 1633–5.

Haan, C.T. 1977 *Statistical methods in hydrology*, IOWA State University Press.

Haas, J.E. 1974 Sociological aspects of weather modification, in Hess, W.N. (ed.) *Weather and climate modification*, J. Wiley, London, pp. 787–811.

Hagan, R.M. 1988 Water issues, in Whitehead, E.E. *et al.* (eds) *Arid lands today and tomorrow*, Westview Press, Boulder, pp. 9–10.

Hagan, R.M. & Stewart, J.I. 1972 Water deficits: irrigation design and programming, *Proc. American Society of Civil Engineers: Irrigation and Drainage Division (June)*, pp. 215–37.

Halcrow, W. 1975 Survey investigations for land and water resources development in the Dhofar, Annex 1, Soils and land capability.

Hall, A.L. 1978 *Drought and irrigation in N.E. Brazil*, Cambridge University Press, London.

Haramata 1989 Bulletin of drylands: people, policies, programmes, No. 5, International Institute for Environment and Development, London.

Hare, F.K. 1983 Climate and desertification, Report for WMO and UNEP.

Hare, F.K. 1984a Recent climatic experience in the arid and semi-arid lands, Desertification Control Bulletin 10, UNEP.

Hare, F.K. 1984b Climate, drought and desertification, Nature and Resources 20(1), UNESCO, 2–8.

Hare, F.K. 1985 Climatic variability and change, in Kates, R.W. *et al.* (eds) *Climate impact assessment*, J. Wiley, Chichester, pp. 37–68.
Harrison, P. 1987 *The greening of Africa*, Paladin, London, 380 pp.
Harrington, J. 1987 Climate change, Canadian Journal of Forestry Research 17: 1327.
Hastenrath, S. 1985 *Climate and circulation in the tropics*, D. Reidel, Boston.
Hastenrath, S., Wu, Ming-Chin and Chu, Pao-Shin 1984 Towards the monitoring and prediction of north east Brazil droughts, Quarterly Journal of the Royal Meteorological Society 110: 411–25.
Hatfield, J.L. 1983 Evapotranspiration obtained from remote sensing methods, in Hillel, D. (ed.) *Advances in irrigation* vol. 2, Academic Press, London, pp. 395–416.
Hayami, Y. 1984 Assessment of the green revolution, in Eicher, C.K. & Staatz, J.M. (eds) *Agricultural development in the third world*, Johns Hopkins University Press, Baltimore, pp. 389–95.
Hayward, D.F. & Oguntoyinbo, J.S. 1987 *The climatology of West Africa*, Hutchinson, London.
Hearn, A. & Ward, R. 1964 Irrigation control experiments on dry season crops in Nyasaland, Empire Journal Experimental Agriculture 32(125): 1–7.
Heathcote, R.L. 1983 *The arid lands: their use and abuse*, Longman, London, 323 pp.
Hegazin, S. 1988 Groundwater monitoring near Aqaba wastewater plant, in Biswas, A.K. & Arar, A. (eds) *Treatment and reuse of wastewater*, Butterworths, London, pp. 180–2.
Heller, J. & Bresler, E. (eds) 1973 Trickle irrigation, in Yaron, B., Danfors, E. and Vaadia, Y. (eds) *Arid zone irrigation*, Chapman and Hall, London, pp. 339–51.
Henderson-Sellers, A. & McGuffie, K. 1987 *A climate modelling primer*, J. Wiley, Chichester.
Hewitt, K. & Burton, I. 1971 *The hazardousness of a place: a regional ecology of damaging events*, Toronto Press, Toronto.
Hewlett, J.D. & Hibbert, A.R. 1963 Moisture and energy conditions within a sloping soil mass during drainage, Journal of Geophysical Research 68: 1081–7.
Hignett, C.T. 1976 A cereal crop water balance model, Notes on Soil Techniques 3: 34–49.
Hillel, D. 1971 *Soil and water*, Academic Press, London.
Hillel, D. (ed.) 1982 *Advances in irrigation*, Academic Press, London, 302 pp.
Hillman, P.F. 1987 Health aspects of irrigation developments, in Rydzewski, J.R. (ed.) *Irrigation development planning*, J. Wiley, London, pp. 117–99.
Hills, E.S. 1966 *Arid lands*, Methuen, London, 641 pp.
Hills, R.C. 1970 The determination of the infiltration capacity of field soils using the cylinder infiltrometer, British Geomorphological Research Group Technical Bulletin 3.
Al-Himyari, A.H. 1985 Managing water resources in the Tigris and Euphrates drainage basins, unpublished PhD Thesis, North Texas State University 1984, University Microfilms International, Ann Arbor, Michigan.
Hoag, R.B. & Ingari, J.C. 1985 Modern approach to ground water exploration in arid and semiarid lands, BCI Geonetics, Santa Barbara. Paper presented at *International Workshop on Sand Transport and Desertification in Arid Lands, Khartoum, November 18.*
Hobbs, J.E. 1988 Recent climatic change in Australia, in Gregogy, S. (ed.) *Recent climatic change*, Belhaven Press, London, pp. 285–97.
Hodges, C.N., Collins, W.L. & Riley, J.J. 1987 Direct seawater irrigation as a major food production technology for the Middle East. Presentation to the Center for Strategic International Studies, June 25, Georgetown University, Washington, DC.
Hodges, C.N., Collins, W.L. & Riley, J.J. 1988 Direct seawater irrigation as a major food production technology for the Middle East, in Starr, J.R. & Stoll, D.C. (eds) *The poli-*

tics of scarcity, Westview Press, Boulder.

Holdworth, P.M. 1971 Users testing schedule for the Wallingford probe system, Institute of Hydrology Report 10, Wallingford.

Holm, L.G., Weldon, L.W. & Blackburn R. 1971 Aquatic weeds, in Detwyler, T. (ed.) *Man's impact on environment*, McGraw-Hill, New York, pp. 246–65.

Hoogmoed, W.B. 1986 Analysis of rainfall characteristics relating to soil management from some selected locations in Niger and India, Wageningen Agricultural University Report 86–3.

Hoogmoed, W.B., Klaij, M.C. & Brouwer, J. 1991 Infiltration, runoff and drainage in the Sudano–Sahelian Zone, in Sivakumar, M.V.K. *et al.* (eds) *Proceedings of Workshop on Soil Water Balance in the Sudano–Sahelian Zone, February 18–23*, IAHS Publication 199, Wallingford, pp. 85–98.

Hoornaert, P. 1984 *Reverse osmosis*, Pergamon Press, Oxford, 212 pp.

Hopkins, A.G. 1975 *An economic history of West Africa*, Longman, London, 337 pp.

Hornburg, C.D. 1987 Desalination for remote areas, *Conference on Water and Engineering for Developing Countries, Developing World Water*, Grosvenor Press International, Hong Kong, pp. 230–2.

Horton, R.E. 1933 The role of infiltration in the hydrological cycle, Transactions of the American Geophysical Union 14: 446–60.

Horton, R.E. 1940 An approach towards a physical interpretation of infiltration capacity, *Proceedings in Soil Science of America* 5: 399–417.

Horton, R.E. 1945 Erosional development of streams and their drainage basins: hydrophysical approach to quantitative morphology, Geological Society of America Bulletin 56: 275–370.

Houston, C.E. 1978 Irrigation development in the world, in Worthington, E.B. (ed.) *Arid land irrigation in developing countries*, Pergamon Press, Oxford, pp. 425–32.

Houston, J. 1987 Groundwater recharge assessment, *Conference on Developing World Water*, Grosvenor International, London, pp. 100–3.

Howe, C.W. & Easter, K.W. 1971 *Interbasin transfers of water: economic issues and impacts*, Johns Hopkins University Press, Baltimore.

Huisman, L. & Kop, J.H. 1986 Artificial recharge of ground water, UNESCO: Nature and Resources XXII (1 & 2): 6–12.

Huisman, L. & Olsthoorn, T.N. 1983 *Artificial groundwater recharge*, Pitman Books, London.

Hulsin, H. 1976 Measurement of peak discharge at dams by indirect methods, USGS.

Hult, J.L. 1982 Water supply from imported Antarctic icebergs, in *Conference on Alternative Strategies for Desert Development and Management, Vol, III: Water*, Pergamon Press, Oxford, pp. 712–19.

Hulme, M. 1989 Is environmental degradation causing drought in the Sahel? Geography 74: 38–46.

Humlum, J. 1959 *La Geographie de l'Afghanistan*, Copenhagen (source Nir, 1974).

Hunt, B.G. 1985 A model study of some aspects of soil hydrology relevant to climate modelling, Quarterly Journal of the Royal Meteorological Society, 111: 1071–85.

Hunting Technical Services Ltd 1976 Pre-feasibility study to investigate alternative means of establishing commercial farming in the Interior of Oman, Final Report.

ICRISAT 1984 *Proceedings of the International Symposium on the Agrometeorology of Sorghum and Millet in the Semi-Arid Tropics, Hyderabad*, ICRISAT.

INRAN 1977 Themes proposes à la vulgarisation pour la campagne 1977, Institut National de la Recherce Agronomique du Niger, Niamey.

International Geographical Union 1971 *Proceedings of symposium of arid zone, Jodhpur*, National Committee for Geography, Calcutta.

Irvine, F.R. 1969 *West African crops*, Vol. 2, OUP.

IUCN 1989 *Sahel studies*, IUCN, Nairobi.
Jackson, I. 1977 *Climate, water and agriculture in the tropics*, Longman, London, 248 pp.
Jain, J.K. 1968 The climatic conditions of the arid region of Rajasthan, India, in *Proceedings of Symposium of Arid Zone, Jodhpur*, National Committee for Geography, Calcutta, pp. 20–7.
Jarvis, P.G. 1975 Heat and mass transfer in the biosphere, in DeVries, D. & Afgan, N. (eds) *Plant environment*, Halsted Press, Washington, DC.
Jennings, M.E. & Benson, M.A. 1969 Frequency curves for annual flood series with some zero events or incomplete data, Water Resources Research 5(1): 276–80.
Jensen, M.E. 1967 Evaluating irrigation efficiency, *Proc. American Society of Civil Engineers; Irrigation and Drainage Division* 95 (March), pp. 83–98.
Jensen, M.E. 1968 Water consumption by agricultural plants, in Koxlowski, T. (ed.) *Water deficits and plant growth*, Academic Press, London.
Jensen, M.E. (ed.) 1974 Consumptive use of water and irrigation water requirements, New York, Report for American Society of Civil Engineers, 215 pp.
Jensen, M.E. 1980 *Design and operation of farm irrigation systems*, American Society of Agricultural Engineering Monograph 3, St Joseph, Michigan.
JICA 1982 The feasibility report on Wadi al Jizzi agricultural development project, Japanese Cooperation Agency, Appendix 1.
Jiusto, J.E. 1981 Fog structure, in Hobbs, P. & Deepak, A. (eds) *Clouds and their formation, optical properties and effects*, Academic Press, London.
Johnson, A.I. 1986 Education and training needs assessment for the water resources sector, Sultanate of Oman, Checci and Company.
Johnson, D.L. 1969 The nature of nomadism, Dept. of Geography Research, Paper 118, University of Chicago.
Johnson (III), S.H. 1988 Large scale irrigation and drainage schemes in Pakistan, in O'Mara, G.T. (ed.) *Efficiency in irrigation: the conjunctive use of surface and groundwater resources*, A World Bank Symposium, Washington DC, pp. 58–79.
Johnson (III), S.H. 1990 Progress and problems of irrigation investment in South and southeast Asia. Water International 15: 15–26.
Jones, D.K.C., Cooke, R.U. & Warren, A. 1988 A terrain classification of the Wahiba Sands of Oman, Journal of Oman Studies Special Report 3: 19–32.
Jones, G.P. & Rushton, K.R. 1981 Pumping test analysis, in Lloyd, J.W. (ed.) *Case studies in groundwater resources evaluation*, Clarendon Press, Oxford, pp. 65–80.
Jones, K.R. (ed.) 1981 Arid zone hydrology for agricultural development, FAO, Rome, 271 pp.
de Jong, E. & Steppuhn, H. 1983 Water conservation: Canadian Prairies, in Dregne, H.E. & Willis, W.O. (eds) *Dryland agriculture*, American Society of Agronomy, Number 23, Madison, WI, pp. 89–104.
Jordan, A.A. & Kilmarx, R.A. 1979 *Strategic mineral dependence: the stockpile dilemma*, Sage, Beverly Hills.
Jurym, W.A., Sinai, G. & Stolzy, I.J. 1979 Future sources of cooling water for power plants in arid regions, Water Resources Bulletin 15(5): 1444–58.
Kabbara, S.A. 1987 Water recycling and re-use plants, *Conference on Water and Engineering for Developing Countries, Developing World Water*, Grosvenor Press International, Hong Kong.
Kadivar, S. 1985 Planning process and principles for establishing national water planning, in Yassin, A.A. (ed.) *Development and rationalisation of water resources in Oman, Oman Water Policy Planning Conference, October 27.*
Kaliappa, R., Selvara, J., Venkatch, S. & Nachappan, K. 1974 Studies on irrigation regimen and nitrogen rate on grain sorghum, Madras Agricultural Journal 61(8): 340–3.

Kay, M. 1983 *Sprinkler irrigation*, Batsford Academic and Educational Ltd, London, 120 pp.

Kay, M. 1986 *Surface irrigation systems and practice*, Cranfield Press, Cranfield, 142 pp.

Keeling, C.D. 1990 Global and national CO_2 emissions from fossil fuel burning, cement production and gas flaring, in Boden, T.A., Kanciruk, P. & Farrell, M.P. (eds) *A compendium of data on global change, ORNL/CDIAC – 36*, Carbon Dioxide Information Analysis Center, Oak Ridge National Laboratory, Oak Ridge, Tennessee, pp. 89–91.

Keeling, C.D., Bacastow, R.B. & Whorf, T.P. 1982 Measurements of concentrations of carbon dioxide at Mauna Loa observatory, Hawii, in Clark, W.C. (ed.) *Carbon dioxide review*, OUP.

Keeling, C.D. & Whorf, T.P. 1990 Atmospheric CO_2 concentrations: Mauna Loa, in Boden, T.A., Kanciruk, P. & Farrell, M.P. (eds) *A compendium of data on global change, ORNL/CDIAC – 36*, Carbon Dioxide Information Analysis Center, Oak Ridge National Laboratory, Oak Ridge, Tennessee, pp. 8–9.

Keen, M. 1985 Cheaper, purer water from the sun, Water and Sewage August, S14–16.

Keen, M. 1990 The groundswell of interest in water saving, *Financial Times* 18 April.

Keenan, J. 1977 *The Tuareg*, Allen Lane, London.

Kellogg, W. 1978 Facing up to climatic change, CERES/FAO 11(6): 13–17.

Kelly, T.J. 1975 Climate and the West African drought, in Newman, J.L. (ed.) *Drought, famine and population movements in Africa*, Syracuse University, Syracuse.

Kerr, R.A. 1988a The weather in the wake of El Nino, Science 240: 883.

Kerr, R.A. 1988b La Nina's big chill replaces El Nino, Science 241: 1037–8.

Khan, A.H. 1986 *Desalination processes and multistage flash distillation practice*, Elsevier, Oxford, 596 pp.

Kidson, J. 1977 African rainfall and its relations to upper air circulations, Quarterly Journal of the Royal Meteorological Society 103: 441–56.

Kilani, A. 1988 Recharge of aquifers by wastewater in Jordan, in Biswas, A.K. & Arar, A. (eds) *Treatment ad reuse of wastewater*, Butterworths, London, pp. 177–8.

Kirby, M.J. & Chorley, R.J. 1967 Throughflow, overland flow and erosion. Bulletin of the IASH 12: 5–21.

Kirschen, P.H. 1979 Water supply planning model for West Africa, *Proceedings of the American Society of Civil engineers*, Journal of the Water Resources and Planning Division 105(No WR2): 413–27.

Kirshna, R. 1986 The Nile river basin. Paper presented at the *Center for Strategic and International Studies Conference on US Foreign Policy on Water Resources in the Middle East: Instrument for Peace and Development, November 25 & 26, Washington, DC.*

Kishi, M. & Inohara, S. 1985 Development of fully automatic MSF plant, Water and Sewage Sept./Oct., 8–10.

Kjellen, R. 1916 *Der Staat als Lebensform.*

Kohnke, H., Dreibelbis, F.R. & Davidson, J.M. 1940 A survey and discussion of lysimeters and a bibliography on their construction and performance, US Dept. of Agriculture Miscellaneous Publication 372.

Kondo, J. & Saigusa, N. 1990 A parameterisation of evaporation from bare soil surfaces, Journal of Applied Meteorology 29: 385–9.

Kondratyev, K. 1988 *Climatic shocks; natural and anthropogenic*, J. Wiley, London.

Köppen, W. 1931 *Die Klimate der Erde*, Berlin.

Kovacs, G. 1987 Estimation of average areal evapotranspiration, Journal of Hydrology 95: 227–40.

Kovda, V.A. 1946 *The origin of saline soils and their regime*, Vol. I and Vol. II, USSR Academy of Sciences Publishing House, translated into English 1971, Israel program

for scientific translation, Jerusalem.
Kovda, V.A. 1980 *Land aridization and drought control*, Westview Press, Boulder, 277 pp.
Kovda, V.A. 1983 Loss of productive land due to salinization, Ambio 12(2): 92.
Kowal, J.M. & Kassam, A.H. 1978 *Agricultural ecology of the Savanna: A study of West Africa*, Clarendon Press, Oxford.
Kozlowski, T.T. (ed.) 1972 *Water deficits and plant growth*, Academic Press, New York.
Kramer, P.J. 1952 Plant and soil water relationships on the watershed, Journal of Forestry 50: 92–5.
Krishnan, A. 1979 Definition of droughts and factors relevant to specification of agricultural and hydrological droughts, *Symp. Hydrological Aspects of Droughts, New Delhi, Indian Natl Comm. for Int. Hydrol. Prog.*
Lamb, H.H. 1973 Is the Earth's climate changing? Courier (FAO) Aug./Sept.: 17–26.
Lamb, H.H. 1974 The Earth's changing climate, The Ecologist 4(1): 10–15.
Lamb, H.H. 1977 *Climate past, present and future*, Methuen, London.
Lamb, H.H. 1981 An approach to the study of the development of climate and its impact in human affairs, in Wigley, T.M.L., Ingram, M.J., & Farmer, G. (eds) *Climate and History: Studies in past climates and their impact on man*, CUP.
Lamb, P.J. 1982 Persistence of Subsaharan drought, Nature 299: 46–7.
Landsberg, H.E. 1982 Climatic aspects of drought, Bulletin of the American Meteorological Society 63: 593–6.
Langmuir, I. 1950 Results of seeding cumulus clouds in New Mexico, Project cirrus occas, Report 24, General Electricity Res. labs, Schenectady, NY.
Larson, M.K. & Pewe, T.L. 1986 Origin of land subsidence and earth fissuring, N.E. Phoenix, Arizona, Bulletin of the Association of Engineering Geologists XXIII(2): 139–45.
Lascano, R.J. 1991 Review of models for predicting soil water balance, in Sivakumar, M.V.T. *et al.* (eds) *Proceedings of Workshop on Soil Water Balance in the Sudano–Sahelian Zone, February 18–23*, IAHS Publication 199, Wallingford, pp. 443–58.
Lassen, L., Lull, H.W. & Frank, B. 1952 Some plant soil water relations in watershed management, US Dept. of Agriculure Circular 910.
Laurie, S. 1988 Water relations and solute content of some perennial plants in the Wahiba Sands, Oman, Journal of Oman Studies Special Report 3: 271–6.
Lavergne, M. 1986 The seven deadly sins of Egypt's high dam, in Goldsmith, E. & Hildyard, N. (eds) The social and environmental effects of large dams; volume 2 case studies, Wadebridge Ecological Centre, Cornwall, pp. 181–3.
Lawson, D. 1990 Let them drink sand as the Moguls did, *Sunday Correspondent*, September 30.
Laylin, J.G. & Bianchi, R.L. 1959 The role of adjudication in international river disputes, American Journal of International Law.
Lean, G., Hinrichsen, D. & Markham, A. 1990 *Atlas of the environment*, Arrow Books, London.
Lee, D.M. 1980 On monitoring rainfall deficiencies in semi desert regions, in Mabbutt, J. & Berkowicz, S.M. (eds) *The threatened dry lands*, School of Geography, Kensington, Australia.
Leeden, V. der 1975 *Water resources of the world*, Water Information Center, New York, 569 pp.
Le Houreou, H.N. & Popov, G.F. 1981 An ecological classification of inter tropical Africa, FAO, Rome.
Lemeur, R. & Zhang, Lu. 1990 Evaluation of three evapotranspiration models in terms of their applicability for an arid region, Journal of Hydrology 114: 395–411.
Letey, J. 1984 Impact of salinity upon the development of soil science, in Shainberg, I. and

Shalhevet, J. (eds.) *Soil salinity under irrigation*, Springer-Verlag, Berlin, pp. 1–11.
Lewis, D. 1991 Drowning by numbers. Geographical LXIII(9): 34–8.
Linacre, E. & Hobbs, J. 1977 *The Australian climatic environment*, J. Wiley, Brisbane.
Lingen, C. 1988 Efficient conjunctive use of surface and groundwater in the People's Victory Canal, in O'Mara, G.T. (ed.) *Efficiency in irrigation: the conjunctive use of surface and groundwater resources*, A World Bank Symposium, Washington, DC, pp. 84–7.
Lloyd, J.W. 1981 Environmental isotopes in groundwater, in Lloyd, J.W. (ed.) *Case studies in groundwater resources evaluation*, Clarendon Press, Oxford, pp. 113–32.
Lockwood, J.G. 1984 The southern oscillation and El Nino, Progress in Physical Geography 8: 102–10.
Logan, R.F. 1968 Causes, climates and distribution of deserts, in Brown, G.W. (ed.) *Desert biology*, Academic Press, London, pp. 21–50.
Long, N. 1980 Planning and politics in urban development, Journal of American Group of Planners 25(6): 168.
Loveday, J. 1984 Amendments for reclaiming sodic soils, in Shainberg, I. and Shalhevet, J. (eds) *Soil salinity under irrigation*, Springer-Verlag, Berlin, pp. 220–37.
Lovelock, J. 1985 Are we destabilising world climate? The Ecologist 15(1/2): 52–5.
Lowes, P. 1987 Half time in the decade, *Conference on Water and Engineering for Developing Countries, Developing World Water*, Grosvenor Press International, Hong Kong, pp. 16–21.
Louw, G. & Seeley, M. 1982 *Ecology of desert organisms*, Longman, London.
Lustig, L.K. 1968 Appraisal of research on geomorphology and surface hydrology of desert environments, in McGinnies, W.G., Goldman, B.J. & Paylore, P., (eds) *Deserts of the world*, University of Arizona Press, pp. 95–286.
L'Vovich, M.I. 1979 *World water resources and their future*, American Geophysical Union, Michigan.
Mabbutt, J. 1977 *Desert Landforms*, Australian National University Press, Canberra.
Mabbutt, J. 1984 A new global assessment of the status and trends of desertification, Environmental Conservation 11: 103–13.
McClearly, J.A. 1968 The biology of desert plants, in Brown, G.W. (ed.) *Desert biology*, Vol. 1, Academic Press, London, pp. 141–94.
Macdonald, L.H. 1986 Natural resources development in the Sahel: the rôle of the UN system, United Nations University, Japan, 95 pp.
McDonald, A.T. & Kay, D. 1980 *Water resources issues and strategies*, Longman Scientific, Harlow.
McElroy, M.B. 1986 Change in the natural environment of the Earth: the historical record, in Clark, W.C. & Munn, R.E. (eds) *Sustainable development of the biosphere*, CUP, pp. 199–211.
McGinnies, W.G. 1968 Appraisal of research on vegetation of desert environments, in McGinnies, W.G., Goldman, B.J. & Paylore, P. (eds) *Deserts of the world*, University of Arizona Press, pp. 381–568.
McGinnies, W.G. 1979 Arid land ecosystems, common features throughout the world, in Goodall, D.W., Perry, R.A. & Howes, K.M.W. (eds) *Arid land ecosystems: structure, functioning and management*, CUP.
McGinnies, W.G. 1988 Climatic and biological conditions of arid lands: a comparison, in Whitehead, E.E. *et al.* (eds) *Arid lands today and tomorrow*, Westview Press, Boulder.
McGinnies, W.G., Goldman, B.J. & Paylore, P. 1968 *Deserts of the world*, University of Arizona Press.
McGowan, M. 1974 Depths of water extraction by roots, in *Isotope and radiation techniques in soil physics and irrigation studies*, IAEA Symposium, Vienna.
McIlveen, R. 1986 *Basic meteorology*, Van Nostrand Reinhold, Wokingham, 457 pp.

McIntosh, D. & Thom, A. 1981 *Essentials of meteorology*, Taylor and Francis, London.
McKie, J. 1984 The race against clock and the thermometer, *The Guardian*, 29 January.
MacKinder, H.J. 1904 The geographical pivot of history, Geographical Journal 23: 412–37.
McKinder, H.J. 1919 *Democratic ideals and reality*, Constable, London.
MacKinder, H.J. 1943 The round world and the winning of peace, Foreign Affairs 21(4): 595–605.
McPherson, H.J. & Gould, J. 1985 Rainwater catchment systems in Botswana, Natural Resources Forum 9(4): 253–63.
Mahan, A.T. 1890 The influence of sea power upon history 1660–1783, Little Brown, Boston.
Maiti, R.K. & Bidinger, F.R. 1981 Growth and development of the pearl millet plant, ICRISAT Research, Bulletin 6, Hyderabad.
Manabe, S. and Stouffer, R.J. 1980 Sensitivity of a global climatic model to an increase of CO_2 concentration in the atmosphere, Journal of Geophysical Research 85: 5529–54.
Mandel, S. 1977 The overexploitation of groundwater resources in dry regions, in Mundlak, Y. & Singer, S.F. (eds) *Arid zone development: potentialities and problems*, Ballinger, Cambridge, Massachusetts.
Mann, H.S. 1984 Natural resource survey and environmental monitoring in arid Rjasthan using remote sensing, in El-Baz, F. (ed.) *Deserts and arid lands*, Martinus Nijhoff, Lancaster, pp. 157–70.
Manwaring, J.F., Lee, R.G. & Hoffbuhr, J.W. 1980 Quenching the thirst for potable water in arid Saudi Arabia, Journal of the American Water Works Association 72(12): 656–65.
Margat, J. & Saad, K.F. 1984 Deep lying aquifers: water mines under the desert? UNESCO: Nature and Resources XX(2): 7–13.
Markar, M.S. & Mein, R.G. 1987 Modelling of evapotranspiration from homogeneous soils, Water Resources Research 23: 2001–7.
Marshall, T.J. & Holmes, J.W. 1979 *Soil physics*, CUP, London, 345 pp.
de Martonne, E. & Aufrere, L. 1927 Maps of interior basin drainage, Geography Review 17(3): 414.
Marwitz, J.D. & Stewart, R.E. 1981 Some seeding signatures in Sierra Storms, Journal of Applied Meteorology 20: 1129–44.
Mason, D.J. 1962 *Clouds, rain and rain making*, CUP.
Mather, J.R. 1954 The measurement of potential evapotranspiration, Publications in Climatology 7(1).
Mather, J.R. 1974 *Climatology: fundamentals and applications*, McGraw-Hill, New York.
Matlock, W.G. 1988 The case for small scale water management systems in developing countries, in Whitehead, E. *et al. Arid lands today and tomorrow*, Westview Press, Boulder, pp. 935–42.
Matveev, L.T. 1984 *Cloud dynamics*, D. Reidel, Boston.
Maull, H.W. 1984 *Raw materials, energy and Western security*, Macmillan, London.
May, P.J. & Williams, W. 1986 *Disaster policy implementation*, Plenum Press, New York.
Mearns, L. 1988 Personal Communication, American Geographical Association Meeting, St Paul, Minnesota.
Meigs, P. 1952 Arid and semiarid climatic types of the world, *Proc. 8th. General Assembly and 7th International Congress, International Geographical Union, Washington, DC.* pp. 135–8.
Meigs, P. 1953 World distribution of arid and semi arid homoclimates, UNESCO Arid Zone Program 1: 203–10.
Micklin, P.P. 1971 Soviet plans to reverse the flow of rivers: the Kama-Vychegda-Pechora

project, in Detwyler, T. (ed.) *Man's impact on environment*, McGraw-Hill, New York, pp. 302–18.

Micklin, P.P. 1986 Soviet river division plans: their possible environmental impact, in Goldsmith, E. & Hildyard, N. (eds) *The social and environmental effects of large dams: volume 2 case studies*, Wadebridge Ecological Centre, Cornwall, pp. 91–106.

Micklin, P.P. 1988 Desiccation of the Aral sea: A water management disaster in the Soviet Union, Science 241: 1170–6.

Middleton, N.J. 1985 Effect of drought on dust production in the Sahel, Nature 316: 431–4.

Milas, S.L. 1983 The years of drought, Desertification Control Bulletin 9: 10–14.

Mitchell, J.F.B. 1987 Simulation of climatic change due to increased atmospheric carbon dioxide, Meteorological Magazine 116: 361–76.

Mitchell, J.F.B. & Warrilow, D.A. 1987 summer dryness in northern mid latitudes due to increased $C0_2$ Nature 330: 238–40.

Mitchell, J.K.N. 1986 The South-east Anatolia project – Commercial opportunities, Preliminary report, British Embassy Ankara and Department of Trade and Industry, August.

Mohan Rao, K.M., Kashyap, D. & Chandra, S. 1990 Relative performance of a soil moisture accounting model in estimating return flow, Journal of Hydrology 115: 231–41.

Monod, R. 1975 *Pastoralism in tropical Africa*, OUP, London.

Monte, J.A. & Cooper, D.I. 1982 Technical manual for groundwater exploration in arid regions, US Department of Interior, Office of Water Research and Technology, Washington DC.

Monteith, J.L. 1957 Dew, Quarterly Journal of the Royal Meteorological Society 83(357): 322–41.

Monteith, J.L. 1980 The development and extension of Penman's formula, in Hillel, D. (ed.) *Applications of soil physics*, Academic Press, London.

Monteith, J.L. 1981 Evaporation and surface temperature, Quarterly Journal of the Royal Meteorological Society 107: 1–27.

Monteith, J.L. 1991 Weather and water in the Sundano–Sahelian zone, in Sivakumar, M.V.K. *et al.* (eds) *Proceedings or Workshop on soil Water Balance in the Sudano–Sahelian zone, February 18–23*, IAHS Publication 199, Wallingford, pp. 11–30.

Montgomery, H.A.C. 1988 Water quality aspects of the recharge of sewage effluents for reuse, in Biswas, A.K. & Arar, A. (eds.) *Treatment and reuse of wastewater*, Butterworths, London, pp. 38–45.

Mooley, D.A. 1971 Independence of monthly and bimonthly rainfall over South East Asia during the summer monsoon season, Monthly Weather Review 99(6): 532–6.

Moorhead, D.L., Reynolds, J.F. & Fonteyn, P.J. 1989 Patterns of stratified water loss in a Chihuahuan desert community, Soil Science 148(4): 244–9.

Morel-Seytoux, H.J. 1976 Derivation of equations for rainfall infiltration, Journal of Hydrology 31: 203–19.

Morgan, R. 1985 Development of early warning drought systems, Disasters 9(1): 44–50.

Moris, J. 1987 Irrigation as a privileged solution in African development, Development Policy Review 5: 99–123.

Morton, F.I. 1983 Operational estimates of areal evapotranspiration and their significance to the science and practice of hydrology, Journal of Hydrology 66: 1–76.

Moses, Y.T. 1986 *Proceedings of the African agricultural development conference, technology, ecology and society, Pamona, CA*, Californian State Polytechnic University.

Motha, R.P., Leduc, S.K., Stayaert, L.T., Sakamoto, C.M. and Strommen, N.D. 1980 Precipitation patterns in West Africa, Monthly Weather Review 108: 1567–78.

Mounier, F. 1986 The Senegal River scheme: development for whom? in Goldsmith, E. & Hildyard, N. (eds) *The social and environmental effects of large dams: volume 2 case*

studies, Wadebridge Ecological Centre, Cornwall, pp. 109–19.
Mundlak, Y. & Singer, S.F. 1977 *Arid zone development; potentialities and problems*, Ballingr, Cambridge, Massachusetts.
Murray, C.S. 1977 Implied reservation claims after Cappaert V. United States, Arizona State Law Journal 3: 647–72.
Myers, N. 1988 Tropical deforestation and climate change, Environmental Conservation 15(4): 293–8.
El-Nadi, A.H. 1969 Efficiency of water use by irrigated wheat in the Sudan, Journal of Agricultural Science 73(2): 261–6.
Naff, T. & Matson, R.C. (eds) 1984) *Water needs in the Middle East*, Westview Press, Boulder.
Nakayama, F.S. & Bucks, D.A. 1986 *Trickle irrigation for crop production*, Elsevier, Oxford, 383 pp.
Namias, J. 1981 Severe drought and recent history, In Rotberg, R.I. & Rabb, T.K. (eds) *Climate and history*, Princeton University Press, 280 pp.
Nasser, M. 1988 Use of sewage effluent for irrigation in Jordan, in Biswas, A.K. & Arar, A. (eds) Treatment and reuse of wastewater, Butterworths, London, pp. 135–8.
Nebbia, G. 1968 Economics of the conversion of saline waters to fresh water for irrigation, in Boyko, H. (ed.) *Saline irrigation for agriculture and forestry*, W. Junk, The Hague, 325 pp.
Nicholson, S.E. 1978 Comparison of historical and recent African rainfall anomalies with late Pleistocene and early Holocene, in van Zindern-Bakker, E. & Coetzee, J.A. (eds) *Paleoecology in Africa*, A.A. Balkema, Rotterdam.
Nicholson, S.E. 1979 Revised rainfall series for the West African Subtropics, Monthly Weather Review 107: 620–3.
Nicholson, S.E. 1981a Rainfall and atmospheric circulation during drought periods and wetter years in West Africa, Monthly Weather Review 109: 2191–208.
Nicholson, S.E. 1981b The historical climatology of Africa, in Wigley, T. *et al.* (eds) *Climate and History*, CUP.
Nicholson, S.E. 1981c Saharan climates in historic times, in Allan, J.A. (ed.) *The Sahara Conference Proc.*, SOAS, London.
Nickum, J.E. 1988 All is not wells in North China, in O'Mara, G.T. (ed.) *Efficiency in irrigation: the conjunctive use of surface and groundwater resources*, A World Bank Symposium, Washington, DC, pp. 88–94.
Nir, D. 1974 *The semi arid world*, Longman, London, 461 pp.
Nizim, B.K. 1968 The Indus, Nile and Jordan: International rivers and factors in conflict potential, Unpublished PhD Thesis, University of Indiana.
Nordensen, T. & Baker, D.R. 1962 Comparative evaluation of evaporation instruments, Journal of Geophysical Research 67: 671–9.
North, R. & Schoon, N. 1989 High level problem facing the world, *The Independent* 2 March.
Norton-Griffiths, M. 1989 Food and agricultural production, IUCN Sahel Studies, Nairobi: 53–82.
Nye, P.H. & Tinker, P.B. 1977 *Solute movement in the soil root system*, Studies in Ecology 4, Blackwell Science, Oxford.
Obeng, L.E. 1975 Too much or too little? CERES 8(4): 18–20.
Obeng, L.E. 1977 Starvation or bilharzia? A rural development dilemma, in Worthington, E.B. (ed.) *Arid land irrigation in developing countries*, Pergamon Press, Oxford, pp. 343–50.
ODI 1987 Coping with African drought, Overseas Development Institute, London, Briefing Paper, July.
Odone, T. 1984 Manmade river brings water to the people, Middle East Digest, 10 August: 39–40.

Odum, E.P. 1971 *Fundamentals of ecology*, W.B. Saunders, London, 574 pp.
van den Oever, P. 1989 *IUCN Sahel studies: Population*, IUCN, Nairobi.
Office of Saline Water 1970 Desalting plants, Inventory report 3, Washington, DC.
Ojiako, G.U. 1988 A study of pollutional effects of irrigation drainage on Anambra river quality and fishing, Water International 13: 66–73.
Ojo, O. 1977 *The climates of West Africa*, Heinemann, London.
Oke, T.R. 1978 *Boundary layer climates*, Methuen, London, 372 pp.
Oliver, H. 1961 *Irrigation and drainage*, Edward Arnold, London.
Oliver, J.E. 1973 *Climate and man's environment*, J. Wiley, London.
Oliver, J.E. 1981 *Climatology: selected applications*, Edward Arnold, London.
Olsson, L. 1983 Desertification or climate?, Lund Studies in Geography No. 60, University of Lund, Sweden.
O'Mara, G.T. (ed.) 1988 *Efficiency in irrigation: the Conjunctive use of surface and groundwater resources*, A World Bank Symposium, Washington, DC, 196 pp.
Oosterbaun, R.J. 1985 Modern water control systems for agriculture in developing countries, in Mock, J.F. (ed.) *Symposium on Man and technology in irrigated agriculture.*
Open University 1974 *Water resources. The Earth's physical resources*, Course S266, Block 5, Open University Press, Milton Keynes.
Ortolano, L. 1979 Environmental assessments in water resources planning, in Golubev, G. & Biswas, A.K. (eds) *Interregional water transfers*, Pergamon Press, Oxford, pp. 159–76.
Orville, H.D. 1986 A review of dynamic mode seeding of summer cumuli, in Braham, Jr, R.R. (ed.) *Precipitation enhancement – A scientific challenge*, Meteorological Monographs of the American Meteorological Society 21(43): 43–62.
Oster, J.D. 1984 Leaching for salinity control, in Shainberg, I. and Shalhevet, J. (eds) *Soil salinity under irrigation*, Springer-Verlag, Berlin, pp. 175–89.
Otterman, J. 1974 Baring high albedo soils by overgrazing, Science 86: 531–3.
Otterman, J. 1975 *Possible rainfall reductions through reduced surface temperature due to overgrazing*, NASA Goddard Space Flight Center, Maryland.
Overman, M. 1976 *Water: solutions to a problem of supply and demand*, Open University Press, Milton Keynes, 204 pp.
Owen, J.A. & Folland, C.K. 1988 Modeling the influence of sea surface temperatures on tropical rainfall, in Gregory, S. (ed.) *Recent climatic change*, Belhaven Press, London, pp. 141–53.
Palmer, W.C. 1965 Meteorological drought, US Weather Bureau Research Paper 45, Washington, DC, 58 pp.
Palmer-Jones, R. 1981 How not to learn from pilot irrigation projects: the Nigerian experience, Water Supply and Management 5: 81–105.
Parker, D.E., Folland, C.K. & Ward, M.N. 1988 Sea surface temperature anomaly patterns and prediction of seasonal rainfall in the Sahel region of Africa, in Gregory, S. (ed.) *Recent climatic change*, Belhaven Press, London 326 pp.
Parker, G. 1985 *Western geopolitical thought in the 20th century*, Croom Helm, London.
Parker, Y. & Smith, P. 1983 Factors affecting the flow of falaj in Oman, Journal of Hydrology 65: 293–312.
Parkes, M. & O'Callaghan, R. 1980 Modelling soil water changes in a well structured freely draining soil, Water Resources Research 16(4): 755–61.
Parry, M.L. 1986 Some applications of climatic change for human development, in Clark, W. and Munn, R. (eds) *Sustainable development of the biosphere*, Cambridge University Press, London, pp. 378–409.
Parry, M. 1990 *Climate change and world agriculture*, Earthscan, London.
PAWR 1983 The hydrology of the Sultanate of Oman, Public Authority for Water Resources, Sultanate of Oman.

Pearce, D. 1988 The sustainable use of natural resources in developing countries, in Turner, R.K. (ed.) Sustainable environmental management, Belhaven Press, London, pp. 102–17.

Pearse, A. 1980 Seeds of plenty, seeds of want, Clarendon Press, Oxford.

Peele, T.C., Beale, O.W. & Lesene, F.F. 1948 Irrigation requirements of South Carolina soils, Agricultural Engineering 29: 157–9.

Pelton, W.L. 1961 The use of lysimetric methods to measure evapotranspiration, *Proc. Hydrological Symp.*, 2, pp. 106–27, Queen's Printer, Ottawa.

Pemberton, B. 1987 Nile water study, *Conference on Developing World Water*, Grosvenor International, London, pp. 301–2.

Penman, H. 1948 Natural evaporation from open water, bare soil and grass, Proceedings of the Royal Society A 193: 120–45.

Penman, H. 1949 The dependence of transpiration on weather and soil conditions, Journal of Soil Science 1: 74–89.

Penman, H. 1963 Vegetation and hydrology, Commonwealth Bureau of Soils, Technical Communication 53, Harpenden.

Pereira, H.C. 1973 *Land use and water resources in temperate and tropical climates*, CUP.

Perry, A.H. 1984 Recent climatic change – is there a signal amongst the noise? Progress in Physical Geography 8: 111–17.

Peterson, J.E. 1978 *Oman in the 20th century*, Croom Helm, London.

Petrov, M.P. 1969 Classification of world deserts, *International Conference on Arid Lands in a Changing World*, American Association for the Advancement of Science.

Petrov, M.P. 1976 *Deserts of the world*, J. Wiley, New York.

Pittock, A.B. 1988 The green house effect and future climatic change, in Gregogy, S. (ed.) *Recent climatic change*, Belhaven Press, London, pp. 306–15.

Pollard, N. 1986 The Sudan Gezira scheme: A study in failure, in Goldsmith, E. & Hildyard, N. (eds) *The social and environmental effects of large dams; volume 2 case studies*, Wadebridge Ecological Centre, Cornwall, pp. 168–80.

Popkin, R. 1969 *Desalination, water for the world's future*, Praeger, New York.

Porteous, A. 1975 *Saline water distillation processes*, Longman, London, 150 pp.

Post, W.S. 1987 Small scale irrigation systems, in Pickford, J. (ed.) *Developing World Water*, Grosvenor Press International, Hong Kong, pp. 261–2.

Postel, S. 1984 Water: rethinking management in an age of scarcity, Worldwatch Paper 62, Washington, DC.

Postel, S. 1986 Effective water use for food production, Water International 11(1): 23–32.

Postel, S. 1989 Water for agriculture: facing the limits, Worldwatch Paper 93, Washington, DC.

Priscoli, J.D. 1983 Water as a political and social tool, in *Water for human consumption, man and his environment*, International Water Resources Association.

Puffet, A.W. 1979 A proposed economical fog sensor configuration for the estimation of slant visual range, RAE Technical memorandum FS 233.

Puffet, A.W. 1980 Vertical fog structure and its application to landing in low visibility, RAE Technical Memorandum FS 301.

Purseglove, J.W. 1972 *Tropical crops – Monocotyledons*, Longman, London.

Ramanathan, V., Cicerone, R.J., Singh, H.B. & Kiehl, J.T. 1985 Trace gas trends and their potential role in climate change, Journal of Geophysical Research 90(V): 5547–66.

Ramdas, D.A. 1960 *Crops and weather in India*, New Delhi, ICAR, 127 pp.

Rantz, S.E. & Eakin, T.E. 1971 A summary of methods for the collection and analysis of basic hydrologic data for arid regions, USGS, Menlo Park, California.

Rasmussen, E.M. 1987 Global climate change and variability: effects on drought and

desertification in Africa, in Glantz, M. (ed.) *Drought and hunger in Africa*, CUP.
Rawitz, E. 1973 Sprinkler irrigation, and gravity irrigation and subirrigation, in Yaron, B., Danfors, E. and Vaadia, Y. (eds) *Arid zone irrigation*, Chapman and Hall, London, pp. 303–22, 323–37 and 353–5.
Reginto, R.J., Jackson, R.D. & Pinter, P.J. 1985 Evapotranspiration from remote multispectral and ground station meteorological data, Remote Sensing of the Environment 18: 75–89.
Reid, I. & Frostick, L.E. 1989 Channel form, flows and sediments in deserts, in Thomas, D. (ed.) *Arid zone geomorphology*, Belhaven Press, London, pp. 87–116.
Reynolds, S.G. 1970 The gravimetric method of soil moisture determination; Parts I and II, Journal of Hydrology 11: 258–87.
Richards, L.A. 1931 Capillary conduction of liquids through porous mediums, Physics 1: 318–33.
Richards, L.A. 1949 Methods for measuring soil moisture tension, Soil Science 68: 95–112.
Richards, L.A. & Weaver, L.R. 1943 Fifteen atmosphere percentage as related to the permanent wilting percentage, Soil Science 56: 331–9.
Richards, S.J. 1957 Time to irrigate? Crops and Soil 9(5): 8–11.
Rind, D. 1984 Global climate in the 21st Century, Ambio 13: 148–51.
Roach, W.T., Brown, R., Caughey, S.J., Garland, J.A. and Readings, C.J. 1976 The physics of radiation fog, Quarterly Journal of the Royal Meteorological Society, 102: 312–33.
Rodda, J.C. 1976 *Facets of hydrology*, J. Wiley, New York.
Rodda, J.C., Downing, R.A. & Law, F.M. 1976 *Systematic hydrology*, Newnes Butterworths, London.
Rodhe, B. 1962 The effects of turbulence on fog formation, Tellus 14: 49–86.
Rodier, J.A. 1985 Aspects of arid zone hydrology, in Rodda, J.C. (ed.) *Facets of hydrology*, Vol. 2, pp. 205–47.
Rodier, J. & Chaperon, P. 1970 Modifications du regime hydrologique du Niger a Niamey depuis 1961, ORSTOM, Paris.
Ron, Z.D. 1985 Development and management of irrigation systems in mountain regions of the Holy Land, Transactions of the Institute of British Geographers 10: 149–69.
Ron, Z.D. 1986 Ancient and modern developments of water resources in the Holy Land and the Israeli–Arab conflict – a reply, Transactions of the Institute of British Geographers 11: 360–9.
Ronghan, H. 1988 Development of groundwater for agriculture in the lower Yellow river alluvial basin, in O'Mara, G.T. (ed.) *Efficiency in irrigation; the conjunctive use of surface and groundwater resources*, A World Bank Symposium, Washington, DC, pp. 80–4.
Roosma, E. & Stakelbeek, A. 1990 Deep well infiltration in North Holland dune area, Water International 15: 151–6.
Rottry, R.M. 1981 Distribution and changes in industrial carbon dioxide production, Carbon Dioxide Assessment Program Contribution 81–23, Oak Ridge Associated Universities, Oak Ridge, Tennessee.
Rowley, G. 1986 Irrigation systems in the Holy Land: a comment, Transactions of the Institute of British Geographers 11: 356–9.
Royal Commission on Pollution 1984 *10th Report: tackling pollution, experience and prospects*, HMSO, London.
Rumney, G.R. 1968 *Climatology and the world's climates*, Macmillan, New York.
Russell, C.S., Arey, D.G. & Kates, R.W. 1970 *Drought and water supply*, Johns Hopkins University Press, Baltimore, 232 pp.
Russell, E.W. 1973 *Soil conditions and plant growth*, Longman, London.

Ruthenberg, H. 1976 *Farming systems of the tropics*, Clarendon Press, Oxford.
Ruttan, V.W. 1986 Assistance to expand agricultural production, World Development 14(1): 39–63.
Rydzewski, J.R. (ed.) 1987 *Irrigation development and planning*, J. Wiley, Chichester, 265 pp.
Saha, S. 1981 River basin planning as a field of study: design of a course structure for practitioners, in Saha, S. & Barrow, C. (eds) *River basin planning*, J. Wiley, New York.
Saha, S. & Barrow, C. 1981 *River basin planning*, J. Wiley, New York.
Sahouri, M.M. 1988 Existing wastewater reuse and its future potential in Jordan, in Biswas, A.K. & Arar, A. (eds) *Treatment and reuse of wastewater*, Butterworths, London. pp.121–7.
Saines, M., Grace, S. & Lysonki, J. 1981 Proposed groundwater exploration program in the Musandam Peninsula, Oman, Public Water Authority for Water Resources, Sultanate of Oman.
Salter, P.J., Berry, G. & Williams, J.B. 1966 The influence of texture on the moisture characteristics of soils, part 3, Soil Science 17(1): 93–8.
Salter, P.J. & Goode, J.E. 1967 Crop responses to water as different stages of growth, Commonwealth Agricultural Bureau.
Salter, P.J. & Williams, J.B. 1965 The influence of texture on the moisture characteristics of soils, parts 1 and 2, Soil Science 16(1): 1–15 & 16(2): 310–17.
Salter, P.J. & Williams, J.B. 1967 The influence of texture on the moisture characteristics of soils, Part IV, Soil Science 18(1): 174–81.
Sandford, S. 1987 Towards a definition of drought, *Symp. on Drought in Botswana, Gaborone*, Clark University Press.
Sargeant, D.H. & Tanner, C.B. 1967 A simple apparatus of Bowen ratio determination, Journal of Applied Meteorology 6: 414–18.
El-Sayed, Y.M. & Silver, R.S. 1980 Fundamentals of Distillation, in Spiegler, K.S. & Laird, A.D.K. (eds) *Principles of desalination, part A*, Academic Press, London, 357 pp.
Schaefer, V.J. 1946 The production of ice crystals in a cloud of supercooled water droplets, Science 104: 457–9.
Schaffer, L.H. & Mintz, M.S. 1980 Electrodialysis, in Spiegler, K.S. & Laird, A.D.K. (eds) *Principles of desalination, part A*, Academic Press, London, 357 pp.
Schick, A.P. 1985 Water in arid lands, in Last, F.T., Hotz, M.C.B. & Bell, B.G. (eds) *Land and its uses – actual and potential*, Plenum Press, London, pp. 213–38.
Schmida, A., Evenari, M. & Noy-Meir, I. 1986 Hot desert ecosystems, in Evenari, M., Noy-Meir, I. & Goodall, D.W. (eds) *Ecosystems of the world: hot deserts and arid shrublands*, Elsevier, Oxford, pp. 379–88.
Schneider, S. 1976 Is there really a food-climate crisis?, in Kopec, R. (ed.) *Atmosphere quality and climatic change*, Studies in Geography 9, University of North Carolina, pp. 106–43.
Schonwiese, C.D. 1988 Volcanism and air temperature variations in recent centuries, in Gregogy, S. (ed.) *Recent Climatic change*, Belhaven Press, London, pp. 20–9.
Schware, R. & Friedman, E.J. 1981 Climate debate heats up, The Bulletin of the Atomic Scientists, December, pp. 31–3.
Schwarz, Y. 1986 Water supply in Israel, Mekorot Company, Tel Aviv.
Scoging, H. 1989 Runoff generation and sediment mobilisation by water, in Thomas, D. (ed.) *Arid zone geomorphology*, Belhaven Press, London, pp. 87–116.
Scroggin, D.G. & Harris, R.H. 1981 Reduction at source, Technology Review, Nov./Dec., 23–8.
Seetharama, N., Mahalakshmi, V., Bidinger, F.R. & Singh, S. 1984 Response of sorghum and pearl millet to drought stress in semi arid India, in *Proc. of Int. Symp. on Agro-*

meteorology of Sorghum and Millet in the semi-arid Tropics, ICRISAT, India, pp. 159–73.

Sellers, W. 1965 *Physical climatology*, University of Chicago Press.

de Seversky, A.P. 1950 *Air power, the key to survival*, Simon and Schuster, New York.

Sewell, W.R.D. 1973 Weather modification: social concerns and public policies, in Sewell, W.R.D. (ed.) *Modifying and weather*, Western Geographical Series 19, University of Victoria, pp. 1–49.

Shalhevet, J. 1973 Irrigation with saline water, in Yaron, B., Danfors, E. and Vaadia, Y. (eds) *Arid zone irrigation*, Chapman and Hall, London, pp. 263–76.

Shalhevet, J., Mantell, A., Bielorai, H. & Shimshi, D. 1976 Irrigation of field and orchard crops grown under semi arid conditions, International Irrigation Information Center Publication 1, Israel.

Shantz, H.L. 1956 History and problems of arid lands development, in White, G.F. (ed.) *The future of arid lands*, American Sociaty for the Advancement of Science Publication 43, pp. 3–25.

Sharon, D. 1972 The spottiness of rainfall in a desert area, Journal of Hydrology 17: 161–75.

Sharon, D. 1981 The distribution in space of local rainfall in the Namib desert, Journal of Climatology 1: 69–75.

Shaw, E. 1983 *Hydrology in practice*, Van Nostrand Rheinhold, Wokingham, 569 pp.

Shimshi, D. 1973 Estimating water status in plants, in Yaron, B., Danfors, E. and Vaadia, Y. (eds) *Arid zone irrigation*, Chapman and Hall, London. pp. 249–59.

Shimshi, D., Yaron, D., Bresler, E., Weisbrod, M. & Strateener, G. 1975 Simulation model for evapotranspiration of wheat: empirical approach, *Proceedings of American Society of Civil Engineers, Irrigation and Drainage Division* 101: 1–12.

Shmueli, E. 1973 Efficient utilisation of water in irrigation, in Yaron, B., Danfors, E. and Vaadia, Y. (eds) *Arid zone irrigation*, Chapman and Hall, London, pp. 411–23.

Shuttleworth, J.W. 1988 Macrohydrology; the new challenge for process hydrology, Journal of Hydrology 100: 31–56.

Shuttleworth, J.W. & Wallace, J.S. 1985 Evaporation from sparse crops; an energy combination theory, Quarterly Journal of the Royal Meteological Society 111: 839–55.

Shuval, J.I. 1977 *Water renovation and reuse*, Academic Press, New York.

Shuval, H.I. 1980 *Water quality management under conditions of scarcity*, Academic Press, New York.

Silver, R.S. 1978 *Steam plant aspects of sea water distillation*, Burlington Press, London 40 pp.

Silverman, B.A. 1986 Static mode seeding of summer cumuli – a review, in Braham, Jr, R.R. (ed.) *Precipitation enhancement – A scientific challenge*, Meteorological Monographs of the American Meteorological Society 21(43): 7–24.

Simpson, J. 1980 Downdrafts as linkages in dynamic cumulus seeding effects. Journal of Applied Meteorology 19: 477–87.

Simpson, R. 1985 Arab water resources; an assessment, Arab–British Chamber of Commerce Report 6, pp. 6–11.

Sivakumar, M.V.K., Wallace, J.S., Renard, C. & Giroux, C. 1991 *Soil Water Balance in the Sudano–Sahelian Zone, Proceedings of Workshop in Niamey February 18–23*, IAHS Publication 199, Wallingford.

Slatyer, R.O. 1956 Evapotranspiration in relation to soil moisture, Netherlands Journal of Agricultural Science 4: 73–6.

Sloan, G.R. 1988 *Geopolitics in the United States strategic policy 1890–1987*, Wheatsheaf, Brighton.

Small, R.J. 1972 *The study of landforms*, CUP.

Smith, E.J. 1974 Cloud seeding in Australia, in Hess, W.N. (ed.) *Weather and climate modification*,

Smith, H.T.U. 1968 Geologic and geomorphic aspects of deserts, in Brown, G.W. (ed.) *Desert biology*, Academic Press, London, pp. 51–100.

Smith, P.L. 1986 an engineer's view on the implementation and testing of seeding concepts, in Braham, Jr, R.R. (ed) *Precipitation enhancement – A scientific challenge*, Meteorological Monographs of the American Meteorological Society 21(43): 151–4.

Smith, R.E. & Schreiber, H.A. 1973 Point processes of seasonal thunderstorm rainfall: distribution of rainfall events, Water Resources Research 9(4): 871–84.

Snowden, C. 1990 Drinking water decade, Geographical Magazine LXII(4): 3–4.

SOAS 1987 Qanats: History, development and utilisation, Unpublished papers on conference organised by Centre of Near and Middle Eastern Studies, SOAS, University of London.

Sofia, S. 1984 Solar variability as a source of climatic change, in Hansen, J.F. and Takahashi, T. (eds) *Climate processes and climate sensitivity*, pp. 202–6.

Sokolov, A.A. Rantz, S.E. and Roche, N. 1976 *Flood flow computation*, UNESCO, Paris.

De Souza, G. 1990 Interrelationships between realtime satellite derived rainfall and vegetation estimates over Africa 1985–1987, Paper presented at *Association of British Climatologists Annual Conference, University of East Anglia.*

Speidel, D.H. & Agnew, A.F. 1988 World water budget, in Speidel, D.H., Ruedisili, L.C. & Agnew, A.F. (eds) *Perspectives on water uses and abuses*, OUP, Oxford.

Speth, J.G. 1988 Arid lands in the global picture, in Whitehead, E. *et al.* (eds) *Arid lands today and tomorrow*, Westview Press, Boulder, pp. 1417–21.

Spiegler, K.S. & Laird, A.D.K. (eds) 1980 *Principles of desalination, parts A. & B*, Academic Press, London, 821 pp.

Spiteri Staines, E. 1987 Aspects of water problems in the Maltese Islands, Paper presented at the *Barcelona Conference*, 19–23 October.

Spykman, N.J. 1944 *The geography of the peace*, Harcourt Brace, New York.

Squire, M.D. 1987 Some recent advances in sprinkler irrigation systems, in Pickford, J. (ed.) *Developing world water*, Grosvenor Press International, Hong Kong, pp. 273–80.

Stanley Price, M.R., Hamoud al-Harthy, A. & Whitcombe, R.P. 1986 Fog moisture and its ecological effects in the Oman Salalah, Planning Committee for Development and Environment in the Southern Region, Sultanate of Oman.

Stewart, B.A. & Musick, J.T. 1982 Conjunctive use of rainfall and irrigation in semi arid regions, in Hillel, D. (ed.) *Advances in irrigation*, Academic Press, London, pp. 1–24.

Stone, L.R., Horton, M.L. & Olson, T.C. 1973 Water loss from an irrigated sorghum field, Agronomy Journal 65(a): 492–7.

Stonehouse, B. 1985 *Philips' pocket guide to the World*, George Philip, London.

Stoner, R. 1990 Future irrigation planning in Egypt, in Howell, P.P. & Allan, J.A. (eds) *The Nile, Conference at RGS, 2–3 May* SOAS, London, pp. 83–92.

Stroosnijder, L., Lascano, R.J., Van Bavel, C.H.M. & Newton, R.W. 1986 Relation between L band soil emittance and soil water content, Remote Sensing of Environment 19: 117–25.

Al-Sudeary, A.M. 1988 Alleviation of rural poverty in arid lands, in Whitehead, E. *et al.* (eds) *Arid lands today and tomorrow*, Westview Press, Boulder, pp. 13–20.

Summer, G. 1988 *Precipitation: process and analysis*, J. Wiley, Chichester.

Sung-Chiao, C. 1984 Analysis of desert terrain in China using Landsat imagery, in El-Baz, F. (ed.) *Deserts and arid lands*, Martinus Nijhoff, Lancaster, pp. 115–32.

Subrahmanyam, V.P. 1976 Incidence and spread of continental drought, WMO Report, Geneva.

Swift, J. (ed.) 1984 Pastoral development in Central Niger: Report of the Niger range and livestock project, Min. du Developpement Rural, Niger.

Tannehill, I.R. 1947 *Drought, its causes and effects*, Princeton University Press, 264 pp.

Taylor, J.G., Downton, M.N. & Stewart, T.R. 1988 Adapting to environmental change; perceptions and farming practices in the Oglalla aquifer region, in Whitehead, E. *et al.* (eds) *Arid lands today and tomorrow*, Westview Press, Boulder, pp. 665–84.

Taylor, S.A. & Ashcroft, G.L. 1972 *Physical edaphology*, W.H. Freeman and Co, San Francisco.

Tennant, D. 1976 Wheat root penetration and total available water on a range of soil types, Australian Journal of Experimental Agriculture & Animal Husbandry 16: 570–2.

Thom, A.S. 1975 Momentum, mass and heat exchange of plant communities in Monteith, J.L. (ed.) *Vegetation and atmosphere*, Academic Press, London, pp. 57–109.

Thom, A.S. & Oliver, H.R. 1977 On Penman's equation for estimating regional evaporation, Quarterly Journal of the Royal Meteorological Society 103: 345–57.

Thom, H. 1958 A note on the gamma distribution, Monthly Weather Review 86(4): 117–22.

Thomas, D.S. (ed.) 1989 *Arid zone geomorphology*, Belhaven Press, London.

Thomas, E.G. 1987 Refugee water supplies in Somalia, *Conference on Developing World Water*, Grosvenor International, London, pp. 27–8.

Thomas, R.G. 1988 Groundwater as a constraint to irrigation, in O'Mara, G.T. (ed.) *Efficiency in irrigation: the conjunctive use of surface and groundwater resources*, A World Bank Symposium, Washington, DC, pp. 168–77.

Thomas, W.A. 1978 Scientific, technological, legal and political uncertainties of weather modification, in Davis, R.J. & Grant, L.O. (eds.) *Weather modification and the law*, American Association for the Advancement of Science 20, Westview Press, Boulder, pp. 1–7.

Thompson, L.M. 1973 Cyclical weather patterns in the middle latitudes, Journal of Soil Water Conservation 28: 87–9.

Thompson, N., Barrie, I.A. & Ayles, M. 1981 The meteorological office rainfall and evaporation calculation system: MORECS, Hydrological Memoir 45, Meteorological Office, Bracknell.

Thompson, R.D. 1975 The climatology of the arid world, Department of Geography Paper No 35, Reading University.

Thorn, D. 1972 A computer program for fitting seasonal probability distributions, Water Research Association, Marlow, Buckinghamshire.

Thorne, W. (ed.) 1963 *Land and water use*, American Association for the Advancement of Science No. 73, Washington, DC.

Thornthwaite, C.W. 1944 A contribution to the report of the committee on transpiration and evaporation, Transactions of the American Geophysical Union 25: 686–93.

Thornthwaite, C.W. 1948 An approach towards a rational classification of climate, Geographical Review 38: 55–94.

Thornthwaite, C.W. 1954 The determination of potential evapotranspiration, in Mather, J. (ed.) *Publications in climatology* 7(1).

Tickell, Sir C. 1986 Drought in Africa: impact and response, Overseas Development 102: 10.

Timberlake, L. 1985 *Africa in crisis*, Earthscan, London.

Tleimat, B.W. 1980 Freezing methods, in Spiegler, K.S. & Laird, A.D.K. (eds) *Principles of desalination, part B*, Academic Press, London, 464 pp.

Tolba, M.K. 1980 Welcoming Address, in Biswas, A.K. *et al.* (eds) *Water management for arid lands in developing countries – training workshop*, Pergamon Press, Oxford, pp. 1–4.

Tolba, M.K. 1988 Heads in the sand: a new appraisal of arid lands management, in Whitehead, E.E. *et al.* (eds) *Arid lands today and tomorrow*, Westview Press, Boulder.

Tolba, M.K. 1990 Climate change and water management, Water International 15: 56–7.

Toulmin, C. 1988a Livestock policy in the Sahel, in Curtis, D. *et al.* (eds) *Preventing famine*, Routledge, London.
Toulmin, C. 1988b Smiling in the Sahel, New Scientist 12 Nov., p. 69.
Touqan, O. 1988 Wastewater irrigation in Western Asia, in Biswas, A.K. & Arar, A. (eds) *Treatment and reuse of wastewater*, Butterworths, London, pp. 91–105.
Trewartha, G.T. 1961 & 1981 *The Earth's problem climates*, University of Wisconsin Press, Madison.
Tsoar, H. 1985 Desert dune sand and its potential for modern agricultural development, in Gradus, Y. (ed.) *Desert development*, D. Reidel, Dordrecht.
Turner, R.K. (ed.) 1988 *Sustainable environmental management*, Belhaven Press, London.
Tyson, P.D. 1980 Climate and desertification in South Africa, in Mabbut, J.A. and Berkowicz, S.M. (eds) *The threatened dry lands symposium*, School of Geography, NSW, pp. 33–44.
Ukayli, M.A. & Husain, T. 1988 Comparative evaluation of surface water availability, waste water reuse and desalination in Saudi Arabia, Water International 13: 215–25.
United Nations 1978a Water development and management, *Proceedings of the UN conference, Mar del Plata, Argentina, March 1977*, Vol. 1, Parts 1, 2 & 3, Pergamon Press, Oxford.
United Nations 1978b Register of international rivers, Department of Economic and Social Affairs, Pergamon Press, Oxford.
United Nations 1985 Waste water reuse and its applications in Western Asia, UN Economic and Social commission for Western Asia, Natural Resources Science and Technology Division.
United Nations 1986 Development guidelines for the economic use of water in the ESCWA (Economic and Social Commission for Western Asia) region.
United Nations Development Programme 1977 The water resources of Qatar and their development, FAO, Rome.
UNESCO 1969 Discharge of selected rivers of the world, Paris.
UNESCO 1977a Development of arid and semi-arid lands: obstacles and prospects, MAB Technical Note 6, Paris.
UNESCO 1977b Map of the world distribution of arid regions, MAB Technical Note 7, Paris.
UNESCO 1978 Management of natural resources in Africa – traditional strategies and modern decision making, MAB Technical Note 9, Paris.
Unger, P.W. 1983 Water Conservation: Southern Great Plains, in Dregne, H.E. & Willis, W.O. (eds) *Dryland agriculture*, American Society of Agronomy, Number 23, Madison, pp. 35–55.
USGS 1984 *National handbook of recommended methods for water data acquisitions*, US Department of Interior.
US Office Technology Assessment 1984 Arica tomorrow: issues in technology, Washington, DC.
US Soil Conservation Service 1972 *National engineering handbook*, US Department of Agriculture, Washington, DC.
Utton, A.E. 1973 International water quality law, Natural Resources Journal 13 April.
Valet, S. 1978 Essai de determination des deficits relatifs d'alimentation maximum en eau du mil à Tillabery et maradi, INRAN, Niger.
Van Bavel, C. 1953 Chemical composition of tobacco leaves as affected by soil moisture conditions, Agronomy Journal 45: 611–14.
Van Wijk, W.R. & Borghorst, A.J.W. 1963 Turbulent transfer in air, in Van Wijk, W.R. (ed.) *Physics of the plant environment*, North-Holland, Amsterdam, pp. 236–76.
Varley, M.E. 1974 *Ecology: Climatic factors*, Unit 10 Course S323, Open University, Milton Keynes.

Veihmeyer, F.J. & Hendrickson, A.H. 1948 The permanent wilting percentage as a reference for the measurement of soil moisture, Transactions of the American Geophysical Union 29: 887–96.

Veihmeyer, F.J. & Hendrickson, A.H. 1955 Does transpiration decrease as the soil moisture decreases? Transactions of the American Geophysical Union 36(3): 425–48.

Vereecken, H. Maes, J., Freyen, J. & Darius, P. 1989 Estimating the soil moisture retention characteristic from texture, bulk density and carbon content, Soil Science 148(6): 389–403.

Vincent, L. 1987 Coordinating grass root improvements in water and health, *Conference on Water and Engineering for Developing Countries, Developing World Water*, Grosvenor Press International, Hong Kong, pp. 70–1.

Vines, R.G. 1986 Rainfall patterns in India, International Journal of Climatology 6: 135–48.

Visser, W.C. 1966 Progress in the knowledge about the effect of soil moisture content on plant production, Institute of Land Water Management Research Technical Bulletin 45, Wageningen.

Vohra, B.B. 1975 No more gigantism, CERES 8(4): 33–5.

Voit, F. Ya., Kornienko, E.E., Kirejko, I.A., Furman, A.I. & Husid, S.B. 1974 Results of seeding cumulonimbus clouds aimed at the modification of precipitation in the Steppe part of the Ukraine, *Proc. of WMO/IAMAP Scientific Conference on Weather Modification*, WMO Report 399, Geneva, pp. 51–4.

Vonnegut, B. 1947 The nucleation of ice formation by silver iodide, Journal of Applied Physics 18(7): 593–5.

Vonnegut, B. 1949 Nucleation of supercooled water clouds by silver iodide smokes, Chemical Review 44: 277.

Vonnegut, B. & Chessin, H. 1971 Ice nucleation by coprecipitated silver iodide and silver bromide, Science 174: 945–6.

Voropaev, G.V. 1979 The scientific principles of large scale areal redistribution of water resources in the USSR, in Golubev, G. & Biswas, A.K. (eds) *Interregional water transfers*, Pergamon Press, Oxford, pp. 90–101.

Wade, N. 1985 Choice of desalination and power plants, Water and Sewage 9(5): 3–6.

Wallace, J.S. 1991 The measurement and modelling of evaporation from semiarid land, in Sivakumar, M.V.K. *et al.* (eds) *Proceedings of Workshop on Soil Water Balance in the Sudano–Sahelian zone, February 18–23*, IAHS Publication 199, Wallingford, pp. 131–48.

Wallace, J.S., Gash, J.H.C. & Sivakumar, M.V.K. 1990 Preliminary measurements of net radiation and evaporation over bare soil and fallow bushland in the Sahel, International journal of Climatology 10: 201–10.

Wallen, C.C. 1967 Aridity definitions and their applicability, Geografiska Annaler 49a: 367–84.

Waller, D.H. 1989 Rain water – an alternative source in developing and developed countries, Water International 14: 27–36.

Walker, H.O. 1958 The monsoon in West Africa, Ghana Meteorological Reports, Departmental Note No 9.

Walker, J. & Rowntree, P.R. 1977 The effect of soil moisture on circulation and rainfall in a tropical model, Quarterly Journal of the Royal Meteorological Society 103: 29–46.

Walpole, R.E. & Myers, R.H. 1972 *Probability and statistics for engineers and scientists*, Collier Macmillan, London.

Walsh, J. 1986 Project born of hope, desperation, Science 232 (May): 1081–3.

Walters, M.O. 1989 A method of estimating design food hydrograph shape in an arid region, Water International 14: 2–5.

Walters, R.E. 1974 *The nuclear trap: an escape route*, Penguin, Harmondsworth.

Walton, K. 1969 *The arid zones*, Hutchinson, London.

Walton, W.C. 1970 *Groundwater resource evaluation*, McGraw-Hill, New York.

Wang, J.R. 1985 Effect of vegetation on soil moisture sensing observed from orbiting microwave radiometers, Remote sensing of the Environment 17: 141–51.

Ward, R.C. 1975 *Principles of hydrology* (2nd edn), McGraw-Hill, London, 367 pp.

Ward, R.C. 1978 *Floods*, Macmillan, London.

Warner, J. 1974 Rain enhancement – A review, in *Proc. WMO/IAMAP Scientific Conference on Weather Modification*, WMO Report 399, Geneva, pp. 43–50.

Warren, A. & Agnew, C. 1988 An assessment of desertification and land degradation in arid and semi-arid areas, International Institute for Environment and Development, paper 2, 30 pp.

Warrick, R.A. 1975 Drought hazard in US: A research assessment, National Technical Information Service, Springfield, Virginia (NSF/RA/C-75-004).

Warrick, R.A. 1983 Drought in the US Great Plains: shifting social consequences? in Hewitt, K. (ed.) *Interpretations of calamity*, Allen & Unwin, London, pp. 67–82.

Water and Engineering for Developing Countries 1987 *Developing World Water*, Pickford, J. (ed.) Grosvenor Press International, Hong Kong, 588 pp.

Water Resources Council 1979 Water resources assessment and appraisal, Sultanate of Oman.

Waterbury, J. 1979 *The hydrology of the Nile valley*, Syracuse University Press, New York.

Weisner, C.J. 1970 *Climate, irrigation and agriculture: a guide to the practice of irrigation*, Angus Robertson, London.

Welbank, P.J. & Williams, E.D. 1968 Root growth of a barley crop estimated by sampling with portable powered soil coring equipment, Journal of Applied Ecology 5: 477–81.

Wengert, N. 1976 The political allocation of benefits and burdens: Economic externalities and due process in environmental protection, Institute of Governmental Studies, University of California, Berkley.

Westing, A.H. (ed.) 1986 *Global resources in international conflict*, OUP, Oxford.

Wetherald, R.T. and Manabe, S. 1979 Sensitivity studies of climate involving changes in CO_2 concentration, in Bach, W. *et al.*, *Man's impact on environment*, pp. 57–64.

Wheater, H.S. & Bell, N.C. 1983 Northern Oman flood study, Proceedings of the Institution of Civil Engineers part 2 75: 453–73.

White, C. 1987 Borehole siting using geophysics, *Conference on Water and Engineering for Developing Countries, Developing World Water*, Grosvenor Press International, Hong Kong, pp. 106–13.

White, G.F. 1960 Science and the future of arid lands, UNESCO, Paris (quoted in McGinnies, 1988).

White, G.F. 1977 The main effects and problems of irrigation, in Worthington, E.B. (ed.) *Arid land irrigation in developing countries*, Pergamon Press, Oxford, pp. 1–71.

White G.F. 1978 Environmental effects of arid land irrigation in developing countries, UNESCO, Paris.

White, G.F. 1986 The changing role of water in arid lands, in Kates, R.W. & Burton, I. (eds) *Geography, resources and environment: The selected writings of Gilbert White*, University of Chicago Press.

Whitear, J. 1982 Irrigation scheduling, how the computer can help, *UK Irrigation Association Conference*, pp. 20–2.

Whitehead, E.E., Hutchinson, C.F., Timmermann, B.N. & Vardy, R.G. (eds) 1988 *Arid lands today and tomorrow*, Westview Press, Boulder, 1,435 pp.

Whitlach, E.E. & de Velle, C.S. 1990 Regionalisation in water resource projects, Water International 15: 70–9.

Whittaker, R.H. 1975 Communities and ecosystems (2nd edn), Collier Macmillan, London.

Wigley, T.M.L. and Raper, S.C.B. 1987 Thermal expansion of sea water associated with global warming, Nature 330: 127–31.

Wijkman, A. & Timberlake, L. 1985 Is the African drought an act of God or of Man?, The Ecologist 15(1/2): 9–18, 34.

de Wilde, J.C. 1967 *Experiences with agricultural development in tropical Africa: Mali – the office du Niger*, John Hopkins University Press, Baltimore.

Wilhite, D. & Glantz, M. 1985 Understanding the drought phenomenon: the role of definitions, Water International 10: 111–20.

Willet, H. 1976 Climatic trends and the ice age, in Kopec, R. (ed.) *Atmosphere quality and climatic change*, Studies in Geography 9, University of North Carolina, pp. 4–37.

Willis, W.O. 1983 Water conservation, in Dregne, H.E. & Willis, W.O. (eds) *Dryland agriculture*, American Society of Agronomy, Number 23, Madison, pp. 21–4.

Willis, W.O., Bauer, A. & Black, A.L. 1983 Water conservation: Northern Great Plains, in Dregne, H.E. & Willis, W.O. (eds) *Dryland agriculture*, American Society of Agronomy, Number 23, Madison, pp. 73–88.

Wilson, E.M. 1984 *Engineering hydrology*, Macmillan, London, 309 pp.

Winstanley, D. 1973a Rainfall patterns and general atmospheric circulation, Nature 245: 190–4.

Winstanley, D. 1973b Drought in the Sahel Zone, Symposium on Drought in Africa, SOAS, University of London.

Winstanley, D. 1983 Desertification: a climatological perspective, in Wells, S.G. and Haragan, D.R. (eds) *Origin and evolution of deserts*, University of New Mexico Press, Albuquerque.

Withers, B. & Vipond, S. 1974 and 1988 *Irrigation design and practice*, Batsford, London, 306 pp.

WMO 1974 *Guide to hydrological pracitces* (3rd edn), World Meteorological Organization, No 168, Geneva.

Woldesmiate, T. & Cox, R. 1987 The food crisis in Kenya, in Yesilada, B.A., Brockett, C.D. and Drury, B. (eds) *Agrarian reform in reverse; the food crisis in the 3rd world*, Westview Press, Boulder, pp. 181–202.

Woodwell, G.M. & Ramakrishna, K. 1989 Will there be a global greenhouse warming? Environmental Conservation 16(4): 289–91.

World Bank 1984 Toward sustained development in Sub-Saharan Africa, Washington, DC.

World Resources Institute 1986 *World Resources 1986*, Basic Books Inc., New York.

World Resources Institute 1987 *World Resources 1987*, Basic Books Inc., New York.

World Resources Institute 1989 *World Resources 1988–89*, Basic Books Inc., New York.

World Resources Institute 1990 *World Resources 1990*, Basic Books Inc., New York.

World Water 1984 The international drinking water supply and sanitation decade directory, Liverpool.

Worthington, E.B. 1977 *Arid land irrigation in developing countries*, Pergamon Press, Oxford, 463 pp.

Wright, J.L. 1981 Crop coefficients for estimates of daily crop evapotranspiration, *Proceedings of Irrigation Scheduling Conference, Chicago*, American Society of Agricultural Engineers, pp. 18–26.

Wright, J.L. 1982 New evapotranspiration crop coefficients, *Proceedings of the American Society of Civil Engineers, Irrigation and Drainage Division* 108: 57–74.

Wrigley, G. 1969 *Tropical agriculture the development of production*, Faber and Faber, London.

Yacoob, M., Brieger, W.R. & Watts. S. 1990 What happened to guinea worm control? Water International 15: 27–34.

Yair, A. & Berkowicz, S.M. 1989 Climatic and nonclimatic controls of aridity: The case

of the Northern Negev of Israel, Catena Supplement 14; Arid and semi-arid environments.
Yiassin, A.A. (ed.) 1985 *Development and rationalization of water resources in Oman, Water Policy Planning Conference, October 27, Muscat.*
Yevjevich, V., Hall, W.A. & Salas, E.D. (eds) 1978 Drought research needs, *Conference Proceedings, Fort Collins*, 264 pp.
Youngs, E.G. 1988 Soil Physics and hydrology, Journal of Hydrology 100: 411–31.
Yousif, A.R.M. & Shajahan, N.K. 1985 Water Consumption for the industrial sector in Oman, in Yassin, A.A. (ed.) *Development and rationalization of water resources in Oman, Water Policy Planning Conference, October 27, Mascat.*
Ziman, K.F. 1979 The carbon cycle, the missing sink and future CO_2 levels in the atmosphere, in Bach, W. *et al., Man's impact on environment*, Elsevier, Oxford, pp. 129–37.
Zoppo, C.E. & Zorgbibe, C. (eds) 1985 *On geopolitics: classical and nuclear*, Martinus Nijhoff, Dordrecht.

INDEX

Printed in the United States
by Baker & Taylor Publisher Services